T0134964

Knowledge Engineering Tools and Techniques for AI Planning

Mauro Vallati • Diane Kitchin
Editors

Knowledge Engineering Tools and Techniques for AI Planning

Editors
Mauro Vallati
Department of Computer Science
University of Huddersfield
Huddersfield, UK

Diane Kitchin
Department of Computer Science
University of Huddersfield
Huddersfield, UK

ISBN 978-3-030-38563-7 ISBN 978-3-030-38561-3 (eBook)
https://doi.org/10.1007/978-3-030-38561-3

This Springer imprint is published by the registered company Springer Nature Switzerland AG.
The registered company address is: Gewerbestrasse 11, 6330 Cham, Switzerland

Preface

Automated planning plays a central role in artificial intelligence (AI), as it focuses on investigating approaches for producing plans, in terms of sequences of actions that need to be performed. These plans allow an agent to achieve identified and clearly stated goals. Being able to plan for the future is a pivotal characteristic of any intelligent system and is an essential capability of autonomous systems. Automated planning has been successfully applied for decades in several areas, including space exploration, machine tool calibration, control of unmanned robots, and urban traffic control to mention a few.

Undoubtedly, the intensive development of domain-independent planners has contributed to the advancement of planning technology, as planning engines can be exploited as embedded components within a larger framework. Since they accept the domain and problem instance in a well-defined interface language and return plans using the same syntax, they can be interchanged without any changes to the rest of the system. On the other hand, the efficiency of plan generation remains one of the most prominent challenges in artificial intelligence. Domain-independent planning engines have to deal with the complexity issues inherent in plan generation, which are exacerbated by the separation of planner logic from domain knowledge.

Knowledge Engineering in Planning and Scheduling (KEPS) was defined in the 2003 PLANET Roadmap, specifically for domain-independent planners, as the collection of processes involving (i) the acquisition, validation and verification, and maintenance of planning domain models; (ii) the selection and optimisation of appropriate planning machinery; and (iii) the integration of (i) and (ii) to form automated planning and scheduling applications.

KEPS can be seen as a special case of knowledge engineering, where the need for methodologies for acquiring, domain modelling, and managing formally captured knowledge has long been accepted. It is also related to the area of capturing conceptual knowledge and developing domain models for qualitative reasoning in the general modelling and simulation area. However, the peculiarities of automated planning and scheduling applications distinguish KEPS from general knowledge-based and simulation systems. Firstly, KEPS is concerned with the acquisition and representation of knowledge about actions, resources, processes, events, and the effect these have on a state. Secondly, this knowledge is to be used to create a system that synthesises plans, rather than performing the more common functions of knowledge systems such as classification, diagnosis, or decision-making. Thirdly, the knowledge is often acquired in two parts: a specification of persistent knowledge (in the literature this part is called the "domain" or "domain model") and a specification of particular scenarios representing the planning problem instance.

Studies on KEPS have led to the creation of several tools and techniques to support the design of domain knowledge structures and the use of planners for real-world problems. Most of these tools have been presented in specialised workshops such as the Knowledge Engineering for Planning and Scheduling workshop and the Verification and Validation of Planning Systems workshop, as well as competitions such as the International Competition on Knowledge Engineering for Planning and Scheduling (ICKEPS).

This book provides an overview of the state of the art of knowledge engineering for planning and scheduling, bringing together work from leading researchers in the field, and covering a variety of tools and approaches. Aiming at being as inclusive as possible, the focus is not given to a specific formalism or paradigm but to the variety of planning approaches and languages that are exploited in real-world applications, as well as in research.

Besides the clear aim of providing a holistic vision of the KEPS field, we hope that this book can foster discussions around KEPS approaches and techniques and highlight areas in which future effort has to be focused in order to advance the field.

Huddersfield, UK

Mauro Vallati
Diane Kitchin

Contents

Part I
Knowledge Capture and Encoding

Chapter 1
Explanation-Based Learning of Action Models

Diego Aineto, Sergio Jiménez, and Eva Onaindia

Abstract The paper presents a classical planning compilation for learning STRIPS action models from partial observations of plan executions. The compilation is flexible to different amounts and types of input knowledge, from learning samples that comprise partially observed intermediate states of the plan execution to samples in which only the initial and final states are observed. The compilation accepts also partially specified action models and it can be used to validate whether an observation of a plan execution follows a given STRIPS action model, even if the given model or the given observation is incomplete.

Keywords Learning action models · Classical planning

1 Introduction

Action models in planning are not only required for plan synthesis [10] but also for other tasks like plan/goal recognition [19, 20]. In both cases, automated planners are required to reason about action models that correctly and completely capture the possible world transitions [9]. Unfortunately building planning action models is complex, even for planning experts, and this knowledge acquisition task is a bottleneck that limits the potential of AI planning [14].

Machine Learning (ML) techniques have shown to be suitable to learn a wide range of different kinds of models from examples [17]. The application of inductive ML to learning STRIPS action models, the vanilla action model for planning [6], is not straightforward though:

– The input to ML algorithms (the learning/training data) is usually a finite vector that represents the value of some fixed object features. The input for learning

D. Aineto · S. Jiménez (✉) · E. Onaindia
Departamento de Sistemas Informáticos y Computación, Universitat Politècnica de València, Valencia, Spain
e-mail: dieaigar@dsic.upv.es; serjice@dsic.upv.es; onaindia@dsic.upv.es

M. Vallati, D. Kitchin (eds.), *Knowledge Engineering Tools and Techniques for AI Planning*, https://doi.org/10.1007/978-3-030-38561-3_1

planning action models is, however, the observation of plan executions, where each plan has a possibly different length (plan length is not a priori bounded) and refers to a different number of objects.
- The output of ML algorithms is usually a scalar value (an integer, in the case of classification tasks, or a real value, in the case of regression tasks). The learning of action models outputs a declarative definition of the preconditions and effects of the modeled actions.

Learning STRIPS action models is a well-studied problem with sophisticated algorithms such as ARMS [27], SLAF [2], or LOCM [4]. All of these learning systems are capable of dealing with partial or null observability of the intermediate states traversed along the plan execution but they also require a full specification of the sequence of actions of the learning examples. Motivated by recent advances on the synthesis of different kinds of generative models with classical planning [3, 22–24], this paper describes a classical planning compilation approach for learning STRIPS action models. The compilation approach is appealing by itself, because it opens up the door to the bootstrapping of planning action models, but also because it is flexible to different amounts and types of available input knowledge:

1. *Learning examples* can range from plans that comprise partially observed intermediate states of the plan execution to samples in which no intermediate state/action is observed, that is, only the initial and final states are observed.
2. *Partially specified action models*, expressing prior knowledge about the structure of actions, can also be provided to the compilation. In the extreme, the compilation can validate whether an observed plan execution is consistent with a given STRIPS action model, even if the model is not fully specified or the input observation is incomplete.

2 Background

In this section we formalize the *classical planning* model, for the *observation* model to represent the execution of a classical plan and the model for the *explanation of a given observation*.

2.1 Classical Planning with Conditional Effects

F is the set of *fluents* or *state variables* (propositional variables). A *literal* l is a valuation of a fluent $f \in F$, i.e., either $l = f$ or $l = \neg f$. L is a set of literals that represents a partial assignment of values to fluents, and $\mathcal{L}(F)$ is the set of all literals sets on F, i.e., all partial assignments of values to fluents. A *state* s is a full assignment of values to fluents. We explicitly include negative literals $\neg f$ in states and so $|s| = |F|$ and the size of the state space is $2^{|F|}$.

A *planning frame* is a tuple $\Phi = \langle F, A\rangle$, where F is a set of fluents and A is a set of *actions*. An action $a \in A$ is defined with *preconditions*, $\mathsf{pre}(a) \in \mathcal{L}(F)$, and *effects* $\mathsf{eff}(a) \in \mathcal{L}(F)$. The semantics of actions $a \in A$ is specified with two functions: $\rho(s, a)$ denotes whether action a is *applicable* in a state s and $\theta(s, a)$ denotes the *successor state* that results of applying action a in a state s. Therefore $\rho(s, a)$ holds iff $\mathsf{pre}(a) \subseteq s$ and the result of applying a in s is $\theta(s, a) = \{s \setminus \neg\mathsf{eff}(a)) \cup \mathsf{eff}(a)\}$, with $\neg\mathsf{eff}(a) = \{\neg l : l \in \mathsf{eff}(a)\}$.

A *planning problem* is defined as a tuple $P = \langle F, A, I, G\rangle$, where I is the initial state in which all the fluents of F are assigned a value true/false and G is the goal set. A *plan* π for P is an action sequence $\pi = \langle a_1, \ldots, a_n\rangle$, and $|\pi| = n$ denotes its *plan length*. The execution of π in the initial state I of P induces a *trajectory* $\tau = \langle s_0, a_1, s_1, \ldots, a_n, s_n\rangle$ such that $s_0 = I$ and, for each $1 \leq i \leq n$, it holds $\rho(s_{i-1}, a_i)$ and $s_i = \theta(s_{i-1}, a_i)$. A plan π solves P if G holds in the last state of the induced trajectory τ; i.e., $G \subseteq s_n$. A solution plan is *optimal* iff its length is minimal.

Now we define actions with *conditional effects* because they allow us to compactly define our compilation. An action $a_c \in A$ with conditional effects is defined as a set of preconditions $\mathsf{pre}(a_c) \in \mathcal{L}(F)$ and a set of *conditional effects* $\mathsf{cond}(a_c)$. Each conditional effect $C \rhd E \in \mathsf{cond}(a_c)$ is composed of two sets of literals: $C \in \mathcal{L}(F)$, the *condition*, and $E \in \mathcal{L}(F)$, the *effect*. An action $a_c \in A$ is applicable in a state s if and only if $\mathsf{pre}(a_c) \subseteq s$, and the *triggered effects* resulting from the action application are the effects whose conditions hold in s:

$$triggered(s, a_c) = \bigcup_{C \rhd E \in \mathsf{cond}(a_c), C \subseteq s} E.$$

The result of applying a_c in state s follows the same definition of successor state, $\theta(s, a)$, but applied to the conditional effects in $triggered(s, a_c)$.

2.2 The Observation Model

Given a planning problem $P = \langle F, A, I, G\rangle$, a plan π that solves P, and the corresponding trajectory τ induced by the execution of π in I, $\tau = \langle s_0, a_1, s_1, \ldots, a_n, s_n\rangle$; there exist as many observations of τ as combinations of observable actions and observable fluents of the states of τ. *The observation model of the trajectory* τ comprises all possible combinations of observable elements of τ. We will refer to the set of observations of τ as $Obs(\tau)$.

Formally, one observation in $Obs(\tau)$ is defined as $\mathcal{O} = \langle s_0^o, s_1^o \ldots, s_m^o\rangle$, $s_0^o = I$, a sequence of possibly *partially observable states*, except for the initial state s_0^o which is fully observable. A partially observable state is one in which $|s_i^o| < |F|$, $1 \leq i \leq m \leq n$; i.e., a state in which at least a fluent of F is not observable. It may be also the case that $|s_i^o| = 0$ when an intermediate state is fully unobservable.

The *minimal observation* needed by our model is $\mathcal{O} = \langle s_0^o, s_1^o \rangle$, where s_0^0 is the fully observable initial state and s_1^o is a partially observable final state.

The observation model can also include *observed actions* as fluents indicating the applied action in a given state. This means that a sequence of observed actions $\langle a_1^o, \ldots, a_l^o \rangle$ is a sub-sequence of $\pi = \langle a_1, \ldots, a_n \rangle$ such that $a_i^o \in s_{i-1}^o$, $0 \leq i \leq l$. Consequently, the number of fluents that represent observed actions, l, can range from 0 (in a fully unobservable action sequence) to $|\pi| = n$ (in a fully observed action sequence).

Given $\mathcal{O} \in Obs(\tau)$, the number of observed states of $\mathcal{O} = \langle s_0^o, s_1^o \ldots, s_m^o \rangle$ ranges from 2 (at least the initial and final states, as explained above) to $|\pi|+1$. The number of fluents of the full observable state s_0^o will be $|F|$, or $|F| + 1$ in case the fluent of the applied action in s_0 is also observed. Every observable intermediate state will comprise a number of fluents between $[1, |F| + 1]$, where a single fluent may represent a sensing fluent of the state or the observation of the applied action.

This observation model can also distinguish between *observable state variables*, whose value may be read from sensors, and *hidden* (or *latent*) *state variables*, that cannot be observed. Given a subset of fluents $\Gamma \subseteq F$ we say that $\mathcal{O}$ is a Γ-observation of the execution of π on P iff for every observed state s_i^o, $1 \leq i \leq m$, s_i^o only contains fluents in Γ.

2.3 Explaining Observations with Classical Planning

In this section we will explore the relationship between a trajectory τ and an observation $\mathcal{O}$. Particularly, we are interested in determining the necessary conditions for $\mathcal{O}$ to belong to $Obs(\tau)$. When the membership $\mathcal{O} \in Obs(\tau)$ is established, we say that $\mathcal{O}$ is *consistent* with τ or that τ *explains* $\mathcal{O}$.

For the sake of simplicity, and given that our observation model encodes the observed applicable actions as fluents in the corresponding state, we will denote a trajectory as $\tau = \langle s_0', s_1', \ldots, s_n' \rangle$, where s_i' comprises a fluent representing the applicable action a_{i+1} in s_i'.

Given an observation $\mathcal{O} = \langle s_0^o, s_1^o \ldots, s_m^o \rangle$ and a trajectory $\tau = \langle s_0', s_1', \ldots, s_n' \rangle$, where $m \leq n$, $s_0^o = s_0'$, and $s_m^o \subseteq s_n'$, it holds that $\mathcal{O} \in Obs(\tau)$ iff τ embeds $\mathcal{O}$; i.e., if there is a monotonic function f mapping the observation indices $j = 1, 2, \ldots, m$ into the trajectory indices $i = 1, 2, \ldots, n$ such that $s_j^o \subseteq s_{f(j)}'$. This definition is a generalization of the one introduced in [19], which states the conditions under which an action sequence satisfies an observation sequence. Since all the elements (sets) of $\mathcal{O}$ are associated to an element (set) of τ, but not vice versa, the fluents of a set of $\mathcal{O}$ are all included in the corresponding set of τ, we can say that τ is a superset of $\mathcal{O}$. All this means that transiting between two consecutive observed states in $\mathcal{O}$ may require the execution of more than a single action ($\theta(s_i^o, \langle a_1, \ldots, a_k \rangle) = s_{i+1}^o$, where $k \geq 1$ is unknown but finite. In other words, the information of $\mathcal{O}$ does not imply knowing the actual length of the trajectory τ.

Given a planning frame $\Phi = \langle F, A \rangle$ and an observation of a plan execution, $\mathcal{O} = \langle s_0^o, s_1^o \ldots, s_m^o \rangle$, we define $P_{\mathcal{O}}$, within the given planning frame, as the planning problem that is built as follows: $P_{\mathcal{O}} = \langle F, A, s_0^o, s_m^o \rangle$.

Definition 1 (Explanation) A plan π (or the trajectory τ) **explains** $\mathcal{O}$ iff π is a solution for $P_{\mathcal{O}}$ and $\mathcal{O} \in Obs(\tau)$.

There may exist more than one solution plan for $P_{\mathcal{O}}$, one or more of which will be optimal solutions if their plan length is minimal. Additionally, other solutions longer than the optimal plan can also be found.

Definition 2 (Best Explanation) A plan π (or the trajectory τ) that solves $P_{\mathcal{O}}$ is the **best explanation** for $\mathcal{O}$ iff $|\pi| = n$ and for every other τ_i s.t. $\mathcal{O} \in Obs(\tau_i)$, $|\pi_i| > n$.

That is, in case that π is optimal, we say that π is the **best explanation** for the input observation $\mathcal{O}$.

The observation $\mathcal{O}$ can also be regarded as a sequence of ordered *landmarks* for the planning problem $P_{\mathcal{O}}$ [11] since all the fluents of the sets in $\mathcal{O}$ must be achieved by any plan that solves $P_{\mathcal{O}}$ and in the same order as defined in the observation $\mathcal{O}$.

3 Explanation-Based Learning of Strips Action Models

The task of learning action models by explaining the observation of a plan execution is defined as a tuple $\Lambda = \langle \mathcal{M}, \mathcal{O} \rangle$, where

- $\mathcal{M}$ is the *initial empty model* that contains only the *header* (i.e., the *name* and *parameters*) of each action model to be learned.
- $\mathcal{O} = \langle s_0^o, s_1^o \ldots, s_m^o \rangle$ is a sequence of partially observed states, except for the initial state s_0^o which is fully observable.

A *solution* to a $\Lambda = \langle \mathcal{M}, \mathcal{O} \rangle$ learning task is a model $\mathcal{M}'$ that is consistent with the headers of $\mathcal{M}$ and that explains $\mathcal{O}$. We say that a model $\mathcal{M}'$ explains an observation $\mathcal{O}$ iff there exists a solution plan for $P_{\mathcal{O}} = \langle F, A, s_0^o, s_m^o \rangle$, where the semantics of the set of actions A are given by $\mathcal{M}'$, such that π explains $\mathcal{O}$. The set of fluents $F \in P_{\mathcal{O}}$ is induced from $s_0^o \in \mathcal{O}$ since it represents a full state.

3.1 *The Space of Strips Action Models*

We analyze here the solution space of the addressed learning task; in this case the space of STRIPS action models.

A STRIPS *action model* is defined as $\xi = \langle name(\xi), pars(\xi), pre(\xi), add(\xi), del(\xi)\rangle$, where $name(\xi)$ and parameters, $pars(\xi)$, define the header of ξ; and $pre(\xi)$, $del(\xi)$, and $add(\xi)$) are sets of fluents that represent the *preconditions*, *negative effects*, and *positive effects*, respectively, of the actions induced from the action model ξ.

Let Ψ be the set of *predicates* that shape the fluents F (the initial state of an observation is a full assignment of values to fluents, $|s_0^o| = |F|$, and so the predicates Ψ are extractable from the observed state s_0^o). The set of propositions that can appear in $pre(\xi)$, $del(\xi)$, and $add(\xi)$ of a given ξ, denoted as $\mathcal{I}_{\xi,\Psi}$, are FOL interpretations of Ψ over the parameters $pars(\xi)$. For instance, in a four-operator *blocksworld* [25], the $\mathcal{I}_{\xi,\Psi}$ set contains five elements for the `pickup`(v_1) model, $\mathcal{I}_{pickup,\Psi}$={`handempty`, `holding`($v_1$),`clear`($v_1$),`ontable`($v_1$), `on`($v_1, v_1$)} and eleven elements for the model of `stack`(v_1,v_2),$\mathcal{I}_{stack,\Psi}$={`handempty`, `holding`(v_1), `holding`(v_2), `clear`(v_1),`clear`(v_2),`ontable`(v_1),`ontable`(v_2), `on`(v_1, v_1),`on`(v_1, v_2), `on`(v_2, v_1), `on`(v_2, v_2)}. Hence, solving a $\Lambda = \langle \mathcal{M}, \mathcal{O}\rangle$ learning task is determining which elements of $\mathcal{I}_{\xi,\Psi}$ will shape the preconditions, positive effects, and negative effects of the corresponding action model.

In principle, for a given STRIPS action model ξ, any element of $\mathcal{I}_{\xi,\Psi}$ can potentially appear in $pre(\xi)$, $del(\xi)$, and $add(\xi)$. In practice, the actual space of possible STRIPS schemata is bounded by:

1. **Syntactic constraints**. The solution $\mathcal{M}'$ must be consistent with the STRIPS constraints: $del(\xi) \subseteq pre(\xi)$, $del(\xi) \cap add(\xi) = \emptyset$, and $pre(\xi) \cap add(\xi) = \emptyset$. *Typing constraints* are also a type of syntactic constraint that reduce the size of $\mathcal{I}_{\xi,\Psi}$ [16].
2. **Observation constraints**. The solution $\mathcal{M}'$ must be consistent with these *semantic constraints* derived from the input observation $\mathcal{O}$. Specifically, the states induced by plans computable with $\mathcal{M}'$ must comprise the observed states of the sample, which further constrains the space of possible action models.

Considering only the syntactic constraints, the size of the space of possible STRIPS models is given by $2^{2\times|\mathcal{I}_{\Psi,\xi}|}$ because one element in $\mathcal{I}_{\xi,\Psi}$ can appear both in the preconditions and effects of ξ. Given $p \in \mathcal{I}_{\Psi,\xi}$, the belonging of p to the preconditions, positive effects, or negative effects of ξ is handled with a propositional encoding that uses fluents of two types, $pre_{p,\xi}$ and $eff_{p,\xi}$. The four possible combinations of these two fluents are summarized in Fig. 1.1. This compact encoding allows for a more effective exploitation of the syntactic constraints, and also yields the solution space of $\Lambda = \langle \mathcal{M}, \mathcal{O}\rangle$ to be the same as its search space.

To illustrate better this encoding, Fig. 1.2 shows the PDDL encoding of the `stack(?v1,?v2)` schema and our propositional representation for this same schema with $pre_{p,stack}$ and $eff_{p,stack}$ fluents ($p \in \mathcal{I}_{\Psi,stack}$).

Encoding	Meaning
$\neg pre_{p,\xi} \wedge \neg eff_{p,\xi}$	p belongs neither to the preconditions nor effects of ξ $(p \notin pre(\xi) \wedge p \notin add(\xi) \wedge p \notin del(\xi))$
$pre_{p,\xi} \wedge \neg eff_{p,\xi}$	p is only a precondition of ξ $(p \in pre(\xi) \wedge p \notin add(\xi) \wedge p \notin del(\xi))$
$\neg pre_{p,\xi} \wedge eff_{p,\xi}$	p is a positive effect of ξ $(p \notin pre(\xi) \wedge p \in add(\xi) \wedge p \notin del(\xi))$
$pre_{p,\xi} \wedge eff_{p,\xi}$	p is a negative effect of ξ $(p \in pre(\xi) \wedge p \notin add(\xi) \wedge p \in del(\xi))$

Fig. 1.1 Combinations of the propositional encoding and their meaning

```
(:action stack
   :parameters (?v1 ?v2)
   :precondition (and (holding ?v1) (clear ?v2))
    :effect (and (not (holding ?v1)) (not (clear ?v2))
                 (clear ?v1) (handempty) (on ?v1 ?v2)))

(pre_holding_v1_stack) (pre_clear_v2_stack)
(eff_holding_v1_stack) (eff_clear_v2_stack)
(eff_clear_v1_stack) (eff_handempty_stack) (eff_on_v1_v2_stack)
```

Fig. 1.2 PDDL encoding of the `stack(?v1,?v2)` schema and our propositional representation for this same schema

```
(:predicates (on ?x ?y) (ontable ?x) (clear ?x) (handempty) (holding ?x))
(:objects blockA blockB blockC)
(:init (ontable blockA) (on blockB blockA) (clear blockB) (handempty))
(:observation (on blockA blockB))
```

Fig. 1.3 Example of a two-state observation for the learning of STRIPS action models in the *blocksworld* domain

3.2 *The Sampling Space*

According to our *observation model* the minimal expression of an observation must comprise at least two state observations $\mathcal{O} = \langle s_0^o, s_m^o \rangle$, a fully observable initial state s_0^o and a partially observed final state s_m^o. Figure 1.3 shows an example of $\mathcal{O} = \langle s_0^o, s_m^o \rangle$ observation that contains only two states. An initial state of the blocksworld where the robot hand is empty and there are two blocks (`blockB` on top of `blockA`). The observation represents also a partially observable final state in which `blockA` is on top of `blockB`.

On the other hand, the maximal expression of an observation corresponds to a fully observed trajectory $\mathcal{O} = \tau$, meaning that all traversed states, and applied actions, are fully observed. Between our minimal and maximal expressions of observation, there exists a whole range of possible degrees of observability. For example, the majority of learning systems such as ARMS [27] or SLAF [2] use observations that comprise the initial state and all the actions of the executed plan.

4 Learning STRIPS Action Models with Classical Planning

Our approach to address a learning task $\Lambda = \langle \mathcal{M}, \mathcal{O} \rangle$ is to compile it into a classical planning problem P_Λ. The intuition behind the compilation is that when P_Λ is solved, the solution plan π_Λ is a sequence of actions that build the output model $\mathcal{M}'$ and verify that $\mathcal{M}'$ explains the observation $\mathcal{O}$.

A solution plan π_Λ includes then two differentiated blocks of actions: a plan prefix with a set of actions, each defining the **insertion** of a fluent as a precondition or an effect of an action model and a plan postfix with a set of actions that determine the **application** of the learned modes while successively **validating** the effects of the action application in every partial state of $\mathcal{O}$. Roughly speaking, in the *blocksworld*, the format of the first block of actions of π_Λ looks like `(insert_pre_stack_holding_v1)`, `(insert_eff_stack_clear_v1)`, `(insert_eff_stack_holding_v1)`..., where the first effect denotes a positive effect and the second one a negative fluent to be inserted in $name(\xi)$ = `stack`; and the format of the second block of actions of π_Λ is like `(apply_unstack blockB blockA)`,`(apply_putdown blockB)`, and `(validate_1)`, `(validate_2)`, where the last two actions denote the points at which the states generated through the action application must be validated with the observed states in $\mathcal{O}$.

4.1 Compilation

Given a learning task $\Lambda = \langle \mathcal{M}, \mathcal{O} \rangle$ the compilation outputs a classical planning task $P_\Lambda = \langle F_\Lambda, A_\Lambda, I_\Lambda, G_\Lambda \rangle$ such that:

- F_Λ extends the set of fluents F (obtained from s_0^o) with the model fluents that are used to represent the preconditions and effects of each $\xi \in \mathcal{M}$ as well as some other fluents to keep track of the validation of $\mathcal{O}$. Specifically, F_Λ contains also:
  - Fluents $pre_{p,\xi}$ and $eff_{p,\xi}$, defined as explained in Sect. 3.1.
  - A set of fluents $\{test_j\}_{0 \leq j \leq m}$ to point at the state observation $s_j^o \in \mathcal{O}$ where the action model is validated. In the example of Fig. 1.3 two tests are required to validate the programmed action model, one corresponding to the initial state and the second one corresponding to the final state.

- A fluent, $mode_{prog}$, to indicate whether action models are being programmed or validated and a fluent *invalid* to indicate that the programmed action model is inconsistent with the input observation.

– I_Λ encodes s_0^o and the following fluents set to true: $mode_{prog}$, $test_0$. Our compilation assumes that action models are initially programmed with no precondition, no negative effect, and no positive effect.
– G_Λ includes the positive literal $test_m$ and the negative literal $\neg invalid$. When these goals are achieved by the solution plan π_Λ, we will be certain that the action models of $\mathcal{M}'$ are validated in the input observation.
– A_Λ includes three types of actions that give rise to the actions of π_Λ.

1. Actions for *inserting* a precondition or effect into $\xi \in \mathcal{M}$ following the syntactic constraints of STRIPS models. These actions will form the prefix of the solution plan π_Λ. Among the *inserting* actions, we find:

   - Actions for inserting a *precondition* $p \in \mathcal{I}_{\xi,\Psi}$ into ξ.

$$\begin{aligned}\text{pre}(\text{insertPre}_{\text{p},\xi}) =& \{\neg pre_{p,\xi}, mode_{prog}\},\\ \text{cond}(\text{insertPre}_{\text{p},\xi}) =& \{\emptyset\} \rhd \{pre_{p,\xi}\}.\end{aligned}$$

   - Actions for inserting an *effect* $p \in \mathcal{I}_{\xi,\Psi}$ into ξ.

$$\begin{aligned}\text{pre}(\text{insertEff}_{\text{p},\xi}) =& \{\neg eff_{p,\xi}, mode_{prog}\},\\ \text{cond}(\text{insertEff}_{\text{p},\xi}) =& \{\emptyset\} \rhd \{eff_{p,\xi}\}\end{aligned}$$

   For instance, given $name(\xi)=$ `stack` and { `(pre_stack_holding_v1)`, `(pre_stack_holding_v2)`,`(pre_stack_on_v1_v2)`,`(pre_stack_clear_v1)`, `(pre_stack_clear_v1)`,...}, the insertion of each item $p \in \mathcal{I}_{\xi,\Psi}$ in ξ will generate a different alternative in the search space when solving P_Λ. The same applies to effects {`(eff_stack_holding_v1)`, `(eff_stack_holding_v2)`, `(eff_stack_on_v1_v2)`, `(eff_stack_clear_v1)`, `(eff_stack_clear_v1)`,...}.
   Note that executing an insert action, e.g., `(insert_pre_stack_holding_v1)`, will add the corresponding model fluent `(pre_stack_holding_v1)` to the successor state. Hence, the execution of the insert actions of π_Λ yields a state containing the valuation of the model fluents that shape every $\xi \in \mathcal{M}$. For example, executing the insert actions that shape the action model $name(\xi) =$ `putdown` leads to a state containing the positive literals `(pre_putdown_holding_v1)`, `(eff_putdown_holding_v1)`, `(eff_putdown_clear_v1)`, `(eff_putdown_ontable_v1)`, `(eff_putdown_handempty)`.
2. Actions for *applying* the action models $\xi \in \mathcal{M}$ built by the insert actions and bounded to objects $\omega \subseteq \Omega^{ar(\xi)}$. These actions will be part of the postfix of the plan π_Λ and they determine the application of the learned action models

according to the values of the model fluents in the current state configuration. Since action headers are known, the variables $pars(\xi)$ are bounded to the objects in ω that appear in the same position.

$$\begin{aligned}
\mathsf{pre}(\mathsf{apply}_{\xi,\omega}) &= \{\},\\
\mathsf{cond}(\mathsf{apply}_{\xi,\omega}) &= \{pre_{p,\xi} \wedge eff_{p,\xi}\} \rhd \{\neg p(\omega)\}_{\forall p \in \Psi_\xi},\\
&\{\neg pre_{p,\xi} \wedge eff_{p,\xi}\} \rhd \{p(\omega)\}_{\forall p \in \Psi_\xi},\\
&\{pre_{p,\xi} \wedge \neg p(\omega)\}_{\forall p \in \Psi_\xi} \rhd \{invalid\},\\
&\{mode_{prog}\} \rhd \{\neg mode_{prog}\}.
\end{aligned}$$

Figure 1.4 shows the PDDL encoding of `(apply_stack)` for applying the action model of the *stack* operator. Let us assume the action `(apply_stack blockB blockA)` is in π_Λ. Executing this action in a state s implies activating the preconditions and effects of `(apply_stack)` according to the values of the model fluents in s. For example, if { `(pre_stack_holding_v1)`, `(pre_stack_clear_v2)`} $\subset s$, then it must be checked that positive literals `(holding blockB)` and `(clear`

```
(:action apply_stack
  :parameters (?o1 - object ?o2 - object)
  :precondition (and )
  :effect (and (when (and (pre_stack_on_v1_v1) (eff_stack_on_v1_v1)) (not (on ?o1 ?o1)))
               (when (and (pre_stack_on_v1_v2) (eff_stack_on_v1_v2)) (not (on ?o1 ?o2)))
               (when (and (pre_stack_on_v2_v1) (eff_stack_on_v2_v1)) (not (on ?o2 ?o1)))
               (when (and (pre_stack_on_v2_v2) (eff_stack_on_v2_v2)) (not (on ?o2 ?o2)))
               (when (and (pre_stack_ontable_v1) (eff_stack_ontable_v1)) (not (ontable ?o1)))
               (when (and (pre_stack_ontable_v2) (eff_stack_ontable_v2)) (not (ontable ?o2)))
               (when (and (pre_stack_clear_v1) (eff_stack_clear_v1)) (not (clear ?o1)))
               (when (and (pre_stack_clear_v2) (eff_stack_clear_v2)) (not (clear ?o2)))
               (when (and (pre_stack_holding_v1) (eff_stack_holding_v1)) (not (holding ?o1)))
               (when (and (pre_stack_holding_v2) (eff_stack_holding_v2)) (not (holding ?o2)))
               (when (and (pre_stack_handempty) (eff_stack_handempty)) (not (handempty)))
               (when (and (not (pre_stack_on_v1_v1)) (eff_stack_on_v1_v1)) (on ?o1 ?o1))
               (when (and (not (pre_stack_on_v1_v2)) (eff_stack_on_v1_v2)) (on ?o1 ?o2))
               (when (and (not (pre_stack_on_v2_v1)) (eff_stack_on_v2_v1)) (on ?o2 ?o1))
               (when (and (not (pre_stack_on_v2_v2)) (eff_stack_on_v2_v2)) (on ?o2 ?o2))
               (when (and (not (pre_stack_ontable_v1)) (eff_stack_ontable_v1)) (ontable ?o1))
               (when (and (not (pre_stack_ontable_v2)) (eff_stack_ontable_v2)) (ontable ?o2))
               (when (and (not (pre_stack_clear_v1)) (eff_stack_clear_v1)) (clear ?o1))
               (when (and (not (pre_stack_clear_v2)) (eff_stack_clear_v2)) (clear ?o2))
               (when (and (not (pre_stack_holding_v1)) (eff_stack_holding_v1)) (holding ?o1))
               (when (and (not (pre_stack_holding_v2)) (eff_stack_holding_v2)) (holding ?o2))
               (when (and (not (pre_stack_handempty)) (eff_stack_handempty)) (handempty))
               (when (and (pre_stack_on_v1_v1) (not (on ?o1 ?o1))) (invalid))
               (when (and (pre_stack_on_v1_v2) (not (on ?o1 ?o2))) (invalid))
               (when (and (pre_stack_on_v2_v1) (not (on ?o2 ?o1))) (invalid))
               (when (and (pre_stack_on_v2_v2) (not (on ?o2 ?o2))) (invalid))
               (when (and (pre_stack_ontable_v1) (not (ontable ?o1))) (invalid))
               (when (and (pre_stack_ontable_v2) (not (ontable ?o2))) (invalid))
               (when (and (pre_stack_clear_v1) (not (clear ?o1))) (invalid))
               (when (and (pre_stack_clear_v2) (not (clear ?o2))) (invalid))
               (when (and (pre_stack_holding_v1) (not (holding ?o1))) (invalid))
               (when (and (pre_stack_holding_v2) (not (holding ?o2))) (invalid))
               (when (and (pre_stack_handempty) (not (handempty))) (invalid))
               (when (modeProg) (not (modeProg)))))
```

Fig. 1.4 PDDL action for applying an already programmed model for *stack*

`blockA)` hold in s. Otherwise, a different set of precondition literals will be checked. The same applies to the conditional effects, generating the corresponding literals according to the values of the model fluents of s.
Note that executing an apply action, e.g., `(apply_stack blockB blockA)`, will add the literals `(on blockB blockA)`, `(clear blockB)`, `(not(clear blockA))`, `(handempty)`, and `(not(clear blockB))` to the successor state if $name(\xi) =$ `stack` has been correctly programmed by the insert actions. Hence, while **insert actions** add the values of the **model fluents** that shape ξ, the **apply actions** add the values of the **fluents of** F that result from the execution of ξ.
When the input plan trace contains observed actions extra preconditions have to be added to ensure that actions are applied in the same order as they appear in $\mathcal{O}$ [1].

3. Actions for *validating* partially observed states $s_j^o \in \mathcal{O}$. These actions are also part of the postfix of the solution plan π_Λ and they are aimed at checking that the observation $\mathcal{O}$ follows after the execution of the apply actions.

$$\mathsf{pre}(\mathsf{validate_j}) = s_j^o \cup \{test_{j-1}\},$$

$$\mathsf{cond}(\mathsf{validate_j}) = \{\emptyset\} \rhd \{\neg test_{j-1}, test_j\}.$$

There will be a validate action in π_Λ for every observed state in $\mathcal{O}$. The position of the validate actions in π_Λ will be determined by the planner by checking that the state resulting after the execution of an apply action comprises the observed state $s_j^o \in \mathcal{O}$.

In some contexts, it is reasonable to assume that some parts of the action model are known and so there is no need to learn the entire model from scratch [28]. In our compilation approach, when an action model ξ is partially specified, the known preconditions and effects are encoded as fluents $pre_{p,\xi}$ and $eff_{p,\xi}$ set to true in the initial state I_Λ. In this case, the corresponding insert actions, $\mathsf{insertPre_{p,\xi}}$ and $\mathsf{insertEff_{p,\xi}}$, become unnecessary making the classical planning task P_Λ easier to be solved.

So far we explained the compilation for learning from a single input trace. However, the compilation is extensible to the more general case $\Lambda = \langle \mathcal{M}, \mathcal{O}_1, \ldots, \mathcal{O}_k \rangle$ where there is an input set of k observations. Taking this into account, a small modification is required in our compilation approach. In particular, the actions in P_Λ for *validating* the last state $s_{m,t}^o \in \mathcal{O}_t,\ \ 1 \leq t \leq k$ of an observation $\mathcal{O}_t$ reset the current state. These actions are now redefined as follows:

$$\mathsf{pre}(\mathsf{validate_j}) = s_{m,t}^o \cup \{test_{j-1}\} \cup \{\neg mode_{prog}\},$$

$$\mathsf{cond}(\mathsf{validate_j}) = \{\emptyset\} \rhd \{\neg test_{j-1}, test_j\} \cup$$
$$\{\neg f\}_{\forall f \in F, f \notin s_{0,t+1}^o} \cup \{f\}_{\forall f \in s_{0,t+1}^o}.$$

Fig. 1.5 Plan for programming the *stack* action model and for validating the programmed *stack* action model with previously specified action models for *pickup*, *putdown*, and *unstack*

```
00 : (insert_pre_stack_holding_v1)
01 : (insert_pre_stack_clear_v2)
02 : (insert_eff_stack_clear_v1)
03 : (insert_eff_stack_clear_v2)
04 : (insert_eff_stack_handempty)
05 : (insert_eff_stack_holding_v1)
06 : (insert_eff_stack_on_v1_v2)
07 : (apply_unstack blockB blockA i1 i2)
08 : (apply_putdown blockB i2 i3)
09 : (apply_pickup blockA i3 i4)
10 : (apply_stack blockA blockB i4 i5)
11 : (validate_1)
```

Finally, we will detail the composition of a solution plan π_Λ to a planning task P_Λ and the mechanism to extract the action models of $\mathcal{M}'$ from π_Λ. The plan of Fig. 1.5 shows a solution to the task P_Λ that encodes a learning task $\Lambda = \langle \mathcal{M}, \mathcal{O} \rangle$ for obtaining the action models of the *blocksworld* domain, where the models for `pickup`, `putdown`, and `unstack` are already specified in $\mathcal{M}$. Therefore, the plan shows the insert actions and validate action for the action model `stack`. Plan steps 00–01 insert the preconditions of the `stack` model, steps 02–06 insert the action model effects, and steps 07–11 form the plan postfix that applies the action models (only the `stack` model is learned) and validates the result in the input observation.

Given a solution plan π_Λ that solves P_Λ, the set of action models $\mathcal{M}'$ that solves $\Lambda = \langle \mathcal{M}, \mathcal{O} \rangle$ learning task is computed in linear time and space. In order to do so, π_Λ is executed in the initial state I_Λ and the action model $\mathcal{M}'$ will be given by the fluents $pre_{p,\xi}$, and $eff_{p,\xi}$ that are set to true in the last state reached by π_Λ, $s_g = \theta(I_\Lambda, \pi_\Lambda)$. For each $\xi \in \mathcal{M}'$, we build the sets of preconditions, positive effects, and negative effects as follows:

$$pre(\xi) = \{p \mid pre_{p,\xi} \in s_g\}_{\forall p \in \Psi_\xi},$$

$$del(\xi) = \{p \mid pre_{p,\xi} \in s_g \wedge eff_{p,\xi} \in s_g\}_{\forall p \in \Psi_\xi},$$

$$add(\xi) = \{p \mid \neg pre_{p,\xi} \in s_g \wedge eff_{p,\xi} \in s_g\}_{\forall p \in \Psi_\xi}.$$

An optimally solved learning task will learn the minimum set of required preconditions; i.e., those that are at the same time negative effects. Optionally, it is possible to infer the maximum set of preconditions that is consistent with the observation and the learned model. This is done via a post-process based on the one proposed by the LOUGA system [15]. The intuition is going through every action counting the number of cases where a literal is present before the action is executed. If a literal is present in all the cases before the action, the literal is considered to be a precondition. This is done by traversing the actions/states found in the validation part of the solution plan π_Λ. For instance, in the example of Fig. 1.5, the used sequence of actions is `(unstack blockB blockA)`, `(put-down blockB)`, `(pick-up blockA)`, and `(stack blockA blockB)`.

4.2 Properties of the Compilation

Lemma 1 *Soundness. Any classical plan π that solves P_Λ produces a model $\mathcal{M}'$ that solves the $\Lambda = \langle \mathcal{M}, \mathcal{O} \rangle$ learning task.*

Proof According to the P_Λ compilation, once a given precondition or effect is inserted into the domain model $\mathcal{M}$ it cannot be undone. In addition, once an action model is applied it cannot be modified. In the compiled planning problem P_Λ, only $(\texttt{apply})_{\xi,\omega}$ actions can update the value of the state fluents F. This means that a state consistent with an observation s_m^o can only be achieved executing an applicable sequence of $(\texttt{apply})_{\xi,\omega}$ actions that, starting in the corresponding initial state s_0^o, validates that every generated intermediate state s_j $(0 < j \leq m)$, is consistent with the input state observations. This is exactly the definition of the solution condition for model $\mathcal{M}'$ to solve the $\Lambda = \langle \mathcal{M}, \mathcal{O} \rangle$ learning task.

Lemma 2 *Completeness. Any model $\mathcal{M}'$ that solves the $\Lambda = \langle \mathcal{M}, \mathcal{O} \rangle$ learning task can be computed with a classical plan π that solves P_Λ.*

Proof By definition $\mathcal{I}_{\xi,\Psi}$ fully captures the set of elements that can appear in an action model ξ using predicates Ψ. In addition the P_Λ compilation does not discard any model $\mathcal{M}'$ definable within $\mathcal{I}_{\xi,\Psi}$. This means that, for every model $\mathcal{M}'$ that solves the $\Lambda = \langle \mathcal{M}, \mathcal{O} \rangle$, we can build a plan π that solves P_Λ by selecting the appropriate $(\texttt{insertpre})_{\mathrm{p},,\xi}$ and $(\texttt{inserteff})_{\mathrm{p},\xi}$ actions for programming the precondition and effects of the corresponding action models in $\mathcal{M}'$ and then, selecting the corresponding $(\texttt{apply})_{\xi,\omega}$ actions that transform the initial state observation s_0^o into the final state observation s_m^o.

The size of the classical planning problem P_Λ depends on the arity of the predicates in Ψ, that shape variables F, and the number of parameters of the action models, $|pars(\xi)|$. The larger these arities, the larger $|\mathcal{I}_{\xi,\Psi}|$. The size of $\mathcal{I}_{\xi,\Psi}$ is the most dominant factor of the compilation because it defines the $pre_{p,\xi}/eff_{p,\xi}$ fluents, the corresponding set of `insert` actions, and the number of conditional effects in the $(\texttt{apply})_{\xi,\omega}$ actions. Note that *typing* can be used straightforward to constrain the FOL interpretations of Ψ over the parameters $pars(\xi)$, which will significantly reduce $|\mathcal{I}_{\xi,\Psi}|$ and hence the size of P_Λ output by the compilation.

Classical planners tend to prefer shorter solution plans, so our compilation may introduce a bias to $\Lambda = \langle \mathcal{M}, \mathcal{O} \rangle$ learning tasks preferring solutions that are referred to action models with a shorter number of preconditions/effects. In more detail, all $\{pre_{p,\xi}, eff_{p,\xi}\}_{\forall e \in \mathcal{I}_{\xi,\Psi}}$ fluents are false at the initial state of our P_Λ compilation so classical planners tend to solve P_Λ with plans that require a smaller number of `insert` actions.

This bias can be eliminated defining a cost function for the actions in P_Λ (e.g., `insert` actions have *zero cost*, while $(\texttt{apply})_{\xi,\omega}$ actions have a *positive constant cost*). In practice we use a different approach to disregard the cost of `insert` actions since classical planners are not proficient at optimizing plan cost with zero-cost actions. Instead, our approach is to use a SAT-based planner [21] that can apply all actions for inserting preconditions in a single planning step (these actions do not interact). Further, the actions for inserting action effects are also applied in another single planning step. The plan horizon for programming any action model is then always bounded to 2. The SAT-based planning approach is also convenient for its ability to deal with planning problems populated with dead-ends and because symmetries in the insertion of preconditions/effects into an action model do not affect the planning performance.

An interesting aspect of our approach is that when a *fully* or *partially specified* STRIPS action model $\mathcal{M}$ is given in Λ, the P_Λ compilation also serves to validate whether the observation $\mathcal{O}$ follows the given model $\mathcal{M}$:

- $\mathcal{M}$ is proved to be a *valid* action model for the given input data $\mathcal{O}$ iff a solution plan for P_Λ can be found.
- $\mathcal{M}$ is proved to be a *invalid* action model for the given input data $\mathcal{O}$ iff P_Λ is unsolvable. This means that $\mathcal{M}$ cannot be consistent with the given observation of the plan execution.

This validation capacity of our compilation is beyond the functionality of VAL (the plan validation tool [12]) because our P_Λ compilation is able to address *model validation* of a partial (or even an empty) action model with a partially observed plan trace. VAL, however, requires a full plan and a full action model for plan validation.

5 Experimental Results

We have tested the proposed explanation-based learning approach in 12 IPC domains that satisfy the STRIPS requirement [7], taken from the PLANNING.DOMAINS repository [18]. In our experiments, we use a set of 5 observations of length 5–7 as learning examples. Each observation corresponds to plan executions generated via random walks. All experiments are run on an Intel Core i5 3.10 GHz $\times$ 4 with 16 GB of RAM.

The learned models are evaluated using the *precision* and *recall* metrics for action models proposed in [1], which compare the learned models against the reference model. Precision measures the correctness of the learned models, while recall measures their completeness. Formally

$$Precision = \frac{tp}{tp + fp}$$

$$Recall = \frac{tp}{tp + fn},$$

where tp (*true positives*) is the number of predicates that appear in both the learned and reference action models, fp (*false positives*) is the number of predicates that appear in the learned action model but not in the reference model, and fn (*false negatives*) is the number of predicates that should appear in the learned action model but are missing.

Table 1.1 Precision and recall scores for learning tasks from labeled plans

	Pre		Add		Del		Global	
	P	R	P	R	P	R	P	R
Blocks	1.0	1.0	1.0	1.0	1.0	1.0	1.0	1.0
Driverlog	0.9	0.64	0.56	0.71	0.86	0.86	0.78	0.73
Ferry	1.0	0.57	1.0	1.0	1.0	1.0	1.0	0.86
Floortile	0.68	0.68	0.89	0.73	1.0	0.82	0.86	0.74
Grid	0.79	0.65	1.0	0.86	0.88	1.0	0.89	0.83
Gripper	1.0	0.67	1.0	1.0	1.0	1.0	1.0	0.89
Hanoi	0.75	0.75	1.0	1.0	1.0	1.0	0.92	0.92
Miconic	0.89	0.89	1.0	0.75	0.75	1.0	0.88	0.88
Satellite	0.82	0.64	1.0	1.0	1.0	0.75	0.94	0.80
Transport	1.0	0.70	0.83	1.0	1.0	0.80	0.94	0.83
Visitall	1.0	1.0	1.0	1.0	1.0	1.0	1.0	1.0
Zenotravel	1.0	0.64	0.88	1.0	1.0	0.71	0.96	0.79
	0.90	0.74	0.93	0.92	0.96	0.91	0.93	0.86

5.1 *Learning from Labeled Plans*

For our first experiment, we use the setting typically followed by most approaches, that is, learning from observations consisting of initial and final states, and the full sequence of actions between these two. In this setting, the number of trajectories that explain a given observation is bounded by the length of the observation and further constrained by the observed sequence of actions.

The results of this experiment are compiled in Table 1.1. Precision (**P**) and recall (**R**) are computed separately for the preconditions (**Pre**), positive effects (**Add**), and negative effects (**Del**), while the last two columns and the last row report average scores. The table show high scores across all domains, with an average precision of 0.93 and average recall of 0.86. Recall is noticeably lower for preconditions at 0.74, which is to be expected given that any relaxation on the preconditions of the reference model will still be able to generate an explanation for the observation.

5.2 *Learning from Initial/Final State Pairs*

Now, we evaluate our approach when observations are reduced to their minimal expression $\mathcal{O} = \langle s_0^o, s_m^o \rangle$; i.e., only the initial and final states are observed. In contrast to the previous experiment, this setting presents an unbounded number of trajectories consistent with the observation. Moreover, the planner must determine how many "gaps" need to be filled between the two observed states.

Table 1.2 summarizes the results obtained for this experiment. Values for the *Zenotravel* and *Grid* domains are not reported because no solutions were found

Table 1.2 Precision and recall scores for learning tasks from initial and final states

	Pre		Add		Del		Global	
	P	R	P	R	P	R	P	R
Blocks	0.75	0.67	0.86	0.67	0.86	0.67	0.82	0.67
Driverlog	1	0.29	0.5	0.71	0.67	0.29	0.72	0.43
Ferry	1	0.57	1	1	1	1	1	0.86
Floortile	0.57	0.36	1	0.64	0.67	0.36	0.75	0.45
Grid	–	–	–	–	–	–	–	–
Gripper	1	0.67	1	1	1	1	1	0.89
Hanoi	1	0.5	1	1	1	1	1	0.83
Miconic	0.5	0.11	0.67	0.5	0.5	0.33	0.56	0.31
Satellite	0.5	0.21	0.57	0.8	0.75	0.75	0.61	0.59
Transport	1	0.3	0.71	1	1	0.6	0.9	0.63
Visitall	–	–	–	–	–	–	–	–
Zenotravel	1	0.29	0.57	0.57	1	0.57	0.86	0.48
	0.83	0.4	0.79	0.79	0.85	0.66	0.82	0.61

under the given timeout of 1000 s. Although the learned models are able to produce explanations for the input observations, we can see that the values of precision and recall are significantly lower than in Table 1.1. This is indicative that the learned models are now considerably different from the reference ones, which is caused by the larger solution space originated from the removal of some observation constraints.

6 Conclusions

We presented a classical planning compilation for learning STRIPS action models from partial observations of plan executions. To the best of our knowledge, this is the first approach on learning action models that is exhaustively evaluated over a wide range of domains and uses exclusively an *off-the-shelf* classical planner. The work in [26] proposes a planning compilation for learning action models from plan traces following the *finite domain* representation for the state variables. This is a theoretical study on the boundaries of the learned models and no experimental results are reported.

When example plans are available, we can compute accurate action models from small sets of learning examples (five examples per domain) in little computation time (less than a second). When action plans are not available, our approach still produces action models that are compliant with the input information. In this case, since learning is not constrained by actions, operators can be reformulated changing their semantics, in which case the comparison with a reference model turns out to be tricky.

An interesting research direction related to this issue is *domain reformulation* to use actions in a more efficient way, reduce the set of actions identifying dispensable information or exploiting features that allow more compact solutions like the *reachable* or *movable* features in the *Sokoban* domain [13].

Generating *informative* examples for learning planning action models is still an open issue. Planning actions include preconditions that are only satisfied by specific sequences of actions which have low probability of being chosen by chance [5]. The success of recent algorithms for exploring planning tasks [8] motivates the development of novel techniques that enable to autonomously collect informative learning examples. The combination of such exploration techniques with our learning approach is an appealing research direction that opens up the door to the bootstrapping of planning action models.

Acknowledgements This work is supported by the Spanish MINECO project TIN2017-88476-C2-1-R. Diego Aineto is partially supported by the *FPU16/03184* and Sergio Jiménez by the *RYC15/18009*, both programs funded by the Spanish government.

References

1. Diego Aineto, Sergio Jiménez, and Eva Onaindia. Learning STRIPS action models with classical planning. In *International Conference on Automated Planning and Scheduling, (ICAPS-18)*, pages 399–407, 2018.
2. Eyal Amir and Allen Chang. Learning partially observable deterministic action models. *Journal of Artificial Intelligence Research*, 33:349–402, 2008.
3. Blai Bonet, Héctor Palacios, and Héctor Geffner. Automatic Derivation of Memoryless Policies and Finite-State Controllers Using Classical Planners. In *International Conference on Automated Planning and Scheduling, (ICAPS-09)*. AAAI Press, 2009.
4. Stephen N Cresswell, Thomas L McCluskey, and Margaret M West. Acquiring planning domain models using LOCM. *The Knowledge Engineering Review*, 28(02):195–213, 2013.
5. Alan Fern, Sung Wook Yoon, and Robert Givan. Learning Domain-Specific Control Knowledge from Random Walks. In *International Conference on Automated Planning and Scheduling, ICAPS-04*, pages 191–199. AAAI Press, 2004.
6. Richard E Fikes and Nils J Nilsson. Strips: A new approach to the application of theorem proving to problem solving. *Artificial Intelligence*, 2(3–4):189–208, 1971.
7. Maria Fox and Derek Long. PDDL2.1: An extension to PDDL for expressing temporal planning domains. *Journal of Artificial Intelligence Research*, 20:61–124, 2003.
8. Guillem Francès, Miquel Ramírez, Nir Lipovetzky, and Hector Geffner. Purely declarative action descriptions are overrated: Classical planning with simulators. In *International Joint Conference on Artificial Intelligence, (IJCAI-17)*, pages 4294–4301, 2017.
9. Hector Geffner and Blai Bonet. *A Concise Introduction to Models and Methods for Automated Planning*. Synthesis Lectures on Artificial Intelligence and Machine Learning. Morgan & Claypool Publishers, 2013.
10. Malik Ghallab, Dana Nau, and Paolo Traverso. *Automated Planning: theory and practice*. Elsevier, 2004.
11. Jörg Hoffmann, Julie Porteous, and Laura Sebastia. Ordered landmarks in planning. *Journal of Artificial Intelligence Research*, 22:215–278, 2004.
12. Richard Howey, Derek Long, and Maria Fox. VAL: Automatic plan validation, continuous effects and mixed initiative planning using PDDL. In *Tools with Artificial Intelligence, 2004. ICTAI 2004. 16th IEEE International Conference on*, pages 294–301. IEEE, 2004.

13. Franc Ivankovic and Patrik Haslum. Optimal planning with axioms. In *International Joint Conference on Artificial Intelligence, (IJCAI-15)*, pages 1580–1586, 2015.
14. Subbarao Kambhampati. Model-lite planning for the web age masses: The challenges of planning with incomplete and evolving domain models. In *National Conference on Artificial Intelligence, (AAAI-07)*, 2007.
15. Jirí Kucera and Roman Barták. LOUGA: learning planning operators using genetic algorithms. In *Pacific Rim Knowledge Acquisition Workshop, PKAW-18*, pages 124–138, 2018.
16. Drew McDermott, Malik Ghallab, Adele Howe, Craig Knoblock, Ashwin Ram, Manuela Veloso, Daniel Weld, and David Wilkins. PDDL—The Planning Domain Definition Language, 1998.
17. Ryszard S Michalski, Jaime G Carbonell, and Tom M Mitchell. *Machine learning: An artificial intelligence approach*. Springer Science & Business Media, 2013.
18. Christian Muise. Planning.domains. *ICAPS system demonstration*, 2016.
19. Miquel Ramírez and Hector Geffner. Plan recognition as planning. In *International Joint conference on Artificial Intelligence, (IJCAI-09)*, pages 1778–1783. AAAI Press, 2009.
20. Miquel Ramírez. *Plan recognition as planning*. PhD thesis, Universitat Pompeu Fabra, 2012.
21. Jussi Rintanen. Madagascar: Scalable planning with SAT. In *International Planning Competition, (IPC-2014)*, 2014.
22. Javier Segovia-Aguas, Sergio Jiménez, and Anders Jonsson. Generating Context-Free Grammars using Classical Planning. In *International Joint Conference on Artificial Intelligence, (IJCAI-17)*, pages 4391–4397. AAAI Press, 2017.
23. Javier Segovia-Aguas, Sergio Jiménez, and Anders Jonsson. Computing hierarchical finite state controllers with classical planning. *Journal of Artificial Intelligence Research*, 62:755–797, 2018.
24. Javier Segovia-Aguas, Sergio Jiménez Celorrio, and Anders Jonsson. Computing programs for generalized planning using a classical planner. *Artificial Intelligence*, 2019.
25. John Slaney and Sylvie Thiébaux. Blocks world revisited. *Artificial Intelligence*, 125(1–2):119–153, 2001.
26. Roni Stern and Brendan Juba. Efficient, safe, and probably approximately complete learning of action models. In *International Joint Conference on Artificial Intelligence, (IJCAI-17)*, pages 4405–4411, 2017.
27. Qiang Yang, Kangheng Wu, and Yunfei Jiang. Learning action models from plan examples using weighted MAX-SAT. *Artificial Intelligence*, 171(2–3):107–143, 2007.
28. Hankz Hankui Zhuo, Tuan Anh Nguyen, and Subbarao Kambhampati. Refining incomplete planning domain models through plan traces. In *International Joint Conference on Artificial Intelligence, IJCAI-13*, pages 2451–2458, 2013.

Chapter 2
Automated Domain Model Learning Tools for Planning

Rabia Jilani

Abstract Intelligent agents solving problems in the real world require domain models containing widespread knowledge of the world. Domain models can be encoded by human experts or automatically learned through the observation of some existing plans (behaviours). Encoding a domain model manually from experience and intuition is a very complex and time-consuming task, even for domain experts. This chapter investigates various classical and state-of-the-art methods proposed by the researchers to attain the ability of automatic learning of domain models from training data. This concerns with the learning and representation of knowledge about the operator schema, discrete or continuous resources, processes and events involved in the planning domain model. The taxonomy and order of these methods we followed are based on their standing and frequency of usage in the past research. Our intended contribution in this chapter is to provide a broader perspective on the range of techniques in the domain model learning area which underpin the developmental decisions of the learning tools.

1 Introduction

Automated planning is one of the most prominent AI challenges. It is the process of finding a procedural course of action through explicit deliberation process to reach a pre-stated objective in the form of goals while optimizing overall performance. Planning is a pivotal task that has to be performed by autonomous agents. The planning community is uplifting planning systems from small problems to capture more complex domains that closely reflect real life applications (e.g., planning space missions, fire extinction management and operation of underwater vehicles)—a way to satisfy the aims of autonomic systems.

R. Jilani (✉)
School of Computing and Engineering, University of Huddersfield, Huddersfield, UK
e-mail: R.Jilani@hud.ac.uk

M. Vallati, D. Kitchin (eds.), *Knowledge Engineering Tools and Techniques for AI Planning*, https://doi.org/10.1007/978-3-030-38561-3_2

In order to perform automated reasoning, planning techniques require formal specification of the application knowledge to be encoded in the form of domain models. In the action-centred view of problem representation, a domain model encodes the domain knowledge in the form of actions that can be executed together with relevant action properties and features. A domain model typically includes the action description, the objects involved in actions, a set of logical state space axioms along with the rules of inference and heuristics to accurately define the operators' description of some real-world domain. It includes both a dynamic and a static object-type hierarchy, constant objects (if any) and the declaration of predicates and functions (for hierarchical domains). In short, it is a declarative depiction of domain world functionalities.

In a centralised approach, this domain model is represented as a knowledge base and automated logical reasoning could be used to determine acts in plans. To generate plans, planning engines search the action descriptions in the provided domain model to achieve the goals. Figure 2.1 shows a typical view of plan generation in AI planning. One of the key factors for the correctness of the planner outcome is the quality of the domain knowledge that otherwise can prove catastrophic.

In the complex domain scenarios, planners also use manually encoded or automatically learned domain-specific control knowledge (in addition to domain model) to guide planner search and cater to scalability issues. Most planners define control knowledge separately from domain model to support different representations. This chapter only focuses on the domain model learning aspect.

Synthesising domain models for planning from scratch by hand is time intense, error-prone and challenging. Knowledge engineering for planning domain models using machine learning (ML) techniques is considered as a paramount for empowering the autonomous learning systems with the capacity to fill implicit human knowledge gaps and errors, requiring least human intervention in domain model development. It not only involves acquisition of a new knowledge from the environment but also the refinement of the already available knowledge of the domain under consideration.

As a result of a planning process the successfully generated plans can be used as the solutions to the desired problems in self-learning systems to enable autonomic properties. The area of ML application to domain model learning systems has received active research attention in recent years but did not make as much stride as the learning of control knowledge.

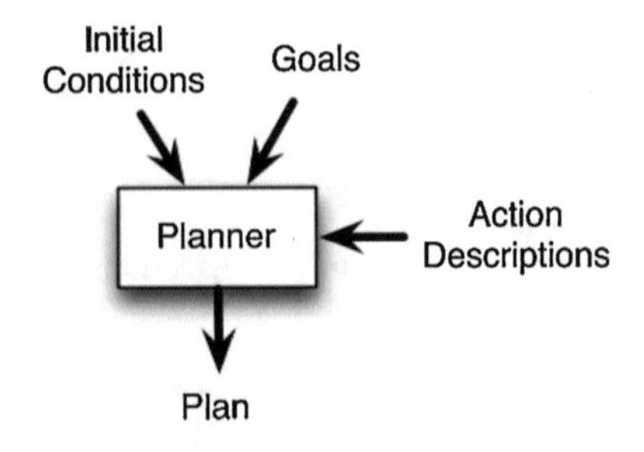

Fig. 2.1 Automated planning as an independent component

1.1 *Knowledge Representation for Knowledge Engineering of Domain Models*

Knowledge representation (KR) is to encoding human knowledge in the form of symbols which can be processed by a computer to obtain intelligent behaviour. Automated reasoning and KR stay in close association with each other as the core aim of the explicit KR is to enable reasoning and inference process. To represent complex problems, KR uses declarative programming for expressing the logic of the computation and behavioural description as compared to describing the control flow as in procedural programming.

In order to effectively engage in the intelligent behaviours, the key attribute of the theories of autonomous agent relies on the agent's internal representation of intelligence. This must be an implicit representation of a domain model to conduct reasoning process. From the point of view of knowledge engineering of this intelligent behaviour, the question to ask is, what KR formalism an agent needs to know to behave intelligently and which computational mechanisms are needed for manipulation of its knowledge, i.e., description of notions, facts, and rules of the world. The knowledge engineering of a domain model is the engineering of a set of sentences/axioms. These sentences are expressed in some KR language which the agent uses to do inference.

For automated reasoning, expressibility and practicability are generally the two main considerations for the KR language of the domain. Riddle et al. [59] empirically proved that a used representation mechanism makes an extensive difference to the planner's ability to solve a problem by exploring six different representations of the blocks world domain. In Brachman and Levesque [7] the authors argue that the expressive power of the representation language is directly proportional to the computational complexity of reasoning with it. In other words, more expressive language makes the reasoning process difficult. The authors demonstrated this by analysing the frame language which later led to an extensive study of the argument put forward by the author in order to search the optimal trade-off.

In automated planning, the key purpose of an explicit KR language for a domain model formulation is for a planner to be able to reason with it and infer new knowledge from it in the form of plans by predicting action outcomes to display some rational behaviour in an explicit way. The essential part of a domain model in the process of reasoning is the representation of the set of actions that a planner can reason with and the elements that include dynamics of the environment that support the specification of actions. It has long been recognised that there can be a variety of encodings and exploitable languages for a domain model formulation. However, the open question is which of these is the best? The choice of encoding language for KR partially depends on the requirements of the planning application itself.

There is no one KR approach just like the reasoning approach that has combined properties for all types and level of deliberation problems. Similarly, there is no single highly specialized KR mechanism to cover a specialized area of learning domain model.

A well-chosen representation language should explicitly model every action effect the system might confront. In addition to that, a domain modelling representation language should have some salient attributes. It should have supporting tools to check its operation and have logically strong inference mechanism to carry out the reasoning. It should be sufficiently expressive to explicitly model complex scenario of the real world. Moreover, it should be customizable and structured to capture every action effect the system might confront in operator definitions. In addition, it should have clear syntax and semantics to support operational aspects of the model.

2 Domain Model Learning Techniques and Tools

Both knowledge acquisition and learning for AI planning systems are essential to improve their effectiveness and to expand the application focus in practice. Most of the literature on learning for AI planning is based upon classical planning and concentrates on the learning of search control rules. For producing a domain model, a general process includes the study of planning application requirements, creating a model that explains the domain and testing it with suitable planning engines. Domain models can be encoded by the human experts or automatically learned through the observation of some existing plans (behaviours). Encoding a domain model manually from experience and intuition is a very complex and time-consuming task, even for domain experts.

Regarding the significance of automatic domain model learning system, the question that arises is why do we need a learning mechanism to learn from data when we can write a program to fulfil the purpose. The significance of learning mechanism becomes apparent when the same program parameters do not fit the new data or when the learning requirements or assumptions of the same data change slightly. The same hard-coded program needs extensive changes to align with the new requirements.

Machine learning is a broad area with a wide variety of sub-fields. It includes various methods starting from sub-symbolic methods like neural networks to high-level symbolic methods like inductive logic programming. Various approaches and techniques have been used by the researchers for the domain model learning task. Inductive learning is the most common technique to expedite learning solutions that are used in the field of supervised learning. Another less common technique outside the scope of supervised learning is the model-based reinforcement learning algorithms that learn model parameters such as probabilities and rewards, but no algorithm yet can produce states and models from observations and action sequences.

For the sake of succinctness, this section describes the commonly used domain model learning techniques in the literature. These techniques are complete in their own capacity and differ from each other in multiple perspectives including the amount and nature of input required, the extent of learning that takes place in the output, environmental characteristics these can work in, etc. Some of these characteristics are discussed in Sect. 4.

2.1 *Inductive Learning*

Inductive methods produce general rules by searching statistical correlations and consistencies in the large set of input training data. The main theory and method behind supervised learning is the inductive learning (IL—Fig. 2.2). Inductive learning can be defined as learning by inferring generalised rules from the training data given in the form of input–output example *pairs* $\mathbf{P(x_i, y_i)}$. By the input–output pairs, we mean the input samples or examples ($\mathbf{x_i}$) and the relevant output observations or the external feedback to the learning system ($\mathbf{y_i = f(x_i)}$). External feedback or output observations are the function of the input samples. The input–output example pairs generally establish the intended relation of input and output values. In simple words, IL is also referred to as learning from examples for function induction. The input–output example pairs can be generated by another system, produced by an instructor or a human expert or could be the traces of expert's behaviour. These do not necessarily need to be numbers and can contain either continuous or discrete values in the form of logical sentences. In the pairs, states are represented by the set of features, i.e., by factored representation.

The *main IL problem* is to generalise the input-to-output mapping *candidate function* or *hypothesis* (h) that satisfies input data so it can estimate the *target function* f [85]. *Hypothesis space* $H = (h_1, h_2, h_3, \ldots h_n)$ is a set of all possible approximations of target function that can exist. A further subset of hypothesis space (H) which is consistent with the given input data is called version space. Extending a hypothesis with every example or in other words generalising a hypothesis which should be in close approximation with the target function f is not an easy task especially when hypothesis space is complex.

A hypothesis (h) is tested for its consistency and correctness of generalised results by using extensive example test set—a set distinct from the example training set $\mathbf{E_T}$. Formally, using inductive learning algorithm **IL**, the problem is to find a hypothesis h which is consistent with the set of example pairs $\mathbf{P(x_i, y_i)}$, such that:

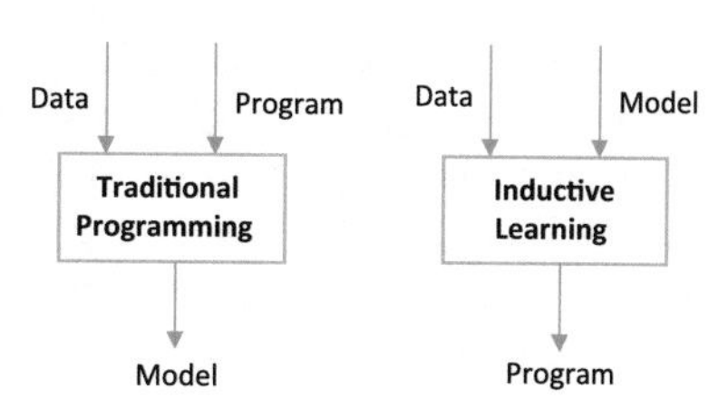

Fig. 2.2 Traditional programming vs. inductive learning

$$\mathbf{IL}\ \Lambda\ \mathbf{P}\,(\mathbf{x_i}, \mathbf{y_i}) \succ \mathbf{h} \qquad \text{(Inductive Learning)}$$

where $\mathbf{y_i} = h(\mathbf{x_i})$ for all P and $\succ$ represents inductive inference. For h to generalise and pass the consistency test, it needs to produce the results like the true target function f by agreeing to the sufficiently large example test set. On getting multiple consistent h in version space, the preference is always given to the simplest h that requires the least speculation and agrees with the data (Ockham's razor principle).

A good example of IL is a classification decision tree algorithm for building prediction models to predict categories from the training dataset. The algorithm has to find a most appropriate tree structure (h) which is consistent with the input data, out of all the available trees with variable attributes in its hypothesis space (H). A decision tree structures like a flow chart where the learning takes place by labelling the variables at nodes and branches of the tree from the training data. Models in which state variables represented in the form of trees can take discrete values are called classification trees. Similarly, trees where state variables can take continuous values are called regression trees. Inducing domain model and its features such as action duration for temporal domains have been well-studied using predictive modelling approaches of a relational decision and regression trees. Inducing regression trees is itself a well-known method for building models for numeric variables.

Artificial neural networks (ANN), reinforcement learning (RL), Bayesian learning (BL) and inductive logic programming (ILP) are some of the most common inductive learning techniques. More common early contributions of these IL techniques since their inception are to learn control knowledge to speed up the classical planning process. An extensive five-dimensional survey of learning-in-planning early work is provided in Zimmerman and Kambhampati [85].

From the domain learning perspective, each of the IL technique produces different types of hypothesis and models based on its expressivity, e.g., BL is used to learn models that allow probabilistic predictions while RL can learn in dynamic and stochastic environmental conditions with no requirement of prior knowledge of transition probabilities. More expressive representation of the learned knowledge which is in close approximation to the target function requires more training examples to narrow down the hypothesis space. Similarly, every technique has its own strengths and weaknesses, e.g., decision trees, ANN and ILP are robust to noisy inputs while RL learns optimal policies even from non-optimal input data.

A number of *properties* that need consideration for developing IL problem include:

- The probability that the training is going to be successful in learning the model. It mainly depends on the quality of training data and also on the inductive bias (discussed ahead in KBIL). Noisy set of training data leads to poor inconsistent models if not given alternate guidance.
- How much input training data should be provided for consistent h to converge? According to computational learning theory, only a few algorithms exhibit this

knowledge through the learning curve with the increase in input examples (PAC learning algorithms).

- Criteria for selection of training data and what presentation medium should be used to learn from available data, i.e., instructions, images, sensory information, environmental perceptions, etc.
- What output examples need to be provided to learn the target function *f* and how it affects the extent and quality of the learnt model in the outcome.
- Estimation of the overall complexity of hypothesis space and the complexity of hypothesis in the space. Complex hypotheses are more susceptible to overfitting.
- What should be the acceptable approximation and estimation error between a consistent hypothesis *h* and a target function *f*.
- What is the size and nature (deterministic or non-deterministic) of the hypothesis space?
- Depending on the nature of the problem, which learning algorithm (e.g., inductive logic programming, Bayesian learning, etc.) and approach to learning will be used (online or batch).
- Approach to use for building hypothesis towards the target function *f*. It could be directly computing the required information to learn the target function, searching for hypothesis from the hypothesis space or the gradual incremental construction of the hypothesis.
- Representation mechanism used to represent learned knowledge. This can include support vector machines, decision trees, graphical models, Bayes networks, finite state machines, logic statements, etc.
- On what time scale learning occurs: eager learners are more common and perform the task up front while lazy learners only learn when needed and are rarely used in machine learning.

Some of the *prominent issues* about the IL that researchers are trying to answer for over a decade include:

- What measures the good hypothesis space and the correctness of the hypothesis *h* if the true function *f* is not known?
- What factors can help reach the trade-off between the complexity of finding consistent *h* and the expressiveness of the hypothesis space? Can we even find *h* in a complex hypothesis space?
- What features and factors build the confidence in the correctness of the output model?
- How to find out computationally complex or intractable problems?

2.1.1 When to Use Inductive Learning

Inductive learning can be used to enhance any essential module or element of the system. It can be used in a variety of situations, some of them include:

1. When the underlying knowledge base or domain model fluctuates frequently and the occurring changes are complex enough that cannot be handled by human efforts every single time in changes, e.g., continuous domains like urban road traffic management, stock market, etc.
2. Situations where the learning only happens with experience and it is not possible to induce learning with a set of instructions, e.g., intelligent self-driving cars [53] where the system requires enormous training data in the form of camera visuals and corresponding steering movements to produce the general rule of driving.
3. Another condition arises when there is no human guidance available to create a reliable domain model, the agent should be intelligent enough to induce the domain model through learning. Building the knowledge base of planetary rovers could be a good example where the lack of reliable, explicit and a priori knowledge can be supported by inductive learning capability of the agent.
4. Last but not the least, situations, where each area of experimentation requires a unique underlying domain model, e.g., all the benchmark domains in International Planning Competition (IPC), require unique domain models and relevant problems to test the efficiency and effectiveness of planning systems.

The strength of the exploited learning approach or algorithm is the key factor to regulate the extent of learning by the system, as it may get stuck in local minima or not be able to capture patterns of the target knowledge within a reasonable time and memory requirements [27]. For example, exploiting reinforcement learning method to learn from a reward-based approach can learn better in a stochastic environment as compared to the inductive learning (which is based on drawing from inference). Similarly, learning for conformant or contingent planning task, the suitable learning approach to adopt is by inference or by inductive generalization to find the best fit for the observed facts. The concept of model-lite [30] planning views a planning problem as an MPE (most plausible explanation) problem. These techniques search for solution plans that are most plausible according to the current domain model, specifically for situations where the first bottleneck is getting the domain model at any level of completeness.

2.2 Knowledge-Based Inductive Learning (KBIL)

Humans learn knowledge with a sequence of experiences and also by reflection on past experiences to facilitate current learning. The challenge for autonomous learning agents is the lack of training examples to reflect on and build the hypothesis. In many learning systems developed since the 1980s, the issue of lacking training data has been covered by the notion of inductive bias (IB) [41]. Mitchell defines inductive bias as the constraint on the hypothesis space (H) of a learning system in addition to the requirement of consistency with the training examples. It is when a learning system prefers one hypothesis over others in hypothesis space. Inductive bias significantly assists the optimal convergence of target function particularly in case of scarce or incorrect training data. This works especially where the new knowledge that needs learning happens with the same set of examples which learned the knowledge earlier.

One kind of inductive bias is the use of background knowledge (B_K) [5] of the domain theory that explains the input training data. In other words, the agent should already know something about the domain it is going to formally induce in the form of accumulated information. Learners use this background knowledge to distinguish useful features from training examples. Learning the background knowledge is itself learning that an agent has to do and is known as a cumulative or incremental learning process. The background knowledge in some cases can be acquired by the relevant domain experts.

KBIL is one of the good examples of inductive learning (IL) which supports the notion of cumulative learning (Fig. 2.3). To reduce the hypothesis space, KBIL induces hypothesis (**h**) with the inductive bias in the form of background knowledge ($\mathbf{B_K}$) and the training examples ($\mathbf{E_T}$), such that:

$$\mathbf{IL} \wedge \mathbf{B_K} \wedge \mathbf{P}(\mathbf{x_i}, \mathbf{y_i}) \succ \mathbf{h} \qquad \text{(KBIL)}$$

where IL is the inductive learning algorithm. The resulting hypothesis h should explain both the background knowledge and training examples. In KBIL, $\mathbf{B_K}$, **h** and $\mathbf{E_T}$ are represented as a set of clauses or as a logical program with predicates (first-order literals) representing the attributes in them. New knowledge learnt for the incremental construction of hypothesis is exploited to improve the background domain knowledge as well. This process is referred to in the literature as the constructive inductive learning [38].

In most of the domain model learning systems, the fundamental motivation models have to solve is model-based planning tasks. One of the prominent inductive

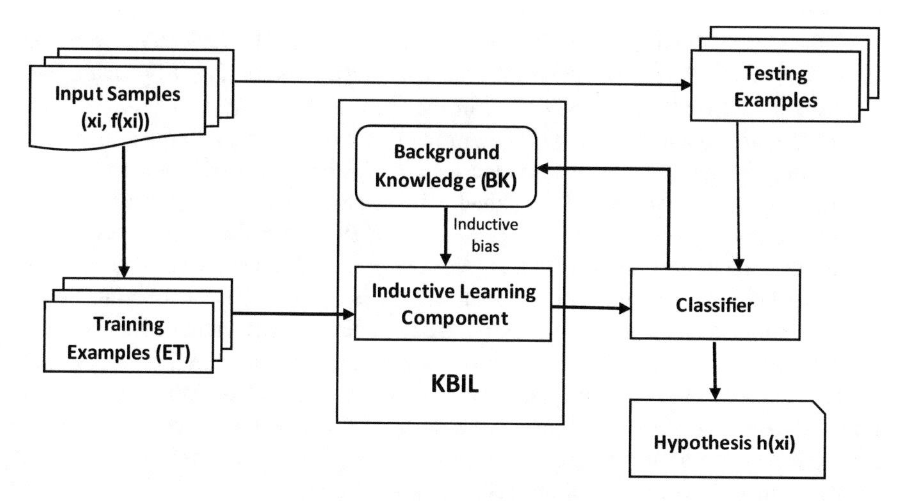

Fig. 2.3 A knowledge-based inductive learning system structure

learning systems is the *LOCM* system [12, 13]. LOCM uses an object-centred representation and performs automated induction of the dynamic aspects of a domain model on FSM representation of object sorts. Each object sort contains objects of the same type that behave in the same way. LOCM requires only a set of fully observable plan traces as the training data with no requirement of background knowledge about the domain. The main assumption which the LOCM relies on is that all objects in the domain go through transitions. This assumption is too strong for some scenarios especially when the domain contains static aspects too (as static constraints are not reflected in the plan traces). Based on this, a drawback of the LOCM process is that it can only induce a domain model which represents the dynamic aspects of objects and not the static aspects. This is problematic since most domains require static predicates to both restrict the number of possible actions and correctly encode real-world constraints.

This LOCM drawback is overcome by the *ASCoL* system [26], an inductive system that exploits graph analysis method to automatically identifying static relations, in order to enhance planning domain models. It uses the same set of input as used by LOCM. *LOP* [21] addresses the same problem of missing static facts in the learned domain model by using optimal goal-oriented plans. LOP compares the optimal input plans with the optimal plans found by using the extended domain model. If the latter is shorter, then some static relations are deemed to be missing.

LOCM2 [86] is an extension of LOCM with the provision of multiple parameterised FSMs to represent each object's separate behavioural aspects. LOCM2 extends the coverage of domains and the captured domain semantics. *NLOCM* [22] extends LOCM to generate fixed action cost numeric domains. It exploits the constraint programming approach to add numeric weights to the states and transitions of the FSMs produced by LOCM. The LOCM family of algorithms is distinct in that it induces the fluents without any additional input support alongside plan traces, i.e., a sequence of actions with no initial, intermediate and goal states mentioned.

Some other inductive leaning systems include *ARMS* [72], *SLAF* [63], *Opmaker* [58], *Opmaker2* [36], *RIMS* [84] and *LSO-NIO* [48]. Zhuo et al. [78] provide two extensions of their work ARMS [72], in the form of a new system *LAMP* which learns domain models written in PDDL, or in other words in terms of quantifiers and logical implications. The other extended system of ARMS learns domain models for hierarchical task networks (HTN), called *HTN-Learner*. *LAMMAS* (learning action models for multi-agent systems) [80] extends ARMS for a multi-agent environment using the same underlying method. *HTNLearn* [81] is another system that induces HTN methods and action models by using as input a collection of plans with partially annotated intermediate state information and a set of annotated tasks.

Hoffmann et al. [23] induce the business process models (BPM) using logs of actions recorded from real life business activity execution. The model then turns into the workflow. The main aim of the developed systems is to use the process mining technique to exploit the sequence of events. The process mining algorithm induces a model in the form of graphs such as Petri Nets.

Framer [34] induces domain models from natural language action description. It uses an estimate of functional similarity, so sentences that describe similar behaviours are represented by the same planning operator. After obtaining planning operators structure, Framer induces formal domain model by using LOCM. Martínez et al. [35] induce a probabilistic relational model including action and exogenous effects. It uses a set of completely observable state transitions as input to learning multiple candidate models using LFIT [25] system. It then uses an optimiser to select the best model out of all the candidates. *LFIT* is a KBIL system that induces a set of propositional rules by understanding the given input transitions in the form of interpretations. In order to learn transition rules of cellular automata, LFIT exploits rules as the background knowledge and conditions on rules as inductive bias.

FAMA [3] induces STRIPS domain model from the observations of plan executions. It demonstrates the ability to learn from partial or totally unobservable actions in plan executions which makes FAMA appropriate to learn from sensory inputs. The least amount of input FAMA requires is the initial and final states of the plan execution along with partial domain model. FAMA also presents two novel model-semantics evaluation metrics that build upon two recognised metrics, precision and recall [14] to evaluate the learned action models with respect to observations of plan executions.

PlanMiner-O2 [62] is an algorithm that uses a classification algorithm, based on inductive rule learning techniques, to learn action models with discrete numerical values (represented as action costs) from incomplete and noisy data. In accepts plan traces with intermediate partially observable states affected by noise as an input.

There exist many more inductive domain model learning systems (with varying level of quality and quantity of input they require) proposed in the last decade. Due to the generality of the topic discussed in this chapter we only include a few here.

2.3 Analytical Learning

KBIL differs from analytical learnings (AL) in the utilisation of training examples. AL mainly uses prior knowledge and exploits a training example just to analyse and discern the relevant features from it. AL uses deductive reasoning method while empirical learning uses inductive reasoning method.

Explanation-based learning (EBL) [15] is the most common type of AL. In EBL the generalised rule logically follows only the prior or background knowledge and does not learn any new facts from examples, such that for EBL the following expression should be valid:

$$\mathbf{B_K} \succ \mathbf{h} \tag{EBL}$$

EBL differentiates from the pure IL in that it looks for only the relevant positive examples to logically justify the background knowledge while IL learns all the true features including relevant to learn from and irrelevant to ignore. This is the reason why IL requires way more training data compared to EBL that can even learn with one relevant example. Mitchell and Thrun [45] quotes a very useful chess example to explain the difference in detail.

Much like the beneficial side of AL, it also suffers from the problem of doggie outcomes if the background knowledge of the domain is incorrect, e.g., in domains with no correct or complete background knowledge available to produce logically justified hypothesis like the stock market. In such situations a reliable source of learning could be the training examples to identify the relevant feature and regularities, i.e., inductive learning to learn statistically justified hypothesis. This leads to a hybrid or mixed inductive-analytical learning approach discussed in the next section.

Some more examples of analytical learning techniques include memorization, static analysis and abstractions learning and case-based reasoning. The analytical learning method is rarely used for domain model learning while the common use of it is in learning of search control knowledge.

One of the most prominent early works that utilise analytical learning is *PRODIGY* [9]. PRODIGY is a planning and learning architecture that integrates a number of learning modules to improve learning and reasoning mechanism. It refines and improves the underlying domain knowledge through experimentation and learns the control rules through experience.

Under a deterministic environment, PRODIGY incrementally learns the domain model actions by a closed-loop integration of observing other agents, learning, planning and executing plans in the environment. It produces operator hypothesis by observing the sequence of changes happening in the environment as the effects a particular action execution. It verifies the correctness of its hypothesised operators during the plan execution stage of planning. In addition to the observations of sequential changes and observed state changes of its learned action, PRODIGY also takes object types and predicate specifications as inputs to learn a domain model.

Among the six main learning modules of the PRODIGY system, the *Experiment* [10] and the *Apprentice* [28] are the two modules used to acquire and improve the underlying domain knowledge through inductive learning.

The remaining four modules of the Prodigy, i.e., *EBL* [39], *Static* [17], *Analogy* [11] *and Alpine* [31], that learn control rules for the PRODIGY planner and assist efficient planning process, use analytical learning.

2.4 Hybrid Learning

Hybrid learning or the multi-strategy learning offers support when either of inductive (statistical) or analytical (logical) knowledge learning cannot individually generalise because of scarce training data or poor background knowledge, respectively. Also, since most machine learning algorithms are custom designed with particular datasets or learning tasks, merging two or more techniques together can improve the overall results and effectiveness of the learner in most of the cases. There are a number of hybrid learning approaches in planning where inductive and analytical learning perform hand-in-hand. These include explanation-based neural network (EBNN), explanation-based learning and inductive logic programming (EBL and ILP) and a combination of explanation-based learning and reinforcement learning (EBL and RL). On literature search, it becomes apparent that the most accepted analytical learning technique in multi-strategy learning is EBL. Among the most common uses of logical learning in EBL and of hybrid approaches is in learning search control rules and heuristics for planning speedup. We only discuss the hybrid techniques which touch the topic of domain model learning.

From the domain model learning perspective, explanation-based neural network (EBNN) blends the two techniques together [44, 45]. It uses NN (backpropagation algorithm) as a form of inductive learning and puts the domain inferences together by updating the NN weights in consistency with training data. For analytical learning, EBL analyses and explains the input training data in terms of the extracted slopes from the prior domain knowledge of already learned NN. The contribution and extent of participation of each technique in EBNN vary depending upon the accuracy of training examples and correctness of the prior knowledge.

PRODIGY, *SOAR* [33] and *THEO* [42] combines both inductive and analytical learning, relying more heavily on the EBL-like deductive methods for acquiring control knowledge.

2.5 *Surprise-Based Learning (SBL)*

Autonomous agents commonly encounter the unknown events which most of the times are realistic and still not engineered in their knowledge base. For example, AUV (autonomous underwater vehicle) that meets an unexpected underwater creature for which it has no model. Ideally, these events can provide the opportunity to learn by experiencing them. Unlike knowledge-based learning methods, SBL is specially designed for autonomous learning and planning in an unforeseen situation with no background knowledge (of the domain available) to the learning agent. SBL works with the notion of prediction rules. Such prediction rules present the agent with the observational model of the environment for pre-action execution time and the predicted observations for post-action execution.

Most of the times in SBL, the learning occurs based on the notion of goal-driven autonomy (GDA) [71]. It is a conceptual model for creating an autonomous agent. In GDA the learning agent continuously monitors and evaluates the activity/plan execution outcome with the already predicted observations [47]. Wherever the action outcome does not match the observed predictions, the algorithm detects and records the discrepancy. The agent then builds explanation by analysing the discrepancy, its cause and effects and updates its hypothesis model and reformulates goals to align with its primary objective. Agents that perform goal reasoning, explicitly model and reason about the goals they try to achieve [2].

The discrepancies in the observations which present themselves as an opportunity to learn are termed as 'surprise' in the title, i.e., situation where the action effects violate its predicted model. SBL is shown to be successful on a modular robot learning and navigating in a small static and the fully observable environment with no prior knowledge available in the knowledge base [55].

FOOLMETWICE [46], the extension of *ARTUE* [47], is a system with a relaxed assumption about domain model completeness. The system implements GDA and presents an algorithm to learn (and apply) environment models of unknown exogenous events. It generalises a new model's preconditions by learning from the states that cause inconsistency.

Nguyen and Leong [51] present *STAR* (surprise triggered adaptive and reactive) system that dynamically learns models of its opponents' strategies in response to surprises. Ranasinghe and Shen [55] present an algorithm for an agent to learn and refine its action models based on the SBL. The model can be used to predict the state changes and identify when surprises occur.

The *LIVE* system [64] enhanced the GPS (general problem-solving) system [16] with the ability to learn models. LIVE learns prediction rules by observing changes in the environment. It assimilates action exploration, experimentation, learning and problem-solving. To create STRIPS-like rules, it requires actions and percept from the environment, a process to provide observation from the environment and the state description of the environment.

EXPO [20] is a learning-by-experimentation system for refining incomplete planning operators. It refines operators, which have missing preconditions and effects by failure-driven experimentation with the environment. It does this by observing plan execution and detecting the inconsistency between observations and predictions. To adjust the inconsistency, it produces a set of hypothesis and empirically tests each. EXPO does incremental learning and also learns conditional effects.

Another example of refinement and incremental learning is the *OBSERVER* [69] system. It induces a STRIPS-like initial model and repairs it continuously during operation by monitoring expert agents. It applies version spaces algorithm [40] to the observations. It repairs plans from incorrect and partial domain knowledge. The framework learns planning operators by observing expert agents and subsequent knowledge refinement in a learning-by-doing paradigm. To refine the learnt operators, it solves practice problems with operators, analyses and learns from the execution traces of the resulting solutions. It uses the pure inductive methodology and does not require background knowledge to do learning. The method is implemented inside PRODIGY [67] that includes a general-purpose planner and several learning modules to improve the planning domain knowledge and also the control knowledge to support the planning algorithm.

To efficiently learn models, there needs to be a quantitative bound on a number of input samples or the required amount of interactions with the environment. Walsh and Littman [68] demonstrate that learning STRIPS operators through raw-experience can require an exponential number of samples, but restricting the size of the precondition lists allows for sample-efficient learning. An external teacher is needed to fulfil the demand of required solution observations in order to eliminate the restriction on the size of the precondition lists.

2.6 Transfer Learning

Through transfer learning (TL), the system exploits data from one or more source domains to improve learning performance in a different target domain in the situations with like and limited training data availability. Knowledge engineering through transfer learning especially for planning domain models has recently received active attention and the resulting learning technique provides a good corpus of work for interested researchers.

Many machine learning methods work well only under a common assumption that the training and test data are taken from the same feature space and the same distribution. In the case of altered feature space and distribution, most statistical models need to be rebuilt from scratch using newly collected training data. In many practical applications, it is expensive or impossible to recollect the needed training data and rebuild the models. It would be nice to reduce the need and effort to recollect the training data especially when data is scarcely available or when it easily becomes outdated. In such cases, knowledge transfer or transfer learning between task domains is desirable. Pan and Yang [52] in a survey on TL explain the benefit of using TL to cover the same feature space assumption of machine learning. The survey also addresses the primary issues of what, when and how to transfer.

Zhuo et al. [76] learn the hypothesis model from plan traces by transferring useful information from other domains whose domain models are already known. The system creates a metric to measure the shared information and transfer this information according to the metric. The larger the metric is, the bigger the information is transferred.

Inspired by the educational psychology of meta-cognitive reflection for better inductive transfer learning practices, a novel L2T framework [73] has been proposed for transfer learning. The system automatically optimizes what and how to transfer between a source and a target domain by leveraging previous transfer learning experiences.

Zhuo et al. [75] present *t-LAMP* (transfer Learning Action Models from Plan traces), which can learn action models in PDDL language with quantifiers. The system exploits plan traces using Markov logic networks to enable knowledge transfer.

Zhuo and Yang [82] proposed *TRAMP* (Transfer learning Action Models for Planning) system, to learn action models with limited training data in the target domain, by transferring as much of the available information from source domains by using web search on top of transfer technique to bridge the transfer gap.

2.7 Policy Learning

A policy is a state-action mapping and policy learning is learning what to do in every possible situation. Based on the assumption that the goals of the learning system are static, instead of learning the domain model of the environment, the system simply learns the reaction or response to common situations that may arise in the environment. The system does not learn the domain model of the environment to take action rather it only learns the policies to respond to the exact situation and this is why this approach is called policy learning. This differs from knowledge-based apprentice systems to learn domain models (discussed in the next section). The planning literature also describes this method as *learning by observation, imitation* or *watching*.

Behavioural cloning [8] also known as learning by imitation is an example of a policy learning approach. In behavioural cloning method, the learning system observes and reproduces the skills of the trainer agent (which is usually human) in carrying out the particular task. It then records the responses of the trainer in every situation along with the cause that gave rise to the response. A sequence of these responses based on the sub-cognitive skills and actions of a trainer is used as input to the learning system. When the learning system itself confronts some situation, it compares this situation with the learnt situation from the trainer and responds to act according to the learnt response of the trainer. It outputs a set of rules which reproduce skilled behaviour. This technique has many advantages including the capability to quickly learn from a few training sessions and to act more reliably than a human trainer that it learnt from Benson [6]. This method is useful for building an automatic control system.

From the perspective of the complex domains, it appears to be a difficult exercise to train the system for all the possible values and situations that can occur inside the environment. However, in cloning the response to take according to a particular situation is usually the same for a big set of possible state space, as the system can then generalise the response and develop a policy to cover more inputs.

ALVINN system [54] is one of the best examples of behavioural cloning. Learning from fully or partially annotated demonstrations by domain expert has been used by several systems for knowledge acquisition in robotics and task modelling [4, 19] but has rarely been used to learn declarative domain model [50] for AI planners. This is partly because even for the moderately complex domains, it is unfeasible for the area expert to specify conjecture for every action, explanations for every inconsistency and all possible effects of the model, as the performance of the learner is affected by the ability of the trainer. In order to cover for trainer's implicit knowledge gap, many systems use reinforcement learning techniques to co-operate with this type of learning such as using Q-learning [70]. Two of the common demonstration approaches include tele-operation and shadowing [4].

2.8 *Other Methods of Knowledge Acquisition*

This section includes potential knowledge acquisition and learning methods that are either not formally classed as the typical learning methods for planning or are in their infancy yet have produced some effective domain learning systems in the past.

Apprentice Systems Mitchell et al. [43] coined the term apprentice system as an interactive knowledge-based system which induce knowledge by observing the users interaction with the system and analysing the problem-solving steps. These systems capture and infer from the training examples of the user's activity and the context on which decisions involved in activity were taken. It then generalizes rules from the training examples that are comparable to the hand-generated rules. The idea has been implemented in a number of application areas. Based on the idea, the same authors created LEAP—a learning apprentice and advice system for digital circuit design. LEAP integrates new knowledge by its experience with the user approval/rejection of its advice about circuit decomposition by observing and analysing the user's problem-solving decisions and steps.

ARMS (acquiring robotic manufacturing schemata) system [61] represents an important first step towards a learning-apprentice system for manufacturing. It learns from a user interface where the user instructs the simulated robots to perform simple tasks. Michele [49]—a groupware toolkit based on a multi-agent model of communication—induces learning mechanism by its interactions with the user and stores learned knowledge in each user's environment. It induces a decision tree for each query to the user and exploits ID4 [60] learning algorithm. Jourdan et al. [29], Tecuci and Dybala [65] and Abbeel et al. [1] are some more examples of the apprentice systems that use various methods of interaction with the user including passive observations or querying the user to record the reason behind the particular decision made.

Crowdsourcing Recently crowdsourcing [24] has been exploited as a novel approach for acquiring planning domain models. Collecting a large amount of training data is not always feasible in terms of reach and cost, e.g., in a situation like a military operation. Instead of collecting training examples, crowdsourcing methods engage different annotators that could include various sources like domain experts, stakeholders, previous data or experience of the general public about the domain to learn. The outcome from various annotators built as the soft constraints can later be solved using max-sat solver to generate a domain model.

While crowdsourcing is comparatively new in learning for planning, it has been used in several planning applications, e.g., Zhang et al. [74] enable a crowd to effectively and collaboratively resolve global constraints to carry out itinerary planning. Gao et al. [18] propose a technique to handle the discrepancy in crowd inputs by first building a set of human intelligence tasks (HITs) for values collection and then estimate the actual values of variables and feed the values to a planner to solve the problem. Raykar et al. [56] label training data for machine learning by crowdsourcing information from experts and non-experts. The system not only evaluates the different experts but also gives an estimate of the actual hidden labels.

Recently, crowdsourcing has also been exploited for acquiring planning domain models [77]. It is worth noting that the problem of encoding domain models is being analysed not only from the point of view of generating models in a specific description language—such as PDDL—but also for generating different sorts of automatically exploitable models. Konidaris et al. [32] proposed a method for constructing symbolic representations for high-level planning by establishing a close relationship between an agent's actions and the symbols required to plan to use them.

MLN To deal with the probability along with imperfect and uncertain knowledge, Markov logic network (MLN) [57] is a dense language to determine very large Markov networks, and has the ability to flexibly and modularly incorporate a wide range of domain knowledge. Many learning systems exploit the technique of MLN that applies the concept of a Markov network to the first-order logic (FOL) and draws the inference from the evidence. In a Markov graph, the vertices are taken as the atomic FOL formulas, and the edges act as the logical connectives used to construct the formulas. Several systems including LAMP—to learn domain model with quantifiers and logical implications [83] and AMAN—a system for action-model acquisition from noisy plan traces [79], learn domain models based on the idea of Markov network as a major driving approach.

LAMP system uses the MLN technique to select the most likely subsets of candidate formulas from all the generated formulas which are later transformed into learned action models. It learns STRIPS action models (with quantifiers and logical implications) for classical planning from plan traces with partially observed states. It learns a domain model for an observable and deterministic environment from training plans with little or no pre-engineered domain knowledge including object types, predicate specifications and action headers.

AMAN builds a graphical model to capture the relations between actions (in plan traces) and states and then learns the parameters of the graphical model. After that, AMAN generates a set of action models according to the learnt parameters. Specifically, the system first exploits the observed noisy plan traces to predict correct plan traces and the domain model based on the graphical model and then executes the correct plan traces to calculate the reward of the predicted correct plan traces according to a predefined reward function. Then, AMAN updates the predicted plan traces and domain model based on the reward. It iteratively performs the above-mentioned steps until a given number of iterations is reached. Finally, the predicted domain model is provided.

TRAMP [82] system conducts MLNs assisted transfer learning to learn domain models.

3 Characteristics of the Domain Model Learning Tools

Using machine learning methods, several tools and techniques have been presented in the recent past to facilitate the transformation process of real planning application requirements into a solver ready PDDL domain model that uses training or observation as inputs. These techniques use various types of knowledge besides plan traces, like general properties and constraints about domain actions, as well as partial knowledge about the kind of domain in which they are operating. To learn expressive domain models, systems tend to require more detailed inputs and substantial a priori knowledge which often include details about initial and goal state information. Some systems also require state information before and after an action execution within each training plan. The main aim of this type of learning is to overcome the knowledge acquisition bottleneck [30], to help planning agents become more autonomous and make them able to adapt and plan for unseen situations and to debug existing domain models.

This section presents a brief overview of the automated tools along with their characteristics that can be exploited to automatically induce planning domain models.

Input Characteristics Input characteristics of the system depend on what the system is trying to learn, the learning method and the extent of learning. Some systems aim to design the complete domain model of a particular world, some refine already built partial domain model and some aim to transfer knowledge from one domain to other. Learning techniques can be supervised or unsurprised learning. It depends if the trainer indicates when something goes wrong in the case of supervised learning. All traditional domain model learning systems accept input in the form of training plans. Based on the learning capability some systems may also have to exploit additional background knowledge B_K or information about the surrounding world. B_K can be in any form like observations, constraints, type structure, initial, intermediate and goal states, fluent, etc. Some systems also require a partial domain model (with missing preconditions and effects in the actions) in the input.

Potential plan traces can be gathered from multiple sources and applications, for example, the sequence of workflow in some process execution, logs of commands for installing a piece of software or the moves or steps captured from game playing, etc.

Obtaining the training plans from sensors, sometimes noise inevitably gets introduced into plan traces when some sensors are occasionally damaged, with unintentional mistakes in the recording of the action sequence, or may be due to the presence of other agents in the same environment. To deal with this, several systems learn domain models with noisy inputs.

Output Characteristics For output, systems can be classified based on the extent and capability to learn in varying world dynamics and the state of observability in the environment. For instance, the characteristics of the surrounding environment can be discrete or continuous, static or dynamic. Action effects can be stochastic or non-deterministic (rolling of dice) compared to a fully deterministic environment. In terms of observability, the learning environment can be fully or partially observable. Learning from deterministic, fully observable, discrete and static environmental characteristic offer lesser challenges than continuous and dynamic environmental features. Jiménez et al. [27] provide a thorough review of techniques based on the learning targets of the systems from various planning paradigms, i.e., learning in varying level of world dynamics and state observability.

The extent of learning by a system is the granularity of the output and the amount of details learnt, for example, full or partial domain model in the output or leaning of domain model with quantifiers and logical implications [83].

No standard evaluation and analysis methods exist to verify the output domain model completeness and quality and like the requirements specification, these characteristics cannot be objectively assessed and proven correct. Learning systems and their output are typically evaluated empirically, based on their divergence from the reference model syntax (which itself can be questionable from multiple perspectives by multiple experts). A step forward in defining the quality of domain models and to improve evaluation method semantically, FAMA [3] presents a method to assess the quality and performance of the learning approaches. The evaluation method alleviates the common limitation of syntactic evaluation methods. A usual limitation of syntactic evaluation methods is when the learned model is semantically correct but syntactically differ from the benchmark model. Unable to evaluate correct but redundant preconditions in the model is another downside of syntax-based assessment methods.

McCluskey et al. [37] use the idea of domain model as a formal specification of a domain and consider what it means to measure the quality of such a specification. To build the notion of quality assessment, they used dynamic and static testing of the domain model. Vallati and McCluskey [66] present a quality framework which aims at representing all the aspects that affect the quality of knowledge in domain models. The framework is based on the interaction between seven different sets that underpin the domain quality.

Some learning systems additionally produce heuristics and various graphical views like finite state machines along with the domain model while some just improve the partial domain model by learning the missing preconditions and effects of operators.

Representational Language and Mechanism A well-chosen representation language should explicitly model every action effect the system might confront. In addition to that, a domain modelling representation language should have some salient attributes. It should have supporting tools to check its operation and have logically strong inference mechanism to carry out reasoning. It should be sufficiently expressive to explicitly model a complex scenario of the real world. Moreover, it should be customizable and structured to capture every action effect the system might confront in operator definitions. In addition, it should have clear syntax and semantics to support the operational aspects of the model.

Different learning systems use different languages to express the output domain model including PDDL, STRIPS, OCL, etc. While choosing the representation language for effective system output, a well-known complexity-expressiveness trade-off of representation is not easy to attain. More expressive languages let the systems produce domains that can express input data in a better way while reducing the expressivity enhances the complexity of the consistent hypothesis. To overcome the KE bottleneck, most systems in the area of domain model learning use propositional and first-order logic to represent domain model for the logical agents and output the model in some variant of PDDL in their initial acquisition phase.

Most learning systems use action-centred representation mechanism where applying an action on a state transforms the state of the system into a new state. These models are represented mostly by PDDL and its variants built on first-order logic. Another mechanism is the object-centred representation that captures the dynamic relationship between objects in state parameters. OCL (object centred language) is used to support this mechanism. Figure 2.4 shows a (non-exhaustive) characteristic that a domain model learning system can cover.

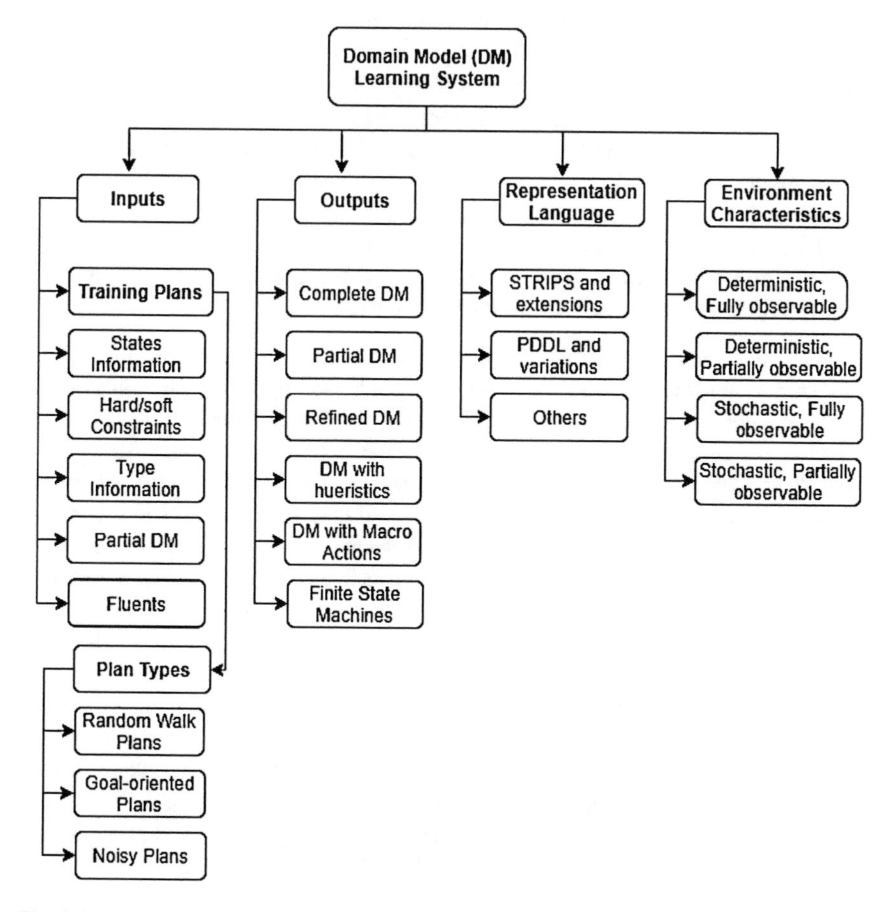

Fig. 2.4 Various characteristics of domain model learning tools

4 Conclusion

Learning is fundamental to autonomic behaviour and it can be defined in many ways. From the point of view of machine learning, it is defined as a change in behaviour through learning to allow improvement in performance. This chapter investigates various classical and state-of-the-art methods to attain such capability for automatically learning domain models from training data. The taxonomy and order of these methods we followed is based on their standing and frequency of usage in the past research. The choice of the learning method and the design of the learning mechanism based on it can be made easy by thinking in terms of the following four factors:

- Firstly, the choice of the learning method depends on the learning goals in a particular domain. For instance, does it need to design the complete domain model of a particular world? Is it trying to transfer knowledge from one domain to other?
- Secondly, it depends upon the availability of the information about the surrounding world. If available, then what is the type and consistency of input assistance or prior knowledge? Is it supervised or unsurprised learning? Is input data consistent or noisy? Sometimes the supervised learning situations become semi-supervised based on the deliberate systematic inconsistencies in the data, e.g., learning the age of people from observing pictures where in the training data some people lied about their age.
- Another factor to consider is what feedback is available. For example, does a trainer indicate when something goes wrong in the case of supervised learning?
- Finally, the characteristics of the world to learn knowledge from also matters, for instance, observability and stochasticity of the environment.

There are several knowledge engineering tools with varying capabilities. These tools support automated planning, not only in the knowledge elicitation process but also for the design, validation and verification of developed models.

There are several relevant aspects that need more focused attention of the planning and learning community in the future. In order to deal with the real-world planning-inherent complexity, learning-augmented planning systems should be able to apprehend the environment, generate corresponding effects and enhance their performance according to the previous experience. This has attracted much research in the recent past but the current state-of-the-art is still a long way from human-level abilities to work in real world. Instead, most of these systems work under classical restrictive environmental assumption with toy domains setup that are more comprehensible and limited proxies of the real-world environment.

Learning systems can be categorized as offline (learns before the planning process starts) and online (learns during the plan search and execution stages). From the domain model learning viewpoint, offline learning is comparatively popular starting from the learning of domain invariants to learning complete domain model from variable sources. Both online and offline domain model learning have pros and cons. Online learning can continuously/incrementally refine the domain model in case an anomaly is detected by improving or adapting to the changes while for offline planning, the planner has to bear with the predefined version till the planning process finishes. Similarly, for online learning of the domain model, the overhead cost incurred for the joint planning-learning process is higher in terms of processing time and efficiency compared to offline learning [85]. This may also explain why online incremental domain model learning has not been very popular in recent years and needs active research attention to effectively reduce the overhead cost.

References

1. Abbeel, P., D. Dolgov, A. Y. Ng and S. Thrun (2008). *Apprenticeship learning for motion planning with application to parking lot navigation*. 2008 IEEE/RSJ International Conference on Intelligent Robots and Systems, IEEE.
2. Aha, D., M. Klenk, H. Munoz-Avila, A. Ram and D. Shapiro (2010). Goal-driven autonomy: Notes from the AAAI workshop, Menlo Park, CA: AAAI Press.
3. Aineto, D., S. J. Celorrio and E. Onaindia (2019). "Learning action models with minimal observability." *Artificial Intelligence*.
4. Argall, B. D., S. Chernova, M. Veloso and B. Browning (2009). "A survey of robot learning from demonstration." *Robotics and autonomous systems***57**(5): 469–483.
5. Baxter, J. (1995). Learning internal representations. *Proceedings of the eighth annual conference on Computational learning theory*. Santa Cruz, California, USA, ACM: 311–320.
6. Benson, S. (1995). *Action model learning and action execution in a reactive agent*. Proceedings of the International Joint Conference on Artificial Intelligence (IJCAI-95).
7. Brachman, R. J. and H. J. Levesque (1984). *The tractability of subsumption in frame-based description languages*. AAAI.
8. Bratko, I. and T. Urbančič (1997). "Transfer of control skill by machine learning." *Engineering Applications of Artificial Intelligence***10**(1): 63–71.
9. Carbonell, J., O. Etzioni, Y. Gil, R. Joseph, C. Knoblock, S. Minton and M. Veloso (1991). "Prodigy: An integrated architecture for planning and learning." *ACM SIGART Bulletin***2**(4): 51–55.
10. Carbonell, J. G. and Y. Gil (1990). Learning by experimentation: The operator refinement method. *Machine learning*, Elsevier: 191–213.
11. Carbonell, J. G. and M. Veloso (1988). *Integrating derivational analogy into a general problem solving architecture*. Proceedings of the First Workshop on Case-Based Reasoning.
12. Cresswell, S. (2009). "LOCM: A tool for acquiring planning domain models from action traces." *ICKEPS 2009*.
13. Cresswell, S., T. L. McCluskey and M. M. West (2009). *Acquisition of Object-Centred Domain Models from Planning Examples*. ICAPS.
14. Davis, J. and M. Goadrich (2006). *The relationship between Precision-Recall and ROC curves*. Proceedings of the 23rd international conference on Machine learning, ACM.
15. DeJong, G. and R. Mooney (1986). "Explanation-based learning: An alternative view." *Machine learning***1**(2): 145–176.
16. Ernst, G. W. and A. Newell (1969). *GPS: A case study in generality and problem solving*, Academic Pr.
17. Etzioni, O. (1991). *STATIC: A Problem-Space Compiler for PRODIGY*. AAAI.
18. Gao, J., H. H. Zhuo, S. Kambhampati and L. Li (2015). *Acquiring Planning Knowledge via Crowdsourcing*. Third AAAI Conference on Human Computation and Crowdsourcing.

19. Garland, A. and N. Lesh (2003). "Learning hierarchical task models by demonstration." *Mitsubishi Electric Research Laboratory (MERL), USA–(January 2002).*
20. Gil, Y. (1992). Acquiring domain knowledge for planning by experimentation, DTIC Document.
21. Gregory, P. and S. Cresswell (2015). *Domain Model Acquisition in the Presence of Static Relations in the LOP System.* ICAPS.
22. Gregory, P. and A. Lindsay (2016). *Domain model acquisition in domains with action costs.* Twenty-Sixth International Conference on Automated Planning and Scheduling.
23. Hoffmann, J., I. Weber and F. Kraft (2009). *Planning@ sap: An application in business process management.* 2nd International Scheduling and Planning Applications woRKshop (SPARK'09).
24. Howe, J. (2008). *Crowdsourcing: How the power of the crowd is driving the future of business*, Random House.
25. Inoue, K., T. Ribeiro and C. Sakama (2014). "Learning from interpretation transition." *Machine Learning***94**(1): 51–79.
26. Jilani, R., A. Crampton, D. Kitchin and M. Vallati (2015). *Ascol: A tool for improving automatic planning domain model acquisition.* Congress of the Italian Association for Artificial Intelligence, Springer.
27. Jiménez, S., T. De la Rosa, S. Fernández, F. Fernández and D. Borrajo (2012). "A review of machine learning for automated planning." *The Knowledge Engineering Review***27**(4): 433–467.
28. Joseph, R. L. (1989). "Graphical knowledge acquisition." *In Proceedings of the Fourth Knowledge Acquisition For Knowledge-Based Systems Workshop, Banff, Canada.*
29. Jourdan, J., L. Dent, J. McDermott, T. Mitchell and D. Zabowski (1993). Interfaces that learn: A learning apprentice for calendar management. *Machine learning methods for planning*, Elsevier: 31–65.
30. Kambhampati, S. (2007). *Model-lite planning for the web age masses: The challenges of planning with incomplete and evolving domain models.* Proceedings of the National Conference on Artificial Intelligence, Menlo Park, CA; Cambridge, MA; London; AAAI Press; MIT Press; 1999.
31. Knoblock, C. A. (1991). Automatically Generating Abstractions for Problem Solving, CARNEGIE-MELLON UNIV PITTSBURGH PA DEPT OF COMPUTER SCIENCE.
32. Konidaris, G., L. P. Kaelbling and T. Lozano-Perez (2014). "Constructing symbolic representations for high-level planning." *Proceedings of the 28th AAAI Conference on Artificial Intelligence.*
33. Laird, J. E., A. Newell and P. S. Rosenbloom (1987). "Soar: An architecture for general intelligence." *Artificial intelligence***33**(1): 1–64.
34. Lindsay, A., J. Read, J. F. Ferreira, T. Hayton, J. Porteous and P. Gregory (2017). *Framer: Planning models from natural language action descriptions.* Twenty-Seventh International Conference on Automated Planning and Scheduling.
35. Martínez, D., G. Alenya, C. Torras, T. Ribeiro and K. Inoue (2016). *Learning relational dynamics of stochastic domains for planning.* Twenty-Sixth International Conference on Automated Planning and Scheduling.
36. McCluskey, T., S. Cresswell, N. Richardson, R. Simpson and M. M. West (2008). "An evaluation of Opmaker2." *The 27th Workshop of the UK Planning and Scheduling Special Interest Group, December 11–12th, 2008, Edinburgh.*: 65–72.
37. McCluskey, T. L., T. S. Vaquero and M. Vallati (2017). *Engineering knowledge for automated planning: Towards a notion of quality.* Proceedings of the Knowledge Capture Conference, ACM.
38. Michalski, R. S. (1993). Learning= inferencing+ memorizing. *Foundations of Knowledge Acquisition*, Springer: 1–41.
39. Minton, S., J. G. Carbonell, O. Etzioni, C. A. Knoblock and D. R. Kuokka (1987). *Acquiring effective search control rules: Explanation-based learning in the PRODIGY system.* Proceedings of the fourth International workshop on Machine Learning, Elsevier.

40. Mitchell, T. M. (1977). *Version spaces: A candidate elimination approach to rule learning*. Proceedings of the 5th international joint conference on Artificial intelligence-Volume 1, Morgan Kaufmann Publishers Inc.
41. Mitchell, T. M. (1980). *The need for biases in learning generalizations*, Department of Computer Science, Laboratory for Computer Science Research ....
42. Mitchell, T. M., J. Allen, P. Chalasani, J. Cheng, O. Etzioni, M. Ringuette and J. C. Schlimmer (1991). "Theo: A framework for self-improving systems." *Architectures for intelligence*: 323–355.
43. Mitchell, T. M., S. Mabadevan and L. I. Steinberg (1990). LEAP: A learning apprentice for VLSI design. *Machine learning*, Elsevier: 271–289.
44. Mitchell, T. M. and S. Thrun (2014). *Explanation based learning: A comparison of symbolic and neural network approaches*. Proceedings of the Tenth International Conference on Machine Learning.
45. Mitchell, T. M. and S. B. Thrun (1996). "Learning analytically and inductively." *Mind matters: A tribute to Allen Newell*: 85–110.
46. Molineaux, M. and D. W. Aha (2014). *Learning unknown event models*. Twenty-Eighth AAAI Conference on Artificial Intelligence.
47. Molineaux, M., M. Klenk and D. Aha (2010). *Goal-driven autonomy in a Navy strategy simulation*. Twenty-Fourth AAAI Conference on Artificial Intelligence.
48. Mourao, K., L. S. Zettlemoyer, R. Petrick and M. Steedman (2012). "Learning strips operators from noisy and incomplete observations." *arXiv preprint arXiv:1210.4889*.
49. Nakauchi, Y., T. Okada and Y. Anzai (1991). *Groupware that learns*. [1991] IEEE Pacific Rim Conference on Communications, Computers and Signal Processing Conference Proceedings, IEEE.
50. Nejati, N., P. Langley and T. Konik (2006). *Learning hierarchical task networks by observation*. Proceedings of the 23rd international conference on Machine learning, ACM.
51. Nguyen, T.-H. D. and T.-Y. Leong (2009). *A Surprise Triggered Adaptive and Reactive (STAR) Framework for Online Adaptation in Non-stationary Environments*. AIIDE.
52. Pan, S. and Q. Yang (2010). A survey on transfer learning. IEEE Transaction on Knowledge Discovery and Data Engineering, 22 (10), IEEE press.
53. Pomerleau, D. A. (1989). *ALVINN: An autonomous land vehicle in a neural network*. Advances in neural information processing systems.
54. Pomerleau, D. A. (1991). "Efficient training of artificial neural networks for autonomous navigation." *Neural Computation***3**(1): 88–97.
55. Ranasinghe, N. and W.-M. Shen (2008). *Surprise-based learning for developmental robotics*. 2008 ECSIS Symposium on Learning and Adaptive Behaviors for Robotic Systems (LAB-RS), IEEE.
56. Raykar, V. C., S. Yu, L. H. Zhao, G. H. Valadez, C. Florin, L. Bogoni and L. Moy (2010). "Learning from crowds." *Journal of Machine Learning Research***11**(Apr): 1297–1322.
57. Richardson, M. and P. Domingos (2006). "Markov logic networks." *Machine learning***62**(1–2): 107–136.
58. Richardson, N. E. (2008). *An operator induction tool supporting knowledge engineering in planning*, University of Huddersfield.
59. Riddle, P. J., R. C. Holte and M. W. Barley (2011). *Does Representation Matter in the Planning Competition?* Ninth Symposium of Abstraction, Reformulation, and Approximation.
60. Schlimmer, J. C. and D. Fisher (1986). *A case study of incremental concept induction*. AAAI.
61. Segre, A. M. (1987). Explanation-Based Learning of Generalized Robot Assembly Plans, ILLINOIS UNIV AT URBANA COORDINATED SCIENCE LAB.
62. Segura-Muros, J. Á., R. Pérez and J. Fernández-Olivares (2018). "Learning Numerical Action Models from Noisy and Partially Observable States by means of Inductive Rule Learning Techniques." *KEPS 2018*: 46.
63. Shahaf, D. and E. Amir (2006). *Learning partially observable action schemas*. Proceedings of the National Conference on Artificial Intelligence, Menlo Park, CA; Cambridge, MA; London; AAAI Press; MIT Press; 1999.

64. Shen, W.-M. and H. A. Simon (1989). *Rule Creation and Rule Learning Through Environmental Exploration.* IJCAI, Citeseer.
65. Tecuci, G. and T. Dybala (1998). *Building Intelligent Agents: An Apprenticeship, Multistrategy Learning Theory, Methodology, Tool and Case Studies*, Morgan Kaufmann.
66. Vallati, M. and T. L. McCluskey (2018). "Towards a Framework for Understanding and Assessing Quality Aspects of Automated Planning Models." *KEPS 2018*: 28.
67. Veloso, M., J. Carbonell, A. Perez, D. Borrajo, E. Fink and J. Blythe (1995). "Integrating planning and learning: The PRODIGY architecture." *Journal of Experimental & Theoretical Artificial Intelligence***7**(1): 81–120.
68. Walsh, T. J. and M. L. Littman (2008). *Efficient learning of action schemas and web-service descriptions.* AAAI.
69. Wang, X. (1995). *Learning by observation and practice: An incremental approach for planning operator acquisition.* ICML.
70. Watkins, C. J. C. H. (1989). *PhD Thesis: Learning from delayed rewards*, University of Cambridge England.
71. Weber, B. G., M. Mateas and A. Jhala (2012). *Learning from demonstration for goal-driven autonomy.* Twenty-Sixth AAAI Conference on Artificial Intelligence.
72. Wu, K., Q. Yang and Y. Jiang (2005). "Arms: Action-relation modelling system for learning action models." *CKE*: 50.
73. Ying, W., Y. Zhang, J. Huang and Q. Yang (2018). *Transfer learning via learning to transfer.* International Conference on Machine Learning.
74. Zhang, H., E. Law, R. Miller, K. Gajos, D. Parkes and E. Horvitz (2012). *Human computation tasks with global constraints.* Proceedings of the SIGCHI Conference on Human Factors in Computing Systems, ACM.
75. Zhuo, H., Q. Yang, D. H. Hu and L. Li (2008). *Transferring knowledge from another domain for learning action models.* Pacific Rim International Conference on Artificial Intelligence, Springer.
76. Zhuo, H., Q. Yang and L. Li (2009). *Transfer learning action models by measuring the similarity of different domains.* Pacific-Asia Conference on Knowledge Discovery and Data Mining, Springer.
77. Zhuo, H. H. (2015). *Crowdsourced action-model acquisition for planning.* Twenty-Ninth AAAI Conference on Artificial Intelligence.
78. Zhuo, H. H., D. H. Hu, Q. Yang, H. Munoz-Avila and C. Hogg (2009). *Learning applicability conditions in AI planning from partial observations.* Workshop on Learning Structural Knowledge From Observations at IJCAI.
79. Zhuo, H. H. and S. Kambhampati (2013). *Action-model acquisition from noisy plan traces.* Twenty-Third International Joint Conference on Artificial Intelligence.
80. Zhuo, H. H., H. Muñoz-Avila and Q. Yang (2011). *Learning action models for multi-agent planning.* The 10th International Conference on Autonomous Agents and Multiagent Systems-Volume 1, International Foundation for Autonomous Agents and Multiagent Systems.
81. Zhuo, H. H., H. Muñoz-Avila and Q. Yang (2014). "Learning hierarchical task network domains from partially observed plan traces." *Artificial intelligence***212**: 134–157.
82. Zhuo, H. H. and Q. Yang (2014). "Action-model acquisition for planning via transfer learning." *Artificial intelligence***212**: 80–103.
83. Zhuo, H. H., Q. Yang, D. H. Hu and L. Li (2010). "Learning complex action models with quantifiers and logical implications." *Artificial Intelligence***174**(18): 1540–1569.
84. Zhuoa, H. H., T. Nguyenb and S. Kambhampatib (2013). *Refining incomplete planning domain models through plan traces.* Proceedings of IJCAI.
85. Zimmerman, T. and S. Kambhampati (2003). "Learning-assisted automated planning: looking back, taking stock, going forward." *AI Magazine***24**(2): 73–73.
86. Cresswell, S. and P. Gregory (2011). *Generalised domain model acquisition from action traces.* Twenty-First International Conference on Automated Planning and Scheduling.

Chapter 3
Formal Knowledge Engineering for Planning: Pre and Post-Design Analysis

Jose Reinaldo Silva, Javier Martinez Silva, and Tiago Stegun Vaquero

Abstract The interest and scope of the area of autonomous systems have been steadily growing in the last 20 years. Artificial intelligence planning and scheduling is a promising technology for enabling intelligent behavior in complex autonomous systems. To use planning technology, however, one has to create a knowledge base from which the input to the planner will be derived. This process requires advanced knowledge engineering tools, dedicated to the acquisition and formulation of the knowledge base, and its respective integration with planning algorithms that reason about the world to plan intelligently. In this chapter, we shortly review the existing knowledge engineering tools and methods that support the design of the problem and domain knowledge for AI planning and scheduling applications (AI P&S). We examine the state-of-the-art tools and methods of knowledge engineering for planning & scheduling (KEPS) in the context of an abstract design process for acquiring, formulating, and analyzing domain knowledge. Planning quality is associated with requirements knowledge (pre-design) which should match properties of plans (post-design). While examining the literature, we analyze the design phases that have not received much attention, and propose new approaches to that, based on theoretical analysis and also in practical experience in the implementation of the system itSIMPLE.

Keywords Planning design · Post-design analysis · Planning automation · Automation by planning

---

J. R. Silva (✉)
Escola Politécnica, Universidade de São Paulo, São Paulo, SP, Brazil
e-mail: reinaldo@usp.br; http://dlab.poli.usp.br

J. M. Silva
Centro Universitario da FEI, São Bernardo do Campo, SP, Brazil
e-mail: jmartinez@fei.edu.br

T. S. Vaquero
Jet Propulsion Laboratory, California Institute of Technology, Pasadena, CA, USA
e-mail: tiago.stegun.vaquero@jpl.nasa.gov

M. Vallati, D. Kitchin (eds.), *Knowledge Engineering Tools and Techniques for AI Planning*, https://doi.org/10.1007/978-3-030-38561-3_3

1 Introduction

The last 20 years witnessed a steady growth in the demand for AI planning technology addressed to real applications. Particularly in the last 5 years we have different applications such as:

- Robotics [20, 30, 48],
- Space exploration [12, 28],
- Traffic management [10, 47],
- Manufacturing [3, 8, 36, 54, 55],
- Workflow generation in business processes, [5, 23, 40],
- Narrative generation in New Media [31],
- Logistics [16, 22], and
- Disaster management [19, 56].

The design of such complex applications requires the engineering of application knowledge into a precise form, so that automated reasoning can derive valid plans that will meet mission goals. This knowledge engineering task, however, is poorly served with both general tools and theoretical underpinnings [27]. At best, these applications are created using in-house application-specific knowledge editing tools, and at worst, a text editor. The immaturity of knowledge engineering techniques in the planning community is demonstrated by the poor use of notation and terminology for the knowledge engineering aspects of AI P&S. For example, in many AI P&S papers the word "domain" is used to refer to a set of formal sentences describing the environment where actions should be executed.

Indeed, for the KEPS researcher or developer, there is currently little published work on how to proceed when developing a knowledge base to be used in a planning application, and on which design steps and tools exist to support that process. An understanding of what makes a good or bad representation is missing, yet it is acknowledged that the representation used makes a large difference to the planners' ability to solve a problem [34]. This conclusion was based on experiments where the use of different, yet logically equivalent, representations can lead to different results in the International Planning Competition (IPC).

The special case of automated planning problems brings an extra challenge: to find a feasible compliance between the automated cycle (the time necessary to sense, process, find the proper control action, and actuate) and the characteristic time of specific applications, that is, the maximum time required to get a feedback from the system.

Applications in manufacturing, traffic management, logistic and, naturally, robotics, are good examples of demand for new applications that raise new services, driven by the digital convergence [42]. In all these applications the automation cycle compliance requires a sound design for AI planning tools that could deliver traceability between requirements model and the planning domain, which we call post-design. That implies also a revision of the terminology to refine the concept of *planning domain* to include properties of the context where the plan should be executed.

In this chapter we revisited available approaches for KEPS and propose a practical one that delivers traceability between the problem definition (requirements) and design. That is a key issue to analyze the plan generated by a planner fed by the design system. itSIMPLE (Integrated Tools and Software Interface for Modeling Plan Environments) will be used as an example for a first attempt to model a design process that is object-oriented and based on UML and Petri Nets, as well as on goal-oriented method with Hierarchical Petri Nets.

2 Knowledge Engineering and Planning

Knowledge engineering for planning & scheduling (KEPS) is concerned with the knowledge modeling for the development a problem solving environment and their respective integration with tools that will synthesize plans. KEPS was defined in the 2003 PLANET Roadmap as the processes involving:

1. The acquisition, validation, verification, and maintenance of planning domain models,
2. The selection and optimization of appropriate planning machinery, and
3. The integration of (1) and (2) to form a planning and scheduling application [25].

The area can be seen as a special case of knowledge-based systems (KBS), where the need for methodologies for acquiring, modeling, and managing knowledge at the conceptual level has long been accepted. However, the peculiarities of AI P&S applications clearly distinguish KEPS from general knowledge-based systems, mainly in the area of the principled acquisition and representation of knowledge about actions [25, 26].

KEPS research focuses on the design process for creating reliable, high quality knowledge models of real domains [24, 50]. It includes the investigation of methods, tools, and representation languages to support and organize the phases of design life cycle. Using a well-structured life cycle to guide design increases the chances of building an appropriate application while reducing the costs of encountering and fixing errors in the future. A simple design life cycle is feasible for the development of a small prototype system, but is likely to fail to produce large, knowledge-intense applications that are reliable and maintainable.

In addition to investigating new techniques to support the design of knowledge models and their integration with planners, research on KEPS has another fundamental purpose. Current P&S technology has limited accessibility to non-experts, such as someone with substantial domain knowledge but no understanding of automated planning and scheduling. The use of planners as an off-the-shelf technology by non-experts is not realistic, given the current state of AI P&S, as a deep understanding of automated planning and scheduling techniques is needed to adequately utilize them.

Research on KEPS has the goal of bridging the gap between P&S non-experts and the AI P&S technology, making it accessible for practical problems in the real world. An extreme of this is the area of learning domain knowledge from observation and/or training. In certain applications where flexibility is important (e.g., embedding an intelligent agent with planning capabilities), it may be desirable to avoid the need to design a knowledge model, but rather to design a learning component that can learn and maintain the knowledge model. Studies on KEPS have led to the creation of tools and techniques to support the design of knowledge models and the use of planners for real-world problems. Most of these tools have been presented in specialized workshops and competitions such as the International Competition on Knowledge Engineering for Planning & Scheduling (ICKEPS). The competition has motivated the development of powerful KEPS systems and the advances in modeling techniques, languages, and analysis approaches. In the next section, we use a design process for knowledge models as a framework for organizing the KEPS literature.

KEPS has not yet reached the maturity of other traditional engineering areas (e.g., software engineering [45]) in having an established standard design process. Nevertheless, research in the KEPS literature has developed design tools and identified the needs and singularities of the design process and life cycle of AI P&S applications [25, 44, 50].

Knowledge Engineering arose initially from the knowledge-based systems (KBSs) community, following by the development of successful expert systems in 1970s, such as Mycin [39] and others. The field's motivation was similar to that of software engineering: "turning the process of constructing KBSs from an art into an engineering discipline" [46].

The focus of all methods and frameworks developed subsequently were the creation of a precise, declarative, and detailed model of the area of knowledge engineering that could be attached to the development life cycle of processes and systems—including its evolution and maintenance. This emphasis was due partly in response to the apparent failure of earlier approaches to knowledge-based system construction and partly to the lack of software tools to support the development life cycle. Systems were composed by a large number of expert-engineered rules without a formal model for the underlying domain knowledge [53].

CommonKADS [37] was considered the first environment to structure a method emphasizing the modeling—together—of expert knowledge and application area. The development of a KBS in CommonKADS involved the construction of the *expertise model*, which contains a mix of knowledge about the problem solving strategies and the declarative knowledge about the application domain.

The field of knowledge engineering is now considerably broader than when the term was coined in the context of expert systems. Nowadays, the key rationale for KEPS is to create declarative representations formalized as requirements specification, instead of relying in implicit code, and potentializing verification and reuse. Replication and restrictions to actions eventually executed with the plan should be directly represented by ontologies. Historically the construction of ontology engineering environments started with Protege [32].

Since 2005 the current discussion about KEPS has focused on the following points:

- The initial model for the process of planning could be directed to the capture of knowledge requirements to domain modeling, which could be later transferred to PDDL or another language understood by planners;
- Using schematic languages such graphs and Petri Nets [14, 49] to analyze and validate/verify planning domain modeling;
- To improve the abstraction of the design process to deal with objectives, intentions, and behavior.

The present work deals with the first two objectives. In what follows, we will describe the domain modeling used in AI planning and specifically the modeling used by itSIMPLE. Therefore we will start discussing the domain modeling process.

3 Domain Modeling in AI Planning

Domain modeling is a term used perhaps with a variety of meanings in computer science and applied mathematics. A domain model is often described as an abstract conceptual description of some application, and is used as an aid to the software development process. It is formed as part of the requirements analysis in order to specify objects, actors, roles, attributes etc., independent of a software implementation. Domain model is often represented imprecisely using diagrams, such as in the Unified Modeling Language (UML), for human consumption—that is, for the benefit of analysts and developers to explore requirements and to subsequently create software in the application area being modeled.

The meaning of domain model for representing knowledge within a planning application is much more specific: it is still an abstract conceptual description of some application area but it is encoded for a different purpose—that is, for the analysis, reasoning, and manipulation by a planning engine in order to solve planning problems.

A planning domain model is a formal description of the application domain part of the requirements specification which represents entities invariant over every planning problem, such as object classes, functions, properties, relations, and actions in the domain. A problem instance is a specific tuple composed by a couple of initial and final state, a set of admissible actions, and a set of rules and heuristics about sequencing such actions. Therefore, domain modeling implies in providing a formal representation to both planning domain and problem instance. It would be even more interesting if an integrated model can be achieved.

However, the process starts with the capture of information about a planning domain and about heuristics that could guide the planning process. Such information should compose a requirements model, in a process we call *pre-design*.

Let us assume that requirements specification for a planning component of some wider project is available. The requirements may be in the minds of domain experts, be described informally in diagrams and textual documents, or described (at least in part) in a formal language (e.g., as in the use of LTL [33]). The requirement specification would naturally contain descriptions of the kind of planning problems

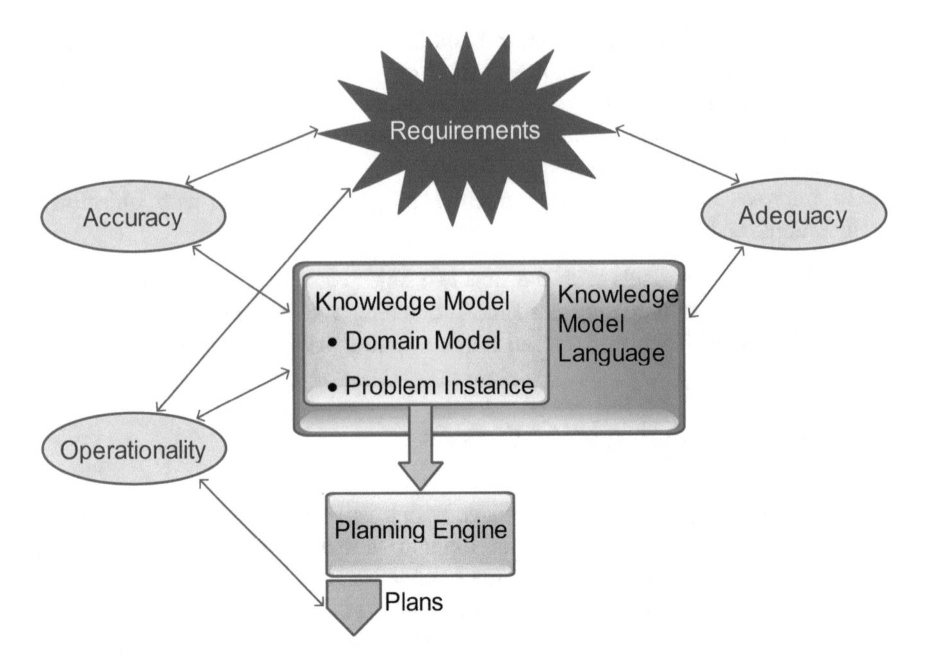

Fig. 3.1 PDM (planning domain modeling properties) basic properties as defined by LeeMc-Cluskey et al. and discussed later in McCluskey [27]

that the planner needs to solve and the kind of plans that need to be output. For example, it might be essential that resource consumption is taken into consideration and so plans need to be generated which achieve goals while minimizing resource consumption. Before a domain-independent planner can be chosen and used, the domain information needs to be conceptualized and formalized. During this process (elaborated in the sections below) the assumptions and features that are essential to represent a domain model are derived from the overall requirements.

Figure 3.1 shows the relationship between requirements and PDM, which should be represented in knowledge model language. Requirements could be modeled in LTL or Petri Nets and verified in a pre-design process to check for completeness and consistency (pre-design mode). PDM is modeled in a knowledge model language and should also be verified to check completeness and consistency. Consistent and sound requirements can then fit PDM design model in a *post-design* mode to verify accuracy and adequacy.

3.1 Accuracy

Accuracy is an attribute of the PDM, related to application domain features considered as part of the requirements specification (often just referred as "requirements" below). Considering the PDM as a logical expression,

A PDM accurately represents the requirement specification if the interpretation given to it by mapping its components to features (objects, relations, etc.) in that specification makes all assertions in the PDM true.

Verifying accuracy is essentially an informal process if the requirements are described informally. If the requirements are already encoded in some formal language, then a PDM is accurate if the requirements provide a model of the PDM. For example, if there is a semantic model encoding the requirements, the accuracy of the PDM equates to interpreting it using the semantic model, and hence accuracy in this case can be proved formally.

If a PDM is encoded in PDDL, then assessing accuracy of the domain model entails,

1. Creating all possible groundings of operator schema, using the objects in the problem file,
2. Mapping the logical expression in the precondition of each grounded schema to a set P of relations and properties in the requirements,
3. Mapping the logical expression in the effects to a set E of relations and properties in the application, and
4. Checking that if P is true in the application, then the action modeled could be executed, and if executed would make E true.

A similar process would be used to assess the accuracy of the problem description. Assessing accuracy of the problem specification is a matter of checking that the initial state and goal map to the problem embedded in the requirements specification.

3.2 *Adequacy*

Adequacy is a relationship between the requirements and the language in which the PDM is encoded.

A language is adequate with respect to a requirements specification if it has the expressive power to represent the requirements within a PDM in sufficient detail so that a complete PDM can be expressed.

Adequacy is related to the level of granularity needed by the requirements and derives from the idea of representational or expressive adequacy of a knowledge representation language.

Completeness of a PDM depends on language adequacy. A PDM could be accurate—all the features present conform to the requirements, in the sense that their interpretation is true—but it may be the case that some requirements cannot be represented at all. Hence, the completeness of a PDM may be prevented because of an inadequate language.

3.3 Operationality

In the AI planning and scheduling literature, the validation of a domain model is often solely based on a test of whether it will lead to acceptable behavior in a P&S system [38], that is, if an acceptable plan can be output. This is a weak form of completeness as defined above. However, there are normally many encodings of any given PDM that would pass this test, but some encodings lead to much more efficiently generated solutions than others. Given a complete model exists, there will always be ways of re-representing the model without compromising completeness. These models may give different results when input to a planner: for example, some may not satisfy some real time constraints in the requirements. More generally, it is also possible that two distinct domain models are complete, but one leads to a more efficient implementation, or better quality plans. Hence, the process of finding an acceptable plan in an application depends not only on the strategy used by the planner, but also the PDM. For example, if the model is not accurate, then the planner will generate flawed plans or no plans at all [25]. Even the planner's speed can be affected under such circumstances. For instance, case studies have shown that fixing and refining the model itself (e.g., adding additional relevant knowledge) can improve the performance of planners, without modifying the planners and their search mechanism [52]. In addition, works like [4, 35] show that adding relevant, redundant constraints (in the form of control knowledge and rules) in the PDM can also speed up planners.

For a given planner and requirements specification, we define operationality as an attribute of a PDM and a planning engine E as follows.

A PDM is operational with respect to planning engine E if E produces a solution S to P within an acceptable time, such that $I(S)$ is an acceptable solution to $I(P)$ according to the requirements.

Note that the definitions do not demand that the planner outputs all the acceptable solutions—this is somewhat unfeasible computationally, hence we have a weaker definition that is more in tune with practice.

Accepted PDM are sent to a planning engine that provide plans which fit PDM operationality—a second level of post-design. In this chapter we will focus only on the first level of post-design.

There are many tools available addressing PDM, and also claiming to do post-design, but up to 2013 only two use requirements specification and therefore could not implement the first level of post-design mentioned above. Figure 3.2 shows a list of tools and its corresponding features concerning KEPS and PDM.

KEPS process is still in use, relying on sound requirement engineering and formal requirement specification (pre-design), which could match a PDM. Those applications use a unified domain model of problem specification where post-design analysis is applied. In what follows, we will present a clear proposal to achieve that.

KEPS tools	**Design Phases and Properties**									
	Design Phases						Domain	Planner	Intended Expert	
	1	2	3	4	5	6	Independent	Independent	Domain	Planning
O-Plan		✓				≈	✓			✓
SIPE		✓					✓			✓
GIPO		✓	✓	✓	✓	✓	✓	≈	≈	✓
itSIMPLE	✓	✓	≈	✓	✓	✓	✓	✓	✓	✓
EUROPA		✓	✓		✓	✓	✓			✓
ModPlan		✓	≈		✓	✓	✓			✓
VIZ		✓		✓			✓		✓	
TIM			✓				✓			✓
DISCOPLAN			✓				✓	≈		✓
RSA			✓				✓			✓
RedOp			✓				✓	✓		✓
VAL						✓	✓	✓		✓
PlanWorks						✓	✓			✓
MrSPOCK						✓			✓	
JABBAH		✓		✓	✓	✓			✓	
PORSCE II				✓	✓			✓	✓	
CoastWatch						✓			✓	
FlowOpt		✓	✓		✓	✓			✓	
MARIO		✓		✓	✓				✓	
SLAF		✓					✓	✓		✓
LAMP		✓					✓	✓		✓
LOCM		✓					✓	≈		✓
ARMS		✓					✓	✓		✓
Bonasso & Boddy, 2010	✓			✓					✓	✓
Bouillet et al., 2007		✓					✓			✓
Fox & Long, 1999			✓				✓			✓
Crawford et al., 1996			✓				✓			✓
Fernández et al., 2009				✓				✓	✓	
Giuliano & Johnston, 2010						✓			✓	
Myers, 2006						✓	✓		✓	
Chrpa et al., 2012a						✓	✓	✓		✓
Chrpa et al., 2012b			✓				✓	✓		✓
Nakhost & Muller, 2010						✓	✓	✓		✓

Fig. 3.2 Summary of available tools and methods in the Knowledge Engineering For Planning and Scheduling literature, Design phases: (1) Requirements, (2) Knowledge Modeling, (3) Model Analysis, (4) Model Preparation, (5) Plan/Schedule Synthesis, (6) Plan/Schedule Analysis and Post-Design. *Checkmark* means that the feature is present in the tool, *approx* means that it is to some degree present, and *blank* means that it is not present

4 A Knowledge Engineering Design Approach for Planning

It is clear that a formal design process must be adopted in P&S—specially for the real applications mentioned in the introduction for this chapter. Figure 3.3 summarizes the main phases of such process highlighting pre-design and the post-design focused in this work. Notice that the last block is the classic post-design, commonly addressed to the output of the plan engine (which is chosen by a corresponding selection of the planning algorithm). Early post-design can reduce the impact of the classic post-design, which has to translate the plan back to a modeling language to be really effective. We will return to that later, now with opening the possibility to use Petri Nets in this process.

The design process of Fig. 3.3 is generic and does not specify either the engineering requirement approach, the modeling representation to requirements, the knowledge model language or the model translation approach. All that should be clearly specified to have a real design process.

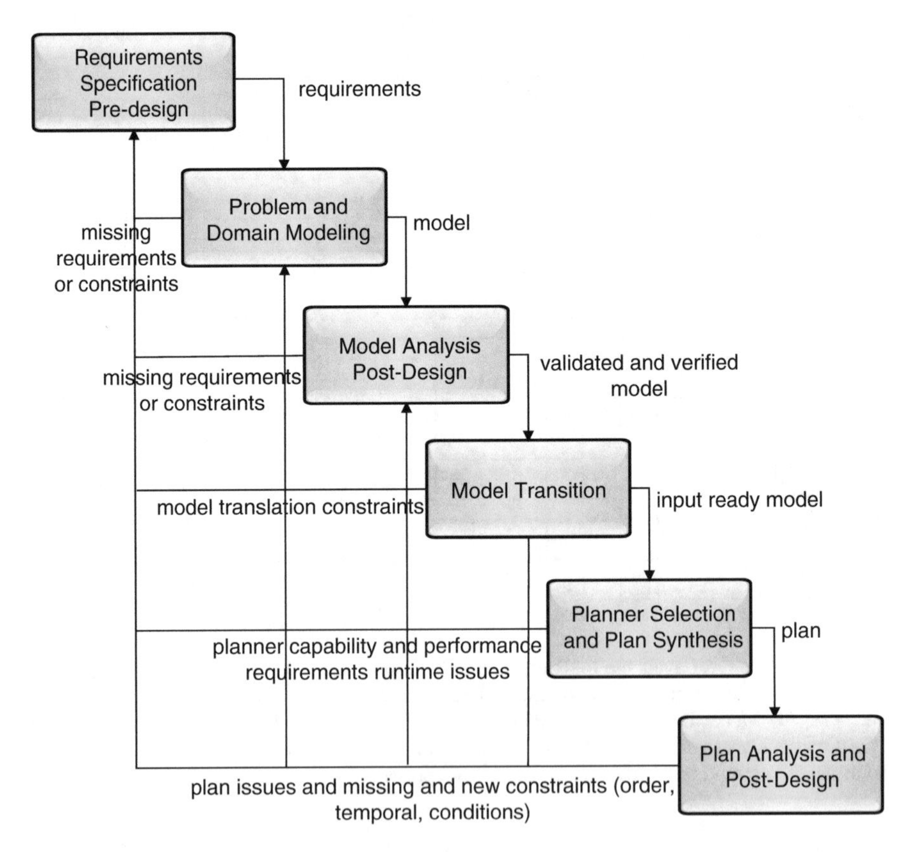

Fig. 3.3 KEPS and PDM design process depicting pre and post-design. First level of post-design is PDM formal verification and traceability with requirements model—pre-design—while a classic post-design approach is directed to the plan engine output

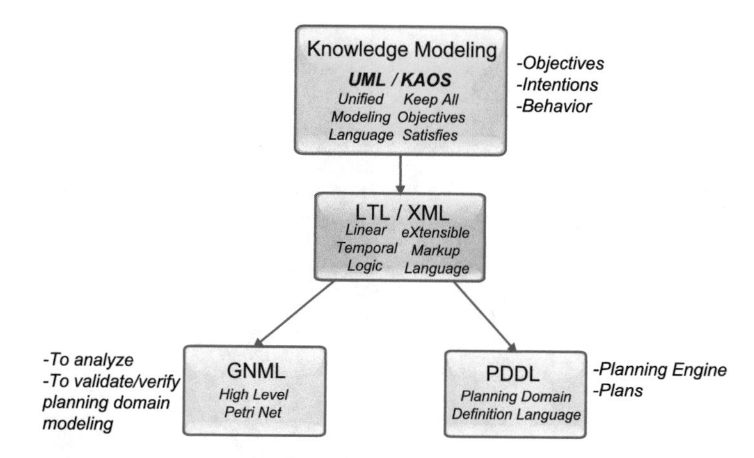

Fig. 3.4 Integration of languages

Starting from last one—model translation—PDDL [17] is a formal representation that collects domain model and problem instance and can feed almost all plan engines, except those that use hierarchical approach. SHOP2 [11] will be the selected representation to hierarchical models.

Both requirements model and knowledge model to PDM will be done in Hierarchical Petri Nets [18, 36, 40]. Requirements will be engineered based on objectives, instead of objects and functionality. Goal oriented requirements engineering is the base for this approach [2, 21], which also uses LTL as a formal representation to requirements. Figure 3.4 shows the choice for language representations proposed to PDM: (1) requirements captured in UML or KAOS diagram [1, 21][1]; (2) model requirements in KAOS can be automatically transferred to LTL or to a based model in XML; (3) formal requirements in XML can be transferred to Petri Nets (using a system called GHENeSys and GNML) or directly to PDDL, and to the plan engine.

There are some reasons to introduce goal-oriented modeling requirements, which are shortly described next.

Using Goal Orientation in Plan Design In the last 5 years the number of tools that give support to requirements modeling grew considerably, many of them using UML. That also means a choice to a functional object-oriented approach which also results in good performance, as in the itSIMPLE tool [51, 52]. However, functional approaches also depend on a complete representation of non-functional requirements, and it is really a challenge to prove such completeness during the acquisition process.

Instead of dozens of diagrams with superposing expressive power, KAOS concentrate the modeling in goals or objectives, which dispense the pair-wising

[1] KAOS (knowledge acquisition in automated specification) is a visual diagram schema proposed in goal oriented requirements engineering that captures knowledge based on goals instead of objects.

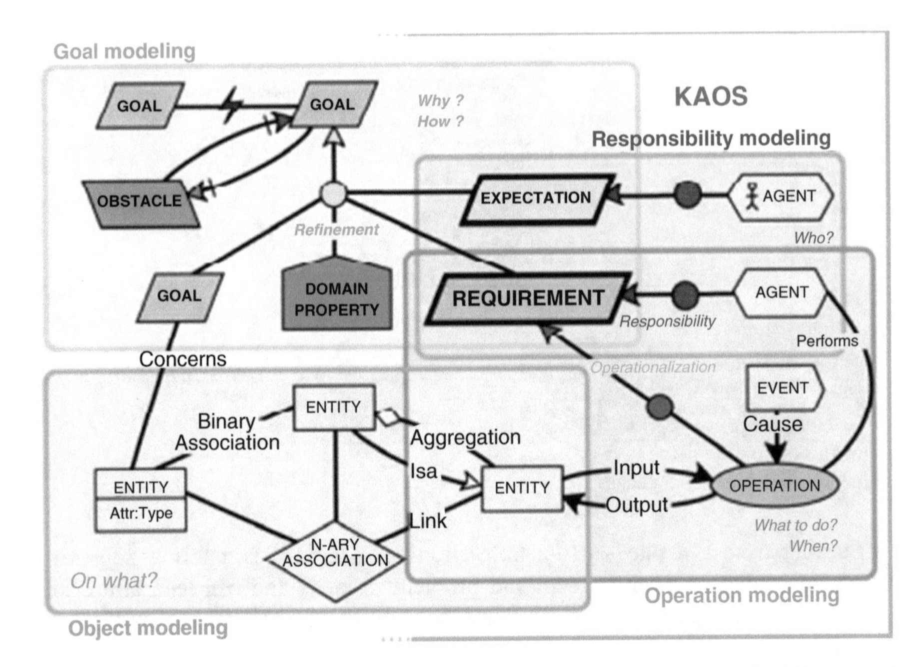

Fig. 3.5 KAOS modeling diagrams (www.objectiver.com/fileadmin/download/documents/KaosTutorial.pdf)

between functional and non-functional requirements to achieve completeness. Incomplete requirements modeling are only due to the lack of knowledge or missing requirements, which is present in any modeling process (Fig. 3.5).

Modeling with KAOS is concentrated in four integrated modeling diagrams: goal model, object model, operation model, and responsibility model. Goal model captures the abstract knowledge about the target and, in the case of planning applications, can capture knowledge about domain and specific problem in an integrated form; object model captures and specifies all objects involved, be it humans or machines, and encapsulates their behavior (which is the main advantage of object-oriented approach); requirements model receives the specification of the planning environment and specifies heuristics, restrictions, and properties that should be used in the planning process; finally, responsibility diagrams map agents and requirements or deliverable expectations providing a base to traceability.

Traceability is a key issue to post-design and also in the guide to fix problems detected in the post-design process. However, it is very hard to implement using object-oriented models. That is the basic reason to look for an alternative representation to knowledge requirements.

Requirements model can be automatically translated to LTL (linear tree logic), which would complete the pre-design process for many short applications. However, to many of the real problems mentioned in the beginning of this chapter that is not enough, and applications require distributed state-transition representation. That is where Petri Nets fit the process.

Petri Nets can be used to model requirements, specially the relationship between the specific problem and the domain environment. The same representation can be refined to fit PDM as shown in Fig. 3.3. Therefore, post-design can be reduced to Petri Net isomorphism and property analysis.

Notice that in what concerns classic post-design, the output of the plan engine should be translated to Petri Nets and can follow the same analysis process, being compared to both nets synthesized from PDM and from requirements. We will not explore this possibility in this work, but from previous experience we can say that is just a matter of adjusting algorithms already developed. A more advanced approach could be achieved by introducing hierarchical Petri Nets into the post-design process, which is what we plan for further work.

Practical problems to translate requirements to Petri Nets are summarized with a realistic example extracted from ROADEF [29]: a sequence of activities in car manufacturing.

5 PDM and Post-Design Modeling Using Petri Nets

Car sequencing is a model problem extracted from a realistic one in the manufacturing process of Renault. It is similar to many similar problems from different car manufacturer that automate partially their sites. The assembling system is divided in three stations: the "white body," where the basic structure is assembled; the painting station; and the assembly, where the final parts (such as doors, windows, etc.) are assembled.

The sequence optimization depends on the proper planning of the demand to reduce setup on stations: changes on the parts available, change of painting gun, etc. It is in fact a classic example where applying AI planning in the manufacturing industry has very good practical results. For that reason it is suitable to show the impact of new planning design approaches (Fig. 3.6).

According to the model of Fig. 3.3, the practical problem to be highlighted is transferring from knowledge modeling representation using in PDM and problem specification to a formal language suitable to model distributed process: Hierarchical Petri Nets (HPN) will be used to formalize post-design analysis.

Automatic transferring from KAOS diagrams to Petri Nets is performed by a tool developed by some of the authors of this work and is called RekPlan [43]. The transference put together an algorithm developed for the tool itSIMPLE which captures object structures in Petri Nets—enhanced by Silva and del Foyo in [41]—

Fig. 3.6 Sequence of stations in car assembling [29]

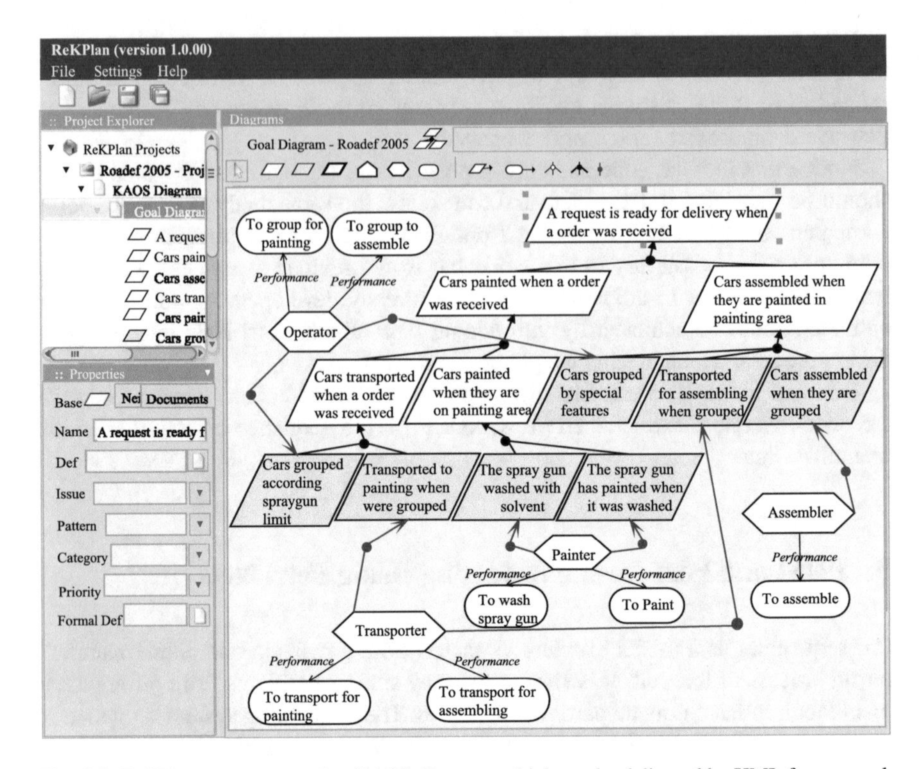

Fig. 3.7 RekPlan system capturing KAOS diagram, which can be delivered by XML from a goal-oriented requirement modeling system

with a new translation algorithm that captures all elements of KAOS diagrams and its context. RekPlan is a stand-alone tool and can be adapted to any other KEPS system, as shown in Fig. 3.7.

Once the KAOS diagram is captured, a Petri Net can be synthesized, as shown in Fig. 3.8.

The Petri Net synthesized is not just a normal place transition net but an extension, called a Unified Net System, which can model classic place/transition net, high level nets and support a transference language—GNML, which stands for GHENeSys Markup Language—and some extensions such as special control elements and hierarchy [41]. All that features fit the standard ISO/IEC 15.909.

The gray circles denote a special place, linked by gates, which marking is not controlled by the system. For instance, the first element is associated with the operator (human) or the availability of a transport system (AGV) from the white body to the painting station. Macro places contains hierarchical subnets and are represented with a rectangle inside, meaning that this an hierarchical place and this place could be replaced by a more detailed network which show, for instance, cars are grouped by special features or according spray gun limitations.

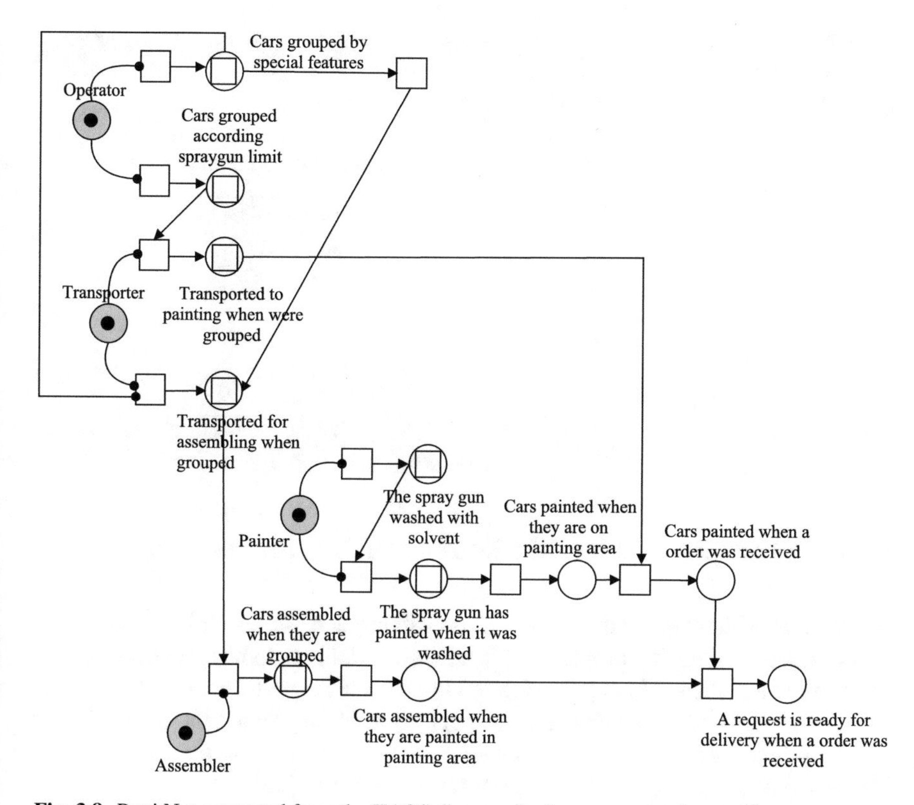

Fig. 3.8 Petri Net generated from the KAOS diagram for the car sequencing problem

Using hierarchical nets makes the transference between requirements modeling to PDM and problem model just by filling up the hierarchical elements with its sub-nets. That guarantees a matching between pre and post-design analysis.

The net models the workflow among the stations without "solving" the planning problem, which will be done by a plan engine. However, all the restrictions and constraints of the domain are present in the net with clear parallelism and distribution. Notice that a plan is not represented just by its shop floor specification. Instead, it is necessary to arrange the demand of cars with defined by the model, color, and, eventually, special features, with the constraints to resource allocation of pieces, such as paint, spray-guns, and others.

Post-design analysis will be done first by property analysis, which is shown in Fig. 3.9.

It is also possible to improve the analysis by introducing analysis of invariants and synchronic distance [36] and timed model checking [41]. That will provide enough elements to face any challenge application as mentioned in the beginning of this chapter.

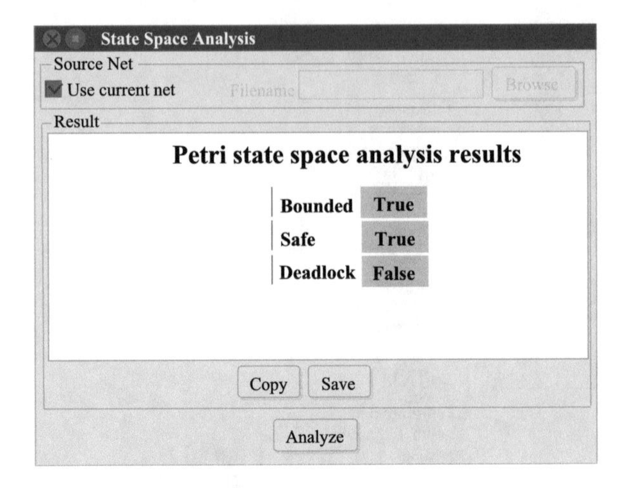

Fig. 3.9 Property analysis generated by GHENeSys tools [41]

6 New Perspectives for AI Planning in Automation Systems

Research in KEPS methods and tools are evolving and taking different routes, as being directed to specific problem applications as roadwork in huge metropolis [47] or complex logistic systems [22], directed to improve the performance of processes [15], to enhance the methods and concepts [13] or by improving the expressibility of methods to model real time systems [7]. Some work rely on machine learning [28], big data [55], or business complex business process [23].

However, there is also a line of work that focuses on the improvement of the abstraction of KEPS design models. In that line of work the goal is to enhance KEPS models to deal with abstract conceptual elements [6, 30] that can provide enough guide to plan engines, but that could also reduce the processing time to synthesize a plan. That will be very important to new collaborative systems composed by humans and robots [9] which will be the new demand, for instance, to new digital manufacturing approaches [42].

Planning systems should face the challenge of fitting the "automated cycle" which stands for the time necessary to sense, process a PDM+problem model, plan the control, and act. Such goal demand more work in abstraction and KEPS modeling but can certainly have a great impact in the use of planning and scheduling systems all over the world.

A step towards a tight automated cycle could be done by using the hierarchical approach showed here to work over a model of behaviors, instead of actions, which can be applied to companion robots and to apprentice machines, that is, machines that work with humans in direct collaboration.

References

1. Almisned, F., Keppens, J.: Requirements Analysis: Evaluating KAOS Models. Journal of Software Engineering and Applications **3**, 869–874 (2010)
2. Ambreen, T., Ikram, N., Usman, M., Niazi, M.: Empirical Research in Requirements Engineering: trends and opportunities. Requirements Engineering **23**(1), 63–95 (2018)
3. Asai, M., Fukunaga, A.: Fully automated cyclic planning for large-scale manufacturing domains. In: 24th. Int. Con. Artificial Planning and Scheduling (June 2014)
4. Bacchus, F.: The AIPS-00 planning competition. AI Magazine **20(3)**, 47–56 (2001)
5. van Beest, N., Russel, N., ter Hofstede, A., Lazovik, A.: Achieving intention-centric BPM through automated planning. In: 7th. IEEE Int. Conf. on Service-oriented Computing and Applications (2014)
6. Bonet, B., Fuentetaja, R., E-Martin, Y., Bonet, B.: Guarantees for Sound Abstractions for Generalized Planning. In: In Proceedings of the 29th. Int. Joint Conference on Artificial Intelligence. AAAI (2019)
7. Cenamor, I.and Vallati, M., Chrpa, L.: On the predictability of domain-independent temporal planners. Computational Intelligence **35**(3) (2019)
8. Cesta, A., Orlandini, A., Umbrico, A.: Fostering Robust Human-Robot Collaboration through AI Task Planning. Procedia CIRP **72**, 1045–1050 (2018)
9. Cesta, A., Finzi, A., Fratini, S., Orlandini, A., Tronci, E.: Validation and Verification Issues in a Timeline-based Planning System. In: Proceedings of the ICAPS 2008 Workshop on Knowledge Engineering for Planning and Scheduling (KEPS). Sydney, Australia (2008)
10. Chen, C., Rickert, M., Knoll, A.: A traffic knowledge aided vehicle motion planning engine based on space exploration guided heuristic search. In: IEEE Intelligent Vehicles Symposium Proceedings. pp. 535–540 (June 2014)
11. Cheng, K., Chen, G., Zhang, R., Wu, L., Wang, Z., Kang, R.: A Method for Unifying the Representation of Domain Knowledge and Planning Algorithm in Hierarchical Task Network. Int. Journal of Pattern Recognition and Artificial Intelligence **31**(8) (2017)
12. Chien, S., Morris, R.: Editorial: Space applications of artificial intelligence. AI Magazine **35**(4), 3–6 (2014)
13. Chrpa, L., Vallati, M., Mccluskey, T.: Inner Entanglements: Narrowing the search in classical planning by problem reformulation. Computational Intelligence **35**(2), 395–429 (2019)
14. Edelkamp, S., Jabbar, S.: Action Planning for Direct Model Checking of Petri Nets. Electronic Notes in Theoretical Computer Science **149**(2), 3–18 (2006)
15. Franco Axela, S., Vallati, M., Mccluskey, T.: Improving Planning performance in PDDL+ Domains via Automated Predicate Reformulation. In: In Proceedings of the International Conference on Computational Science. Springer Verlag (2019)
16. Gath, M., Herzog, O., Edelkamp, S.: Autonomous and flexible multiagent system to enhance transport logistic. In: Proc. of 11th Proc. of the Int. Conf. & Expo on Emergent Technologies for a Smarter World (October 2014)
17. Gerevini, A., Long, D.: Preferences and Soft Constraints in PDDL3. In: Gerevini, A., Long, D. (eds.) Proceedings of ICAPS workshop on Planning with Preferences and Soft Constraints. pp. 46–53. AAAI Press (2006), http://www.plg.inf.uc3m.es/icaps06/preprints/i06-ws1-allpapers.pdf
18. Harie, Y., Mitsui, Y., Fujimori, K., Batajoo, A., Wasaki, K.: HiPS: Hierarchical Petri Nets design, simulation, verification and model checking tool. In: Proceedings of IEEE Global Conference on Consumer Electronics (2017)
19. Kim, S., Shin, Y., Lee, G., Moon, I.: Early stage response problem for post-disaster incidents. Engineering Optimization **50**(7), 1198–1211 (2018)
20. Lallement, R., Silva, L., Alami, R.: HATP: An HTN planner for robotics. In: 24th. Int. Con. Artificial Planning and Scheduling (June 2014)
21. Lamsweerde, A.: Requirements Engineering: from system goals to UML Models to Software Specifications. John Wiley & Sons (2009)

22. Leofante, F., Abraham, E., Tacchela, A.: Task Planning with OMT: An Application to Production Logistic. In: C., F., Winter, K. (eds.) Integrated Formal Methods—Lecture Notes in Computer Science. vol. 11023. Springer (2018)
23. Marrella, A.: Automated Planning for Business Process Management. Journal of Data Semantics pp. 1–20 (2018), https://doi.org/10.1007/s1374
24. McCluskey, T.L.: Knowledge Engineering: Issues for the AI Planning Community. In: Workshop on Knowledge Engineering Tools and Techniques for AI Planning. Sixth International Conference on Artificial Intelligence Planning and Scheduling. pp. 1–4. Toulouse, France (2002)
25. McCluskey, T.L., Aler, R., Borrajo, D., Haslum, P., Jarvis, P., Refanidis, I., Scholz, U.: Knowledge Engineering for Planning Roadmap (2003)
26. McCluskey, T.L., Simpson, R.M.: Knowledge Formulation for AI Planning. In: Knowledge Acquisition, Modeling and Management (EKAW). pp. 449–465 (2004)
27. McCluskey, T.L., V.T.V.M.: Engineering Knowledge for Automated Planning: Towards a Notion of Quality. In: Proceedings of the Knowledge Capture Conference (2017)
28. Mohr, F., Wever, M., Hullermeier, E.: ML-Plan: Automated Machine learning via hierarchical Planning. Machine Learning **107**(8–10) (2018)
29. Nguyen, A.: Challenge ROADEF 2005: Car sequencing problem. Online reference at http://challenge.roadef.org/2005/files/suite_industrielle_2005.pdf, last visited on August of 2016 **23** (2005)
30. Pecora, F., Andreasson, H., Mansouri, M., Peckov, V.: A Loosely-coupled Approach for Multi-Robot Coordination on Automated Planning and Scheduling. In: In Proceedings of the 28th. Int. Joint Conference on Artificial Intelligence. AAAI (2018)
31. Porteous, J., Cavazza, M., Charles, F.: Applying planning to interactive storytelling: Narrative control using state constraints. ACM Transactions on Intelligent Systems and Technology **1**(2), 1–21 (2014)
32. Puerta, A., Egar, J., Tu, S., Musen, M.: A multiple-method knowledge-acquisition shell for the automatic generation of knowledge-acquisition tools. Knowledge Acquisition **4**, 171–196 (1992)
33. Raimondi, F., Pecheur, C., Brat, G.: Verification and Validation of Planning and Scheduling Systems. In: Proceedings of ICAPS 2009. AAAI (2009)
34. Riddle, P., Holte, R.C., Barley, M.: Does representation matter in the planning competition. In: Gnesereth, M.R., Revesz, P.Z. (eds.) SARA. AAAI (2011)
35. de la Rosa, T., McIlraith, S.: Learning Domain Control Knowledge for TLPlan and Beyond. In: Proceedings ICAPS 2011—Workshop on Planning and Learning (2011)
36. Salmon, A., del Foyo, P., Silva, J.: Scheduling real-time systems with periodic tasks using model-checking approach. In: Proc. of 12th IEEE Int. Conf. on Industrial Informatics (July 2014)
37. Schreiber, G., Wielinga, B., Breuker, J. (eds.): KADS: A Principled Approach to Knowledge-Based System Development, Knowledge Based Systems, vol. 11. Academic Press, London (1993)
38. Shah, M., Chrpa, L., Kitchen, D., McClyskey, T.L., V.M.: Exploring Knowledge Engineering Strategies in Designing and Modelling a Road Traffic Accident Management Domain. In: Proceedings IJCAI 2013. AAAI (2013)
39. Shortliffe, E.: MYCIN: A rule-based computer program for advising physicians regarding antimicrobial therapy selection. Ph.D. thesis, Stanford University (1974)
40. Sid, I., Reichert, M., Ghomari, A.: Enabling Flexible task compositions, order and granularities for Knowledge-intensive business process. Enterprise Information System **13**(3), 376–423 (2019)
41. Silva, J., del Foyo, P.: Timed Petri Nets. In: IntechOpen (ed.) Petri Nets—Manufacturing and Computer Science. Springer-Verlag (2012)
42. Silva, J., Nof, S.: Perspectives on Manufacturing Automation Under the Digital and Cyber Convergence. Polytechnica **1**(1–2), 36–47 (2018)

43. Silva, J.M, Silva, J.R.: A New Hierarchical Approach to Requirements Analysis of Problems in Automated Planning. Eng. App. of Artificial Intelligence, **81**, 373–386 (2019).
44. Simpson, R.M.: Structural Domain Definition using GIPO IV. In: Proceedings of the Second International Competition on Knowledge Engineering for Planning and Scheduling. Providence, Rhode Island, USA (2007)
45. Sommerville, I.: Software Engineering. Pearson, 10th edn. (2016)
46. Studer, R., Benjamins, V.R., Fensel, D.: Knowledge Engineering: Principles and Methods. Data and Knowledge Engineering **25**(1–2), 161–197 (March 1998)
47. Vallati, M., Chrpa, L., Kitchin, D.: How to Plan Roadworks in Urban Regions? A Principled Approach Based on AI Planning. In: In Proceedings of the International Conference on Computational Science. Springer Verlag (2019)
48. Vaquero, T.S., Nejat, G., Beck, J.: Planning and scheduling single and multi-person activities in retirement home settings for a group of robots. In: 24th. Int. Con. Artificial Planning and Scheduling (June 2014)
49. Vaquero, T.S., Silva, J.R., Tonidandel, F., Beck, J.C.: itSIMPLE: Towards an Integrated Design System for Real Planning Applications. The Knowledge Engineering Review Journal, special issue on International Competition on Knowledge Engineering for Planning and Scheduling (ICKEPS) (2011)
50. Vaquero, T.S., Romero, V., Tonidandel, F., Silva, J.R.: itSIMPLE2.0: An integrated Tool for Designing Planning Environments. In: Proceedings of the 17th International Conference on Automated Planning and Scheduling (ICAPS 2007). pp. 336–347. AAAI Press (2007)
51. Vaquero, T.S., Sette, F.M., Silva, J.R., Beck, J.C.: Planning and Scheduling of Crude Oil Distribution in a Petroleum Plant. In: Proceedings of ICAPS 2009 Scheduling and Planning Application workshop (2009)
52. Vaquero, T.S., Silva, J.R., Beck, J.C.: Improving Planning Performance Through Post-Design Analysis. In: Proceedings of ICAPS 2010 workshop on Scheduling and Knowledge Engineering for Planning and Scheduling (KEPS). pp. 45–52 (2010)
53. Vaquero, T., Silva, J., Beck, J.: A brief review on tools and methods for knowledge engineering for planning and scheduling. In: Proc. of KEPS Workshop, ICAPS 2011. AAAI Press (2011)
54. Xu, L., Wang, C., Bi, Z., Yu, J.: Object-oriented templates for automated assembly planning of complex products. EEE Trans. on Automation Science and Engineering **11**(2), 492–503 (2014)
55. Xu, Y., S., T., Zeng, X.: AI for Apparel Manufacturing in Big Data Era: A Focus on Cutting and Sewing. In: S., T., Zeng, X. (eds.) Artificial Intelligence for Fashion Industry in the Big Data Era, pp. 125–151. Springer (2018)
56. Yuan, C.C., Chua, F.F.: Autonomic execution of web service composition using AI planning method. Int. J. of Information Technologies and Systems Approach **8**(1), 28–45 (2015)

Chapter 4
MyPDDL: Tools for Efficiently Creating PDDL Domains and Problems

Volker Strobel and Alexandra Kirsch

Abstract The Planning Domain Definition Language (PDDL) is the state-of-the-art language for specifying planning problems in artificial intelligence research. Writing and maintaining these planning problems, however, can be time-consuming and error- prone. To address this issue, we present myPDDL—a modular toolkit for developing and manipulating PDDL domains and problems. To evaluate myPDDL, we compare its features to existing knowledge engineering tools for PDDL. In a user test, we additionally assess two of its modules, namely the syntax highlighting feature and the type diagram generator. The users of syntax highlighting detected 36% more errors than non-users in an erroneous domain file. The average time on task for questions on a PDDL type hierarchy was reduced by 48% when making the type diagram generator available. This implies that myPDDL can support knowledge engineers well in the PDDL design and analysis process.

Keywords PDDL · Planning · Knowledge engineering

1 Introduction

Being a key aspect of artificial intelligence (AI), *planning* is concerned with devising a sequence of actions to achieve a desired goal [7]. AI planning has made remarkable progress in solving planning problems in large state spaces that would be impossible for humans to handle. The International Planning Competition[1] has

[1] http://www.icaps-conference.org/index.php/Main/Competitions/.

V. Strobel (✉)
IRIDIA, Université Libre de Bruxelles, Brussels, Belgium
e-mail: vstrobel@ulb.ac.be

A. Kirsch
Independent Scientist, München, Germany

M. Vallati, D. Kitchin (eds.), *Knowledge Engineering Tools and Techniques for AI Planning*, https://doi.org/10.1007/978-3-030-38561-3_4

led to a number of open source planners that are ready to be used by practitioners and researchers outside the AI planning field.

However, the effectiveness of planning largely depends on the quality of the problem formalization [18]. PDDL (*Planning Domain Definition Language*) [11] is the de facto standard for the description of planning tasks [10]. It divides the description of a planning task into a domain model and problem descriptions: the description of a household with its objects and locations would be a domain, with possible tasks such as making breakfast, cleaning the windows, or changing a light bulb. Someone with a non-planning background, for example, from robotics, has to get used to the PDDL syntax and its possibilities to describe the world. She also has to keep track of the facts in the domain model and the different planning tasks. While automated planning can save vast amounts of time to find a valid solution, creating the planning task specifications is a complex, error-prone, and cumbersome task. An ill-defined problem is often the reason for finding suboptimal plans or no plan at all. The household domain in this example comes with another challenge. For many tasks the distances between objects or other numerical input may be necessary. PDDL is by its very nature as a planning language designed for symbolic specifications. To use numerical data efficiently, it must often be preprocessed.

In this chapter, we describe MYPDDL (Fig. 4.1), a knowledge engineering toolkit that supports knowledge engineers in the entire design cycle of specifying planning

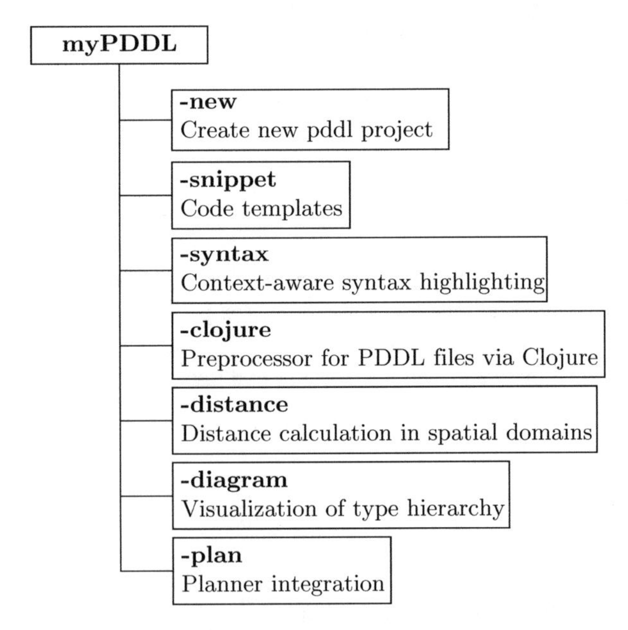

Fig. 4.1 MYPDDL is a highly customizable and extensible modular system, designed for supporting knowledge engineers in the process of writing, analyzing, and expanding PDDL files and thereby promoting the collaboration between knowledge engineers and the use of PDDL in real-world applications. It consists of the parts shown in the figure

tasks without having to become an expert in AI planning. In the initial stages, it allows for the creation of structured PDDL projects that should encourage a disciplined design process. With the help of snippets, that is, code templates, often used syntactic constructs can be inserted into PDDL files. A syntax highlighting feature that speeds up the error detection supports intermediate stages. Understanding the textual representation of complex type hierarchies in domain files can be confusing, so an additional tool enables their visualization. PDDL's limited modeling capabilities were bypassed by developing an interface that converts PDDL code into code of the functional programming language Clojure [8] and vice versa. Within this project, the interface was employed for a feature that calculates distances between objects specified in a problem model, but the interface provides numerous other possibilities and could also be used to further automate the modeling process. A basic planner integration allows for quickly running a desired planner. All of the features were integrated into the customizable and extensible Sublime Text[2] editor.

Since the main aim in the development of the toolkit was for it to be easy to use and maintain, it is evaluated with regard to these criteria. Another aim was to make planning more accessible in real-life tasks and to enable inexperienced users to get started with planning problems. Therefore, MYPDDL's usability was assessed by means of a user test with eight subjects that had no prior experience with AI planning. The results show that MYPDDL facilitates both error detection and the understanding of a given domain.

This chapter is an extended version of work published in a previous paper [20]. The remainder of this chapter is structured as follows. Section 2 compares PDDL knowledge engineering tools to lay a foundation for MYPDDL. Section 3 describes the different modules of MYPDDL and their design principles. Section 4 evaluates MYPDDL via a user test. Section 5 concludes the chapter and outlines future work.

2 Related Work

This section introduces, compares, and discusses knowledge engineering tools that allow text-based editing of PDDL files to set the stage for MYPDDL (Table 4.1).

PDDL STUDIO [15] is an IDE (*integrated development environment*) for creating and managing PDDL projects, that is, a collection of PDDL files. Its main features are syntax highlighting, error detection, context sensitive code completion, code folding, project management, and planner integration. Many of these features are based on a parser, which continuously analyzes the code and divides it into syntactic elements. These elements and the way in which they relate to each other can then be identified. The syntax highlighter is a tool that colors constructs according to their syntactical meaning within the code. In the case of PDDL STUDIO, it colors names, variables, errors, keywords, predicates, types, and brackets each

[2]http://www.sublimetext.com/.

Table 4.1 Comparison of knowledge engineering tools and their features

Feature	Function	PDDL STUDIO	ITSIMPLE	PDDL-mode	Planning.domains	vscode-PDDL	MYPDDL
Latest supp. PDDL version	Considering recent PDDL features	1.2	3.1	2.2	3.1	3.1	3.1
Syntax highlighting	Supporting error detection and code navigation	Yes	Basic	Basic	Basic	Yes	Yes
Semantic error detection	Supporting error detection	Yes	No	No	No	Yes	No
Automatic indentation	Supporting readability and navigation	No	No	Yes	No	Yes	Yes
Code completion	Speeding-up the knowledge engineering process	Yes	No	Yes	Yes	Yes	Yes
Code snippets	Speeding-up the knowledge engineering process externalizing user's memory	No	Yes	Yes	Basic	Yes	Yes
Code folding	Supporting keeping an overview of the code structure	Yes	No	Yes	Yes	Yes	Yes
Domain visualization	Supporting fast understanding of the domain structure	No	No	No	No	No	Yes
Project management	Supporting keeping an overview of associated files	Yes	Yes	No	Yes	Yes	Yes
UML to PDDL translation	Supporting initial modeling	No	Yes	No	No	No	No
Planner integration	Allowing for convenient planner access	Yes	Yes	No	Yes	Yes	Yes
Plan visualization	Supporting understanding and crosschecking the plan	No	Yes	No	No	Yes	No
Dynamic analysis	Supporting dynamic domain analysis	No	Yes	No	No	No	No
Declaration menu	Supporting code navigation	No	No	Yes	No	Yes	No
Interface with programming language	Automating tasks extending PDDL's modeling capabilities	No	No	No	No	No	Yes
Customization features	Acknowledging individual needs and preferences	Basic	No	Yes	Basic	Yes	Yes

in a different customizable color. PDDL STUDIO's error detection can recognize both syntactic (missing keywords, parentheses, etc.) and semantic (wrong type of predicate parameters, misspelled predicates, etc.) errors. This means that PDDL STUDIO can detect errors based on a mismatch between domain and problem file in real time. The code completion feature offers recommendations for standard PDDL constructs as well as for previously used terms. Code folding allows the knowledge engineer to hide currently not needed code blocks. In this case only the first line of the block is displayed. Lastly, a command-line interface allows the integration of planners in order to run and compare different planning software.

Unlike PDDL STUDIO, which provides a text-based editor for PDDL, the ITSIMPLE [22] editor has, as its main feature, a graphical approach that allows designing planning tasks in an object-oriented approach using UML (*Unified Modeling Language*). In the process leading up to ITSIMPLE, UML.P—UML in a planning approach—was proposed, which is a UML variant specifically designed for modeling planning domains and problems [21].

The main purpose of ITSIMPLE is supporting knowledge engineers in the initial stages of the design phase by making tools available which help transform the informality of the real world to formal specifications of domain models. The professed aim of the project is to provide a means to a "disciplined process of elicitation, organization, and analysis of requirements" [22]. However, subsequent design stages are also supported. Once domain and problem models have been created, PDDL representations can be generated from the UML.P diagrams, edited, and then used as input to a number of different integrated planning systems.

With ITSIMPLE, it is possible to directly input the domains and problems into a planner and to inspect the output from the planning system using the built-in plan analysis. This consists of a plan visualization, which shows the interaction between the plan and the domain by highlighting every change caused by an action. ITSIMPLE's modeling workflow is unidirectional as changes in the PDDL domain do not affect the UML model and UML models have to be modeled manually, meaning that they cannot be generated using PDDL.

Starting in version 4.0, ITSIMPLE expanded its features to allow the creation of PDDL projects from scratch, that is, without the UML to PDDL translation process [23]. So far, a basic syntax highlighting feature recognizes PDDL keywords, variables, and comments. ITSIMPLE also provides templates for PDDL constructs, such as requirement specifications, predicates, actions, initial state, and goal definitions.

PDDL-mode[3] for Emacs builds on the sophisticated features of the widely used Emacs editor and uses its extensibility and customizability. PDDL-mode provides syntax highlighting by way of basic pattern matching of keywords, variables, and comments. Additional features are automatic indentation and code completion as well as bracket matching. Code snippets for the creation of domains, problems, and actions are also available. Finally, PDDL-mode keeps track of action and problem

[3]http://rakaposhi.eas.asu.edu/planning-list-mailarchive/msg00085.html.

declarations by adding them to a menu and thus intending to allow for easy and fast code navigation.

PDDL-mode for Emacs supports PDDL versions up to 2.2, which includes derived predicates and timed initial predicates [3], but does not recognize later features like object-fluents.

The online tool editor.planning.domains allows for editing PDDL files in a web browser. Its features comprise syntax highlighting, code folding, PDDL-specific auto-completion, and multi-tab support. The editor is part of the Planning-Domains[4] initiative which aims at providing three pillars to the planning community: (1) an API to access existing PDDL domains and problems; (2) a planner-in-the-cloud service which can be accessed via a RESTful API; and (3) an online PDDL editor. The online editor is also connected to the planner in the cloud.

The PDDL plugin vscode-PDDL[5] for the editor VS Code (*Visual Studio Code*) offers a wide range of editing functions, such as syntax highlighting, code completion, code folding, and code snippets. It offers a mature planner integration and plan visualization. Thanks to a PDDL parser integration, it is possible to detect semantic errors immediately when they are made.

2.1 Critical Review

All the above-mentioned tools provide environments for the creation of PDDL code. Their advantages and disadvantages are reviewed in this section. At the end of each discussed feature, the approach that was used in MYPDDL is introduced.

PDDL STUDIO, ITSIMPLE, and editor.planning.domains for the most part do not build on existing editors and therefore cannot fall back on refined implementations of features, such as selection of tab size, defining custom key shortcuts, customizing the general look and feel, and bracket matching. In contrast, vscode-PDDL and PDDL-mode for Emacs are integrated into mature code editors and can be used in combination with other plugins. To have both basic editor features and a high customizability, it was decided to use an existing, extensible text editor to integrate MYPDDL into.

The tools can also be compared in terms of their syntax highlighting capabilities. In PDDL-mode for Emacs (up to PDDL 2.2), editor.planning.domains (up to PDDL 3.1), and vscode-PDDL (up to PDDL 3.1) keywords, variables, and comments are highlighted. However, this is only done via pattern matching without controlling for context. This means that wherever the respective terms appear within the code they will get highlighted, regardless of the syntactical correctness. Different colors can be chosen by customizing Emacs and Visual Studio Code. editor.planning.domains provides two fixed color schemes. ITSIMPLE's syntax highlighting for PDDL 3.1

[4]http://planning.domains/.

[5]https://marketplace.visualstudio.com/items?itemName=jan-dolejsi.pddl.

is, except for the PDDL version difference, equally as extensive as that of PDDL-mode for Emacs but does not allow for any customization. PDDL STUDIO has advanced syntax highlighting that distinguishes all different PDDL 1.2 constructs depending on the context and allows knowledge engineers to choose their preferred highlighting colors. One of the primary objectives of MYPDDL is to help users in keeping track of their PDDL programs. As a means to this end, it was decided to also implement sophisticated, context-dependent syntax highlighting.

Another useful feature for fast development is the ability to insert larger code skeletons or snippets. PDDL STUDIO does not support the insertion of code snippets. ITSIMPLE features some code templates for predicates, derived predicates, functions, actions, constraints, types, comments, requirements, objects, and metrics. However, the templates are neither customizable nor extensible. PDDL-mode for Emacs provides three larger skeletons: one for domains, one for problems, and one for actions. Further skeletons could be added. Both editor.planning.domains and vscode-PDDL provide many code snippets. MYPDDL aims to combine the best of these latter tools and support customizable and extensible snippets for domains, problems, types, predicates, functions, actions, and durative actions.

PDDL STUDIO, PDDL-mode for Emacs, and editor.planning.domains do not provide visualization options. ITSIMPLE, on the other hand, is based entirely on visually modeling domains and problems. Therefore, since the first version, the focus has mainly been on exporting from UML.P to PDDL and to visualize plans. MYPDDL is to reverse this design approach and enable type diagram visualization of some parts of the PDDL code. vscode-PDDL does not provide domain visualization but is able to visualize a found plan.

Searching for errors can be one of the most time-consuming parts of the design process. Hence, any tool that is able to help detect errors faster is of great value to the knowledge engineer. While PDDL-mode for Emacs, ITSIMPLE, and editor.planning.domains facilitate error detection only by basic syntax highlighting, both PDDL STUDIO and vscode-PDDL are able to detect errors via a PDDL parser. In MYPDDL, a different approach is taken and syntactic errors are *not* highlighted by the syntax highlighting feature, while all correct PDDL code *is* highlighted.

A major drawback of PDDL STUDIO and PDDL-mode for Emacs especially is that they are not updated regularly to support the most recent PDDL versions. PDDL STUDIO's parser is only able to parse PDDL 1.2, while the latest PDDL version is 3.1. PDDL has significantly evolved since PDDL 1.2 and was extended in PDDL 2.1 to include *durative actions* to model time dependent behaviors, *numeric fluents* to model non-binary changes of the world state, and *plan-metrics* to customize the evaluation of plans. PDDL-mode for Emacs is only compatible with PDDL versions up to 2.2, which introduced *derived predicates* and *timed initial predicates* but does not recognize later features like *object-fluents*. It follows that the range of functions specified in the domain file cannot include object types in addition to numbers. ITSIMPLE, editor.planning.domains, vscode-PDDL, and MYPDDL support the latest PDDL version.

PDDL STUDIO falls short of customization options since they are limited to the choice of font style and color of highlighted PDDL expressions. Furthermore,

PDDL STUDIO is written as a standalone program, meaning that there are no PDDL-independent extensions. The same holds true for ITSIMPLE. Since both Emacs and VS Code are established editors, PDDL-mode and vscode-PDDL are highly customizable and extensible. This is the other major reason why it was decided that MYPDDL should be integrated into an existing, extensible, and customizable text editor. These requirements are met by Sublime Text, a text editor that offers a wide variety of features and plugins.

All in all, MYPDDL must be understood as complementary to the other existing knowledge engineering tools. MYPDDL is distributed as a package for Sublime Text and provides context-aware syntax highlighting, code snippets, syntactic error detection, and type diagram visualization. Additionally, it allows for the automation of modeling tasks due to an interface with Clojure that supports the conversion of PDDL code into Clojure code and vice versa. Therefore, MYPDDL is intended to support both the initial design process of creating domains with code snippets, syntax highlighting and the Clojure interface, and the later step of checking the validity of existing domains and problems with the type diagram generator. Lastly, the visualization capabilities of MYPDDL are meant to facilitate collaboration among knowledge engineers.

3 MyPDDL

MYPDDL is a highly customizable and extensible modular system, designed for supporting knowledge engineers in the process of writing, analyzing, and expanding PDDL files and thereby promoting the collaboration between knowledge engineers and the use of PDDL in real-world applications. The modules of MYPDDL are described in the next section.

3.1 Modules

myPDDL-IDE is an integrated development environment (IDE) for the use of MYPDDL in the text and code editor *Sublime Text*.[6] Since MYPDDL-SNIPPET and -SYNTAX (see below) are devised explicitly for Sublime Text, their integration is implicit. The other tools described below (MYPDDL-new, -diagram, -distance, -plan) can be used independently of Sublime Text via the command-line but can also be called from the editor.

myPDDL-new helps to organize PDDL projects. In many cases PDDL domains are created ad hoc [19]. However, each implementation of a PDDL task specification comprises one domain and at least one corresponding problem file. Since

[6]http://www.sublimetext.com.

several team members may be working on these files, keeping PDDL projects organized will facilitate collaboration. An automatically created, standardized project folder structure could facilitate the collaboration between users and the maintenance of consistency across projects. To this end, MYPDDL-NEW creates the following folder structure when creating a new PDDL project:

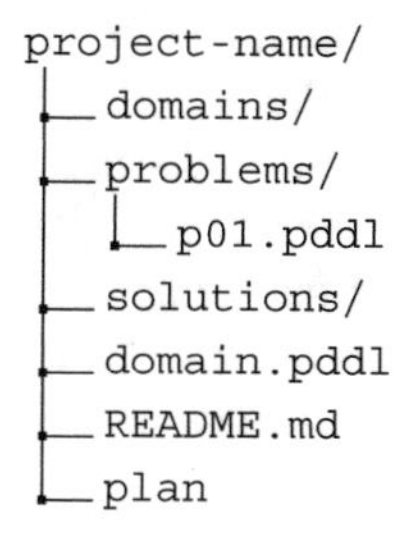

All of the templates to create the files can be customized and new templates can be added. The domain file `domain.pddl` and the problem file `p01.pddl` initially contain corresponding PDDL skeletons. Additionally the project name is used as the domain name within the files `domain.pddl` and `p01.pddl`. All problem files that are associated with one domain file are collected in the folder `problems/`. `README.md` is a Markdown file, which is intended for information about the authors of the project, contact information, informal domain and problem specifications, and licensing information. Markdown files can be converted to HTML by various hosting services like GitHub or Bitbucket. The basic planner integration MYPDDL-PLAN provided by the file `plan` is described below.

myPDDL-snippet provides code skeletons, that is, templates for often used PDDL constructs such as domains, problems, type and function declarations, and actions. They can be inserted by typing a triggering keyword. Table 4.2 displays descriptions of all available snippets and the corresponding trigger.
For example, typing `action` and pressing the tabulator key inserts a skeleton to specify an action. PDDL constructs with a specified arity can be generated by adding the arity number to the trigger (p2 would insert the binary predicate template `(pred-name ?x - object ?y - object)`).

Table 4.2 The snippets that can be inserted into PDDL files by typing the trigger

Snippet description	Trigger
Domain skeleton	`domain`
Problem skeleton	`problem`
Type declaration	`t1, t2, ...`
Typed predicate declaration	`p1, p2, ...`
Typed function declaration	`f1, f2, ...`
Action skeleton	`action`, `durative-action`

Every snippet is stored in a separate file, located in the packages folder of Sublime Text. New snippets can be added and existing snippets can be customized by changing the templates in this folder.

myPDDL-syntax is a context-aware syntax highlighting feature for Sublime Text. It recognizes all PDDL constructs up to version 3.1, such as comments, variables, names, and keywords and highlights them in different colors. Using regular expressions and a sophisticated pattern matching heuristic, it detects both the start and the end of PDDL code blocks and constructs. It then divides them into *scopes*, that is, named regions. Sublime Text colorizes the code elements via the assigned scope names and in accordance with the current color scheme. These scopes allow for a fragmentation of the PDDL files, so that constructs are only highlighted if they appear in the correct context. Thus missing brackets, misplaced expressions, and misspelled keywords are visually distinct and can be identified (Fig. 4.2).

myPDDL-Clojure provides a preprocessor for PDDL files to bypass PDDL's limited mathematical capabilities, thus reducing modeling time without overcharging planning algorithms. Since PDDL is used to create more and more complex domains [5, 6], one might need the square root function for a distance optimization problem or the logarithmic function for modeling an engineering problem. While these mathematical operations are currently not supported by PDDL itself, preprocessing PDDL files in a programming language and then hardcoding the results back into the file seem to be a reasonable workaround. With the help of such an interface, the modeling time can be reduced. We decided to use the functional programming language Clojure [8], a modern Lisp dialect, facilitating input and output of the Lisp-style PDDL constructs. Once a part is extracted and represented in Clojure, the processing possibilities are diverse and the full capacities of Clojure are available. It can be used for generating PDDL

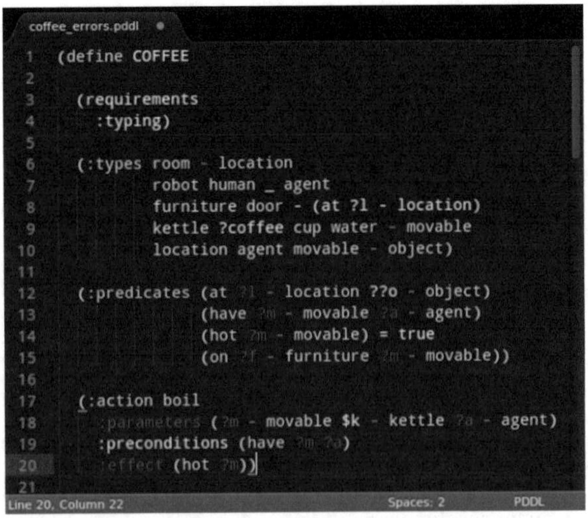

```
coffee_errors.pddl
1  (define COFFEE
2
3    (requirements
4      :typing)
5
6    (:types room - location
7            robot human _ agent
8            furniture door - (at ?l - location)
9            kettle ?coffee cup water - movable
10           location agent movable - object)
11
12   (:predicates (at ?l - location ??o - object)
13                (have ?m - movable ?a - agent)
14                (hot ?m - movable) = true
15                (on ?f - furniture ?m - movable))
16
17   (:action boil
18     :parameters (?m - movable $k - kettle ?a - agent)
19     :preconditions (have ?m ?a)
20     :effect (hot ?m))
21
Line 20, Column 22                                Spaces: 2      PDDL
```

Fig. 4.2 The figure shows the use of MYPDDL in the text editor Sublime Text. Syntax errors in the domain are detected by MYPDDL-SYNTAX's context-aware syntax highlighting feature and displayed in white

constructs, reading domain and problem files, handling, using and modifying the input, and generating PDDL files as output.
The interface is provided as a Clojure library and based on two methods described below.

read-construct(keyword, file) This method allows for the extraction of code blocks from PDDL files. The following code block shows an example in which the goal state `(:goal (exploited magicfailureapp))` is extracted from a PDDL problem file.

Clojure command:

```
(read-construct :goal "garys-huge-problem.pddl")
;;=> ((:goal (exploited magicfailureapp)))
```

add-construct(file, block, part) This method provides a means for adding constructs to a specified code block in PDDL domain and problem files. This is illustrated in the following two code blocks where the predicate `(hungry gisela)` is added to the `(:init ...)` block.

Clojure command:

```
(add-construct "garys-huge-problem.pddl" :init '((hungry gisela)))
```

Updated PDDL file:

```
(:init (hungry gary)
       (in pizza-box big-pepperoni)
       (has-access gisela magicfailureapp))
       (hungry gisela))
```

myPDDL-distance provides special preprocessing functions for distance calculations. In some domains, every object needs a location specified by x and y coordinates. While the location of objects can be implemented using the predicate `(location ?o - object ?x ?y - number)`, with `x` and `y` being the spatial coordinates of an object, calculating the Euclidean distance requires using the square root function. However, PDDL 3.1 supports only the four basic arithmetic operators.
Parkinson and Longstaff [14] describe a workaround for this drawback. By writing an action `calculate-sqrt`, they bypass the missing square root function by making use of the Babylonian root method. Although this method approximates the square root function, it requires many iterations and would most likely have an adverse effect on plan generation [14].
More usable and probably faster results can be achieved by using the interface between PDDL and Clojure as a distance calculator, implemented in the tool MYPDDL-DISTANCE. It reads a problem file into Clojure and extracts all locations, defined in the `(:init ...)` code block. The Euclidean distances between these locations are then calculated and written back into

a new and now extended copy of the problem file, using the predicate `(distance ?o1 ?o2 - object ?n - number)`, which specifies the distance between two objects. The code blocks below show the `(:init ...)` block of a PDDL problem file before and after using MYPDDL-distance.
Before:

```
(:init ...
       (location gary 4 2)
       (location pizza 2 3))
```

After:

```
(:init ...
       (location gary 4 2)
       (location pizza 2 3)
       (distance gary gary 0.0)
       (distance gary pizza 2.2361)
       (distance pizza gary 2.2361)
       (distance pizza pizza 0.0))
```

The calculator works on any arity of the specified location predicate, so that locations can be specified in 1D, 2D, 3D, and even used in higher dimensions. A disadvantage of this method is that the calculated distances have to be stored in the PDDL problem file, potentially requiring many lines of code. If the number of locations is n, the number of calculated distances is n^2, so that every location has a distance to every other location and itself. Therefore, a sensible next step would be to extend PDDL by increasing its mathematical expressivity [14], perhaps by declaring a requirement `:math` that specifies further mathematical operations.

myPDDL-diagram generates a PNG image based on the type hierarchy of a PDDL domain file (Fig. 4.3). The diagrammatic representation of textual information helps to quickly understand the connection of hierarchically structured items and should thus be able to simplify the communication and collaboration between developers. In the diagram, types are represented with boxes, with every box consisting of two parts:

- The header displays the name of the type.
- The lower part displays all predicates that use the corresponding type at least once as a parameter. The predicates are written just as they appear in the PDDL code.

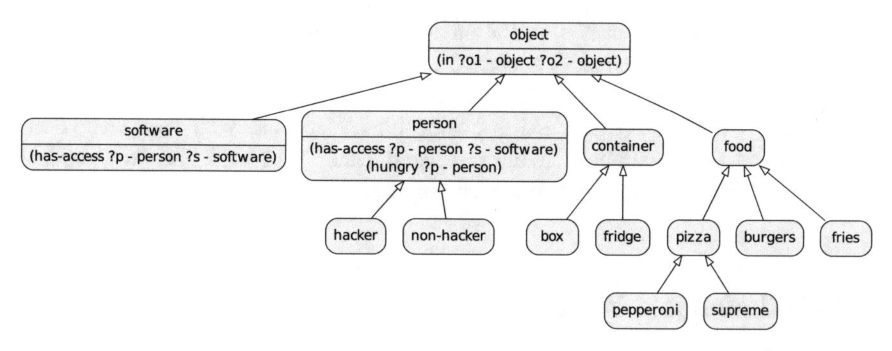

Fig. 4.3 The type diagram generated by MYPDDL-DIAGRAM helps to grasp the relationship between types in the domain file. Additionally, it displays all predicates that use the corresponding type at least once as a parameter

Generalization relationships express that every subtype is also an instance of the illustrated super type (e.g., "a hacker *is a* person). This relationship is indicated in the diagram with an arrow from the subtype (here: *hacker*) to the super type (here: *person*).

In order to create the diagram, MYPDDL-diagram utilizes dot from the Graphviz package [4] and takes the following steps:

1. A copy of the domain file is stored in the folder `domains/`.
2. The `(:types ...)` block is extracted via the PDDL/Clojure interface.
3. In Clojure, the types are split into super types and associated subtypes using regular expressions and stored in a Clojure hashmap.
4. Based on the hashmap, the description of a directed graph in the DOT language is created and saved in the folder `dot/`.
5. The DOT file is passed to dot, creating a PNG diagram and saving it in the folder `diagrams/`.
6. The PNG diagram is displayed in a window.

Every time MYPDDL-diagram is invoked, these steps are executed and, optionally, the names of the saved files are extended by an ascending revision number. Thus, one cannot only identify associated PDDL, DOT, and PNG files, but also use this feature for basic revision control.

myPDDL-plan is a basic planner integration for MYPDDL. After creating a new project with MYPDDL-NEW, the file `plan` in the project folder contains a shell script for executing a planner with the new domain and problem files as input. The desired planner can be specified in the file `plan` or by editing the templates of MYPDDL-NEW. Due to the versatility of shell scripts, any planner can be used and arbitrary command-line options can be specified. The planner can be invoked from Sublime Text or via the command-line.

In order to provide easy installation and maintenance, MYPDDL-IDE can be installed using Sublime Text's Package Control.[7] The project source code is hosted on GitHub,[8] providing the possibility to actively participate in the design process. Additionally, MYPDDL-CLOJURE is hosted on GitHub[9] as well as a standalone version to call the functions from the command- line.[10] The MYPDDL project site[11] provides room for discussing features and reporting bugs.

4 Validation and Evaluation

To assess the utility of MYPDDL, we evaluated its performance in terms of collaboration, experience, efficiency, and debugging in a user test. We analyzed the user performance both with and without using MYPDDL-SYNTAX and MYPDDL-DIAGRAM.

4.1 User Evaluation

The two most central modules of MYPDDL are MYPDDL-SYNTAX and MYPDDL-DIAGRAM, since they support collaboration, efficiency, and debugging independently of the user's experience with PDDL. To evaluate their usability, they were evaluated in a user study. To this end, we compared the user performance regarding several tasks, both with and without using the respective module.

Participants In Usability Engineering, a typical number of participants for user tests is five to ten. Studies have shown that even such small sample sizes identify about 80% of the usability problems [9, 12]. Our study design required eight participants. Three female and five male participants took part in the study (average age of 22.9, standard deviation of age 0.6). All participants were required to have basic experience with at least one Lisp dialect in order not to be confused with the many parentheses, but no experience with PDDL or AI planning in general.

Approach Twenty-four hours before the experiment was to take place, participants received the web link[12] to a 30-min interactive video tutorial on AI planning and

[7]https://sublime.wbond.net/about.

[8]https://github.com/Pold87/myPDDL.

[9]https://github.com/Pold87/pddl-clojure-interface.

[10]https://github.com/Pold87/pddl-clojure-interface-standalone.

[11]http://pold87.github.io/myPDDL/.

[12]Tutorial in German: https://www.youtube.com/watch?v=Uck-K8VnNOU&list=PL3CZzLUZuiIMWEfJxy-G6OxYVzUrvjwuV.

PDDL. This method was chosen in order not to pressure the participant with the presence of an experimenter when trying to understand the material.

Procedure We defined four tasks (Appendix "Tasks"): two debugging tasks for testing the syntax highlighting feature and two type hierarchy tasks for testing the type diagram generator. A within-subjects design was considered most suited due to the small number of participants. Therefore, it was necessary to construct two tasks matched in difficulty for each of these two types to compare the effects of having the tools available. Each participant started either with a debugging or type hierarchy task and was given the MYPDDL tools either in the first two tasks or the second two tasks, so that each participant completed each task type once with and once without MYPDDL. This results in 2 (first task is debugging or hierarchy) $\times$ 2 (task variations for debugging and hierarchy) $\times$ 2 (starting with or without MYPDDL) $=$ 8 individual task orders, one per participant.

- Debugging Tasks
  For the debugging tasks, participants were given 6 min (a reasonable time frame tested on two pilot tests) to detect as many of the errors in the given domain as possible. They were asked to record each error in a table using pen and paper with the line number and a short comment. Moreover, they were instructed to immediately correct the errors in the code if they knew how to, but not to dwell on the correction otherwise. For the type hierarchy task, participants were asked to answer five questions concerning the domains, all of which could be facilitated with the type diagram generator. One of the five questions (Question 4, see Appendix sections "Planet Splisus" and "Store") also required looking into the code. Participants were told that they should not feel pressured to answer quickly, but to not waste time either. Also they were asked to say their answer out loud as soon as it became evident to them. They were not told that the time it took them to come up with an answer was recorded, since this could have made them feel pressured and thus led to more false answers.
- Type Hierarchy Tasks
  The two tasks to test syntax highlighting presented the user with domains that were 54 lines in length, consisted of 1605 characters and contained 17 errors each. Errors were distributed evenly throughout the domains and were categorized into different types. The occurrence frequencies of these types were matched across domains as well, to ensure equal difficulty for both domains. To test the type diagram generator, two fictional domains with equally complex type hierarchies consisting of non-words were designed (five and six layers in depth, 20 and 21 types). The domains were also matched in length and overall complexity: five and six predicates with approximately the same distribution of arities, one action with four predicates in the precondition and two and three predicates in the effect.
- System Usability Scale
  At the end of the usability test the participants were asked to evaluate the perceived usability of MYPDDL using the system usability scale [2].

4.1.1 Analysis

- Debugging Tasks
  To compare differences in the debugging tasks, a paired sample t-test was used; normality was tested with a Shapiro–Wilk test. To compare the arithmetic means (Ms) of detected errors, the test was performed two-tailed, since syntax highlighting might both help or hinder the participants. Arithmetic standard deviations (SDs) were calculated for each condition.
- Type Hierarchy Tasks
  For the type hierarchy tasks, t-tests were performed on the logarithms of the data values to compare the geometric means for the two conditions for each question; normality was tested with a Shapiro–Wilk test on the log-normalized data values. The geometric mean is a more accurate measure of the mean for small sample sizes as task times have a strong tendency to be positively skewed [17]. The geometric standard deviation (GSD) was calculated for each question and condition. Only those task completion times were included in the calculation of the t-values, where the respective participant gave a correct answer for both occurrences of a question. This approach should reduce the influence of random guessing. Again, two-tailed t-tests were used to account for both, improvements and drawbacks, of using MYPDDL-DIAGRAM.
- System Usability Scale
  The arithmetic mean and standard deviation for the score on the system usability scale was calculated.

4.1.2 Results

- Debugging Tasks
  The participants detected more errors using the syntax highlighting feature ($M = 10.3$, $SD = 3.45$) than without it ($M = 7.6$, $SD = 2.07$); $t(7) = 2.68$, $p = 0.03$. That is, approximately 36% more errors were found with syntax highlighting. The arithmetic means are displayed in Fig. 4.4, where each cross ($\times$) represents the data value of one participant.
- Type Hierarchy Tasks
  Figure 4.5 shows the geometric mean of the completion time of successful tasks for each question with and without the type diagram generator. With the type diagram generator participants answered all questions (except Question 4) on average nearly twice as fast ($GM = 33.0$, $GSD = 2.23$) as without it ($GM = 57.8$, $GSD = 2.05$); $t(32) = -3.34$, $p = 0.002$. This difference slightly increases if Question 4 is excluded from the calculations: with type generator: $GM = 31.1$, $GSD = 2.17$, without: $GM = 58.1$, $GSD = 2.07$; $t(30) = -3.68$, $p < 0.001$. Table 4.3 gives an overview of geometric means, geometric standard deviations, t-values, and p-values for each question.
- System Usability Scale

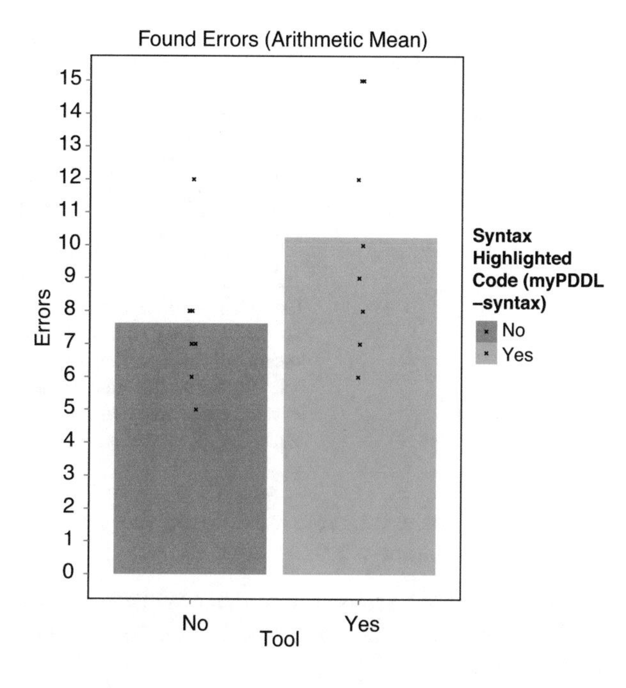

Fig. 4.4 Comparison of detected errors with and without the syntax highlighting feature. Each cross (×) shows the data value of one participant. The bars display the arithmetic mean

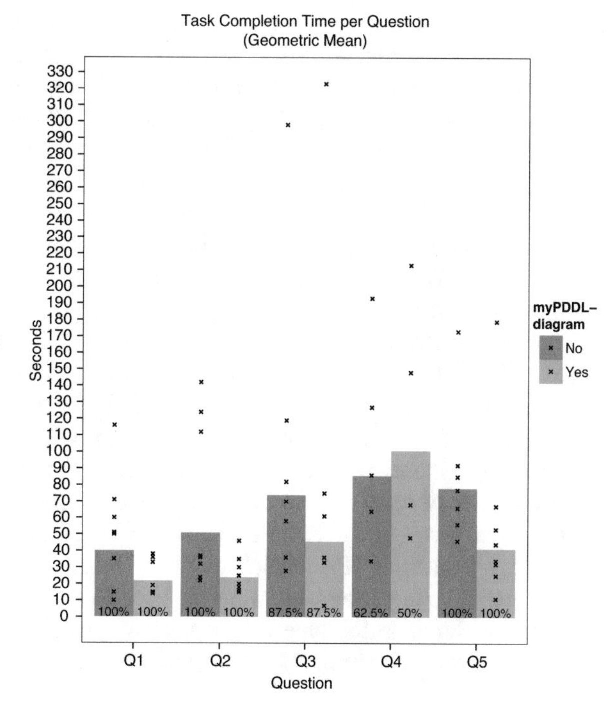

Fig. 4.5 Task completion time for the type hierarchy tasks. The bars display the geometric mean averaged over all participants; each cross (×) represents the data value of one participant. The percent values at the bottom of the bars show the percentage of users that completed the task successfully. The questions can be found in the Appendix sections "Planet Splisus" and "Store"

Table 4.3 Overview of geometric means (GMs), geometric standard deviations ($GSDs$), degrees of freedom (df), t-values, and p-values

	Type diagram generator						
	With		Without				
Question	GM	GSD	GM	GSD	df	t	p
Q1	21.8	1.52	40.0	2.26	7	−1.86	0.11
Q2	23.8	1.49	50.8	2.16	7	−1.91	0.10
Q3	48.0	3.49	83.2	2.20	5	−0.86	0.43
Q4	84.3	2.22	54.1	1.93	1	4.48	0.14
Q5	41.2	2.24	78.0	1.48	7	−2.75	0.03

The calculation for Q4 is based on only two paired data values ($df = 1$). This table only considers paired data values, this means only if a participant answered the question correctly in both domains, the data value is considered (since *paired t-tests* are calculated). In contrast, Fig. 4.5 displays the geometric means for all correct answers

MYPDDL reached a score of 89.6 on the system usability scale [2], with a standard deviation of 3.9.

Discussion The user test shows that MYPDDL-SYNTAX and -DIAGRAM provide useful tools for novices in AI planning and PDDL. Below, we will discuss each part of the user test in turn.

- Debugging Tasks
  While, in general, the syntax highlighting feature was considered very useful, two participants remarked that the used colors confused them and that they found them more distracting than helpful. One of them mentioned that the contrast of the colors was so low that they were hard for her to distinguish. She found the same number of errors with and without syntax highlighting. The other of the two was the only participant who found less errors with syntax highlighting than without it. With MYPDDL-SYNTAX, two participants found all errors in the domain, while none achieved this without syntax highlighting. While every participant had to use the same color scheme in the experiment, colors are customizable in Sublime Text.
- Type Hierarchy Tasks
  In spite of the rather large difference between the GMs for Question 3, a high p-value is obtained ($p = 0.43$). This might be due to the high GSD for the *with* condition and the rather small degrees of freedom ($df = 5$). Testing more participants would probably yield clearer results here. The fact that the availability of tools did not have a positive effect on task completion times for Question 4 can probably be attributed to the complexity of this question (see Appendix sections "Planet Splisus" and "Store"): in contrast to the other four questions, here, participants were required to look at the actions in the domain file in addition to the type diagram. Most participants were confused by this, because they had assumed that once having the type diagram available, it alone

would suffice to answer all questions. This initial confusion costs some time, thus negatively influencing the time on the task.
Visualization tools such as MYPDDL-diagram can improve the understanding of unknown PDDL code and thus support collaboration. But users may be unaware of the limitations of such tools. A possible solution is to extend MYPDDL-diagram to display actions, but this can overload the diagram and, especially for large domains, render it unreadable. Different views for different aspects of the domain or dynamically displayed content could integrate more data, but this also hides functionality, which is generally undesired for usability [13].
- System Usability Scale
Since the overall mean score of the system usability scale has an approximate value of 68 with a standard deviation of 12.5 [16], the score of MYPDDL is well above average with a small standard deviation. A score of 89.6 is usually attributed to superior products [1]. Furthermore, 89.6 corresponds approximately to a percentile rank of 99.8%, meaning that it has a better perceived ease of use than 99.8% of the products in the database used by Sauro [16].

In summary, the user test shows that customizability is important, as not all users prefer the same colors or syntax highlighting at all and their personal preferences seem to correlate with the effectiveness of the tools.

5 Conclusion

We designed MYPDDL to support knowledge engineers in creating, understanding, modifying, and extending planning domains. MYPDDL's code editing features such as syntax highlighting and code snippets, as well as a type diagram generator, an interface with the programming language Clojure, and a planner integration can help in the various stages of working with PDDL domains. MYPDDL's extensible and customizable architecture helps to fulfill the different preferences and requirements of knowledge engineers. In the conducted user test, MYPDDL users were able to grasp the domain structure of a PDDL file more quickly than non-users and also found more errors in a deliberately erroneous domain file. Moreover, the users found the tools easy and pleasant to use.

In future work, MYPDDL's set of features could be extended in several directions. The interface between PDDL and Clojure offers a basis for creating dynamic planning scenarios. Applications could be the modeling of learning and forgetting by adding facts to or retracting facts from a PDDL file or the modeling of an ever changing real world via dynamic predicate lists. Another way of putting the interface to use would be by making the planning process more interactive, allowing for the online interception of planning software in order to account for the needs and wishes of the end user. Since many features of MYPDDL can be called via the command line, interfaces with other editors could be developed. So far, there is a basic integration with the code editor Atom.[13]

All in all, the overall increase of efficiency due to facilitated collaboration and support in maintaining an overview should encourage a shift of focus toward real-world problems in knowledge engineering. The full modeling potential can only be reached with appropriate tools, with MYPDDL hopefully leading to a broader acceptance and use of PDDL for planning problems.

Appendix: Tasks

Deliberately Erroneous Logistics Domain

```
;;;; Logistics domain

(define (domain ?logistics)

  (:requirements
    :types)

  (:typing truck airplane motorboat - vehicle
           package vehicle suitcase furniture - thing
           airport garage station - location
           car1 car 2 car3 - vehicle
           city location thing - object)

  (:predicates (in-city ?l - location ?c - city)
               (at ?obj - thing ?l - location)
               (key ?v - vehicle) = true
               (full ?v - vehicle)
               (in ?p - package ??veh - vehicle))

  (:action drive
    :parameters (?t - truck ?from ?to - location ?c - city)
    :precondition (and (at ?tr ?from)
                       (in-city ?from ?c)
                       (incity ?to ?c))
    :effect (and (not (at ?t ?from))
                 (at ?t ?to)))

  (:action fly
    :parameters (?a - airplane ?from ?to - airport)
    :precondition (at ?a ?from)
    :effect (and (n0t (at ?a ?from))
                 (at ?a ?to)))

  (:action fuel
    :parameters (?v - vehicle ?c - city ?to airport)
    :precondition (and (not (full ?v))
                       (in-city ?to ?c)
                       (at ?v ?to))
    :effect (full ?v))

  (:action load
   parameters: (?v - vehicle ?p - package ?l - location)
   precondition: (and (?v ?l)
                      (at ?p ?l))
    :effect (and (ay ?p ?l)
                 (in ?p ?v)))

  (:action unload
    :parameters (?v - vehicle p - package ?l - location)
    :precondition (and (at ?v ?l)
                            ?p ?v)
    :effects (and (not (in ?p ?v))
                  (at ?p - ?l))))
```

[13] https://github.com/Pold87/myPDDL-Atom.

The original file can be downloaded at http://ipc.informatik.uni-freiburg.de/PddlExtension:

Deliberately Erroneous Coffee Domain

```
(define COFFEE

  (requirements
    :typing)

  (:types room - location
                robot human _ agent
                furniture door - (at ?l - location)
                kettle ?coffee cup water - movable
                location agent movable - object)

  (:predicates (at ?l - location ??o - object)
              (have ?m - movable ?a - agent)
              (hot ?m - movable) = true
              (on ?f - furniture ?m - movable))

  (:action boil
    :parameters (?m - movable \$k - kettle ?a - agent)
    :preconditions (have ?m ?a)
    :effect (hot ?m))

  (:action grip-some
    :parameters (?m - movable ?r - robot ?f - _furniture ?l - location)
    :precondition (and (at ?l ?r)
                       (on ?fu ?m)
                       (at ?l ?f))
    :effect (and (have ?m ?r)))

  (:action move
    :parameters: (?m - movable ?a - agent ?from ?to - location)
    :precondition (or (\"{a}nd (at ?from ?a)
                           (at ?from ?m))
                           (and (at ?from ?m)
                                (location ?from ?a)))
    :effect (and (not (at ?from ?m))
                 (at ?to ?m)))

  (:action change-room
    :parameters (?from-r ?to-r - room ?a - agent)
    :precondition (at ?fromr ?a)
    :effect (and (not (at ?from-r ?a))
                 (at ?tor ?a)))

  (:action prep-coffee
    :parameters (?a - agent ?c - cjp ?w - water ?cof - coffee)
    :precondition (and (have ?c ?a)
                       (hot ?w))
    :effect (have ?cof ?a))

  (:action ?hand-over
    :parameters (?m - movable ?a1 - agent ?a2 - agent)
    :precondition (have ?m ?a1))
    :effect (and (not (have ?m ?a1))
                 (have ?m ?a2))))
```

Planet Splisus

```
(define (domain splisus)

  (:requirements :typing)

  (:types splis - gid
          spleus - splos
          schprok schlok - splus
          rud mekle - lech
          hulpf hurpf - hupf
          sipsi flipsi hupf - splis
          schmok schkok - splus
          gid splos splus - ruffisplisus
          merle - hupf
          ruffisplisus mak lech - object)

  (:predicates (father-of ?r1 - ruffisplisus ?r2 - ruffisplisus)
               (married ?s1 - splos ?s2 - splis)
               (has-weapon ?h - sipsi)
               (dead ?r1 - ruffisplisus)
               (at ?l - lech ?r - ruffisplisus))

  (:action kill
    :parameters (?l - lech ?r1 - ruffisplisus ?s - splis)
    :precondition (and (at ?l ?r1)
                       (at ?l ?s)
                       (married ?r1 ?s)
                       (has-weapon ?s))
    :effect (and (dead ?r1)
            (not (married ?r1 ?s)))))
```

Please answer the following five questions on the society and structure of Planet Splisus:

1. Are all *Flipsis* also of the type *Ruffisplisus*?
2. Are all *Merles* also *Splus*?
3. Can a *Spleus* be married to a *Schlok*?
4. Only theoretically: Could a *Hurpf* murder a *Spleus*?
5. Let us assume there are three categories of object types on Splisus: places, beings, and food. Match the three object types *Ruffisplisus*, *Mak*, and *Lech* with these categories.

Store

```
(define (domain store)

  (:requirements :typing)

  (:types lala lila - zahls
          blisis blusis - ultri
          iltre lula - nulls
          zahls schwinds - knozi
          minis - lala
          ultri sopple schmitzl - lila
          ultres raglos wexis - lola
          kosta - nulls
          nulls spax - minis
          lola - zahls
          knozi schmus - object)

  (:predicates (product ?k - knozi) ; Produkt
               (workplace ?l1 - lola ?l2 - lala)
               (product-at ?l1 - lola ?l2 - lila)
               (cashier ?k - knozi)
               (customer ?s - spax)
               (owns ?l - lila ?s - spax))

  (:action sell
    :parameters (?p - lila ?z - zahls ?l - lola ?w - wexis ?s - spax)
    :precondition (and (product ?p)
                       (cashier ?z)
                       (product-at ?l ?p)
                       (customer ?s))
    :effect (and (product-at ?w ?p)
                 (not (product-at ?l ?p))
                 (owns ?p ?s))))
```

Please answer the following five questions concerning the environment store:

1. Are objects of the type *Lula* also of the type *Minis*?
2. Are *Spax* and *Schmus Zahls*?
3. Is it possible for an Iltre to work at a *workplace* of the type *Knozi*?
4. Only theoretically: Could a *Lala* sell a *Schmitzl* to a *Kosta*?
5. Let us assume our domain *store* models a grocery store. There are three categories: humans, products, and places. Can you match these world terms with the object types *lila*, *lala*, and *lola* from the domain?

References

1. Bangor, A., Kortum, P.T., Miller, J.T.: An empirical evaluation of the system usability scale. Intl. Journal of Human–Computer Interaction **24**(6), 574–594 (2008)
2. Brooke, J.: Sus—a quick and dirty usability scale. Usability evaluation in industry **189** (1996)
3. Edelkamp, S., Hoffmann, J.: PDDL2.2: The language for the classical part of the 4th International Planning Competition. 4th International Planning Competition (IPC-04) (2004)
4. Ellson, J., Gansner, E., Koutsofios, L., North, S.C., Woodhull, G.: Graphviz—open source graph drawing tools. In: Graph Drawing. pp. 483–484. Springer (2002)

5. Goldman, R.P., Keller, P.: "Type problem in domain description!" or, outsiders' suggestions for PDDL improvement. WS-IPC 2012 p. 43 (2012)
6. Guerin, J.T., Hanna, J.P., Ferland, L., Mattei, N., Goldsmith, J.: The academic advising planning domain. WS-IPC 2012 p. 1 (2012)
7. Helmert, M.: Understanding Planning Tasks: Domain Complexity and Heuristic Decomposition, vol. 4929. Springer (2008)
8. Hickey, R.: The Clojure programming language. In: Proceedings of the 2008 symposium on Dynamic languages. ACM (2008)
9. Hwang, W., Salvendy, G.: Number of people required for usability evaluation: the 10±2 rule. Communications of the ACM **53**(5), 130–133 (2010)
10. Ilghami, O., Murdock, J.W.: An extension to PDDL: Actions with embedded code calls. In: Proceedings of the ICAPS 2005 Workshop on Plan Execution: A Reality Check. pp. 84–86 (2005)
11. McDermott, D., Ghallab, M., Howe, A., Knoblock, C., Ram, A., Veloso, M., Weld, D., Wilkins, D.: PDDL—the planning domain definition language (1998)
12. Nielsen, J.: Estimating the number of subjects needed for a thinking aloud test. International journal of human-computer studies **41**(3), 385–397 (1994)
13. Norman, D.A.: The design of everyday things. Basic books (2002)
14. Parkinson, S., Longstaff, A.P.: Increasing the numeric expressiveness of the Planning Domain Definition Language. In: Proceedings of The 30th Workshop of the UK Planning and Scheduling Special Interest Group (PlanSIG2012). UK Planning and Scheduling Special Interest Group (2012)
15. Plch, T., Chomut, M., Brom, C., Barták, R.: Inspect, edit and debug PDDL documents: Simply and efficiently with PDDL Studio. ICAPS12 System Demonstration (2012)
16. Sauro, J.: A practical guide to the system usability scale: Background, benchmarks & best practices. Measuring Usability LLC (2011)
17. Sauro, J., Lewis, J.R.: Quantifying the user experience: Practical statistics for user research. Elsevier (2012)
18. Shah, M., Chrpa, L., Jimoh, F., Kitchin, D., McCluskey, T., Parkinson, S., Vallati, M.: Knowledge engineering tools in planning: State-of-the-art and future challenges. Knowledge Engineering for Planning and Scheduling (2013)
19. Shah, M.M., Chrpa, L., Kitchin, D., McCluskey, T.L., Vallati, M.: Exploring knowledge engineering strategies in designing and modelling a road traffic accident management domain. In: Proceedings of the Twenty-Third International Joint Conference on Artificial Intelligence. pp. 2373–2379. AAAI Press (2013)
20. Strobel, V., Kirsch, A.: Planning in the wild: Modeling tools for PDDL. In: Lutz, C., Thielscher, M. (eds.) KI 2014: Advances in Artificial Intelligence, LNCS, vol. 8736, pp. 273–284. Springer, Cham, Switzerland (2014)
21. Vaquero, T.S., Tonidandel, F., de Barros, L.N., Silva, J.R.: On the use of UML.P for modeling a real application as a planning problem. In: ICAPS. pp. 434–437 (2006)
22. Vaquero, T.S., Tonidandel, F., Silva, J.R.: The itSIMPLE tool for modeling planning domains. Proceedings of the First International Competition on Knowledge Engineering for AI Planning, Monterey, California, USA (2005)
23. Vaquero, T., Tonaco, R., Costa, G., Tonidandel, F., Silva, J.R., Beck, J.C.: itSIMPLE4.0: Enhancing the modeling experience of planning problems. In: System Demonstration–Proceedings of the 22nd International Conference on Automated Planning & Scheduling (ICAPS-12) (2012)

Chapter 5
KEPS Book: Planning.Domains

Christian Muise and Nir Lipovetzky

Abstract In this chapter we describe the main pillars of the Planning.Domains initiative (API, Solver, Editor, and Education), detail some of the current use-cases for them, and outline the future path of the initiative. We further dive into some of the most recent developments of Planning.Domains, and shed light on what is next for the platform.

Keywords PDDL · Modeling · Online services

The inaugural International Planning Competition (IPC) was held in 1998.[1] Since that time, there have been 9 IPCs,[2] each with their own set of benchmarks and problem compilers. While many of the contest websites are still available online, the benchmark problems used by the planning community are scattered and collected only in an ad-hoc manner for specific planners (for example, on the websites for FD[3] and FF [4] planners). In 2016, The Planning.Domains initiative was announced to address this and other key pain points for researchers in the planning community.

The fundamental objective of the Planning.Domains (PD) initiative is to provide a set of resources, repositories, and tools for researchers and educators to discover,

[1] http://ipc98.icaps-conference.org/.

[2] http://icaps-conference.org/index.php/Main/Competitions.

[3] http://hg.fast-downward.org/file/1b5bf09b6615/benchmarks.

[4] https://fai.cs.uni-saarland.de/hoffmann/ff-domains.html.

C. Muise (✉)
School of Computing, Queen's University, Kingston, ON, Canada
e-mail: muise@cs.queensu.ca

N. Lipovetzky
University of Melbourne, Melbourne, VIC, Australia
e-mail: nir.lipovetzky@unimelb.edu.au

M. Vallati, D. Kitchin (eds.), *Knowledge Engineering Tools and Techniques for AI Planning*, https://doi.org/10.1007/978-3-030-38561-3_5

develop, and disseminate planning problems and planning techniques. PD is made up of four principal components:

1. http://api.planning.domains: A programmatic interface to all existing planning problems
2. http://solver.planning.domains: An open and extendable interface to a planner-in-the-cloud service
3. http://editor.planning.domains: A fully featured editor for creating and modifying PDDL
4. http://education.planning.domains: A central source for educational resources for planning.

Development on the PD initiative began in 2014, and the site was announced at ICAPS-2015 in Jerusalem, Israel. It has since garnered the involvement of seven institutions, as well as financial support from the ICAPS organization. Ultimately, the aim for the PD initiative is to be a resource created both for and by planning researchers, as well as an avenue for outreach to other communities unfamiliar with planning technology.

One of the pillars—the online editor—has served as the foundation for several courses in AI that focus on the modeling of planning problems, and in 2016 the editor was extended to include a rich plugin framework. This has subsequently enabled individual contributions from planning researchers throughout the world on the shared platform for engineering planning models.

In this chapter, we will exhibit each of the PD pillars, their capabilities, and general use. We will also explore the future of each, and shed light on the future of the initiative.

1 Planning.Domains Solver

The genesis of Planning.Domains was to represent a canonical source for the existing planning benchmarks. However, there are a variety of issues that arise when one considers the naïve solution of a simple directory containing the existing benchmarks: What naming convention should be used? What is the hierarchy? How do you treat repeated domains over multiple years? and How do you partition the benchmarks? To address these and other concerns, we store the benchmarks in a version controlled publicly accessible repository, and *provide a programmatic access to the domains through an API.* While access to files directly is always available, the structure is flat and not intended to be the first point of contact. Rather, the API and the set of libraries that surround it are the direct entry-point for anyone looking to retrieve or interact with existing planning problems.

Database access is provided at three levels: (1) individual planning Problems; (2) a set of Problem objects for a single Domain; and (3) a set of Domain objects for a single Collection. Collections can correspond to individual IPCs or commonly used benchmark suites. One key advantage is the ability to define canonical collections such as "all STRIPS IPC domains"—a general term that many papers in the field reference, but few align on—that incorporates best practices of picking such a set of domains. Several experts in the field were solicited for advice on building such a set, and the following desiderata is what influences the current collection #12 (All-IPC (STRIPS)):

- Only one copy of each domain is used.
- Benchmarks from the satisficing tracks are preferred over the optimal track.
- Use only the most recent versions of domains when duplicates exist.
- Use bug-free versions if they become available after the IPC.
- Sample uniformly from large domain sets so that all domains have roughly the same number of instances.
- Discard domains that are trivially solved by all modern planning techniques.

Domains represent a set of Problem instances, and naturally have their associated descriptions and origin information. Individual Problems have information about the correct domain and problem PDDL to use (which may be non-standard for some benchmarks), as well as statistics about the individual instances. These include best known lower bounds, upper bounds, classical effective width [9], etc. This enables queries such as "all problems where we do not know the optimal plan, but have effective width 1 and are thus trivial to satisfy suboptimally."

1.1 Libraries

The API component of PD also comes with JavaScript and Python libraries to interface with the database, as well as a command-line utility to fetch and use the stored benchmarks. These three components represent the direct interface for researchers to interact with the API, and we will briefly illustrate their functionality here.

Python API The Python API is meant to mirror the programmatic access to all of the collections, domains, and problems online. A common use-case, shown in the following example, is to build a dictionary mapping a domain name to the list of domain-problem file tuples for testing planners locally. Note that only one line would need to be changed in order to test with a different collection (e.g., a newer IPC).

```
1
2  import sys
3
4  print "Loading domains...",
5  sys.stdout.flush()
6
```

```
7  import planning_domains_api as api
8
9  # 12 is the collection for all STRIPS IPC domains
10 domains = {}
11 for dom in api.get_domains(12):
12
13   # Turn the links into relative paths for this machine
14   probs = api.get_problems(dom['domain_id'])
15
16   # Map the domain name to the list of domain-problem pairs
17   domains[dom['domain_name']] = []
18   for p in probs:
19     domains[dom['domain_name']].append((p['domain_path'], \\
20                                         p['problem_path']))
21
22 print "done!"
```

Example to load all of the domain/problem files

JavaScript API The JavaScript API serves a slightly different purpose than that of the Python API. Rather than mirror the RESTful API directly, it surfaces functionality that allows web developers to quickly build out interfaces around the exploration of the data behind the endpoints. In the example below, two functionalities are shown: (1) a table showing the domains in a particular collection; and (2) an interactive table to explore all of the collections, domains, and problems.

```
1
2  <h2>Showing IPC-2018 Domains:</h2>
3  <div id="domains-table">
4    Loading...
5  </div>
6
7  <h2>Problem Navigator</h2>
8  <div style="min-height:600px" id="problem-navigator">
9      Loading...
10 </div>
11
12 <script type="text/javascript" src="planning-domains.js">
13     fetch_domains('/domains/13', '#domains');
14     insert_navigator('#problem-navigator', 'alert');
15 </script>
```

Examples to (1) create a navigator for collections and (2) display the domains of a particular collection

The function `fetch_domains` used above populates the HTML object “domains-table” with the domains from IPC-2018 collection, and `insert_navigat`… populates the HTML object “problem-navigator,” calling the built-in javascript alert function when a row is selected.

Command-Line Utility For quick access to the database information, packaging of the benchmarks, etc., a command-line utility was also constructed. Figures 5.1

Fig. 5.1 Command-line example to show problem details

```
$ ./planning.domains.py show problem 4154
{u'domain': u'gripper',
 u'domain_id': 119,
 u'lower_bound': 11,
    ...
```

Fig. 5.2 Command-line example to show the plan of a problem

```
$ ./planning.domains.py show plan 4154
['(pick ball1 rooma left)',
 '(pick ball2 rooma right)',
 '(move rooma roomb)',
    ...
```

and 5.2 (respectively) show just two examples of the utility to (1) fetch details on a particular problem and (2) show an incumbent plan for the problem.

1.2 API Future

Moving forward, we have three key objectives for the API: (1) to expand the repository to alternative planning formalisms (FOND, POND, RDDL, RMPL); (2) to open the database to a curated form of statistics submission so that any researcher may contribute to the information on problems; and (3) to introduce a tagging mechanism for Problems, Domains, and Collections to allow for custom categories (e.g., identifying all delete relaxed problems, specifying the requirements used in the modeling language, etc.).

The general setting of having a crowd-sourced database of statistics for academic use poses interesting design challenges for the infrastructure and API. Most crucially, we must address the authenticity and accuracy of the submitted data so that the statistics can be trusted for academic use. For some aspects, such as the best known plan cost, we can independently verify the data using tools such as VAL [7].[5] For other aspects, such as the best known lower bound, we must develop a mechanism for trusted submission and traceability of the results (so any error discovered can be cross-referenced with all potential erroneous entries). This is an area of ongoing work in the development of the API.

[5]This is, in fact, already in place for the (potentially anonymous) submission of new incumbent plans.

2 Solver Planning Domains

For many outside of the automated planning research community, getting a planner to compile and run can be a daunting task. The Solver component of PD offers a planner-in-the-cloud service that can be invoked using a standard RESTful API—the PDDL is sent as raw text, and a planner running remotely returns a plan. The following Python code, for example, operates as an IPC-ready planner that accepts PDDL files and produces a plan using the service.[6]

```
1
2  import urllib2, json, sys
3
4  dom = open(sys.argv[1], 'r').read()
5  prob = open(sys.argv[2], 'r').read()
6  url='http://solver.planning.domains/solve'
7
8  data = {'domain': dom, 'problem': prob}
9  data = json.dumps(data)
10
11 req = urllib2.Request(url)
12 req.add_header('Content-Type',
13                'application/json')
14 resp = urllib2.urlopen(req, data)
15 resp = json.loads(resp.read())
16
17 plan = []
18 for act in resp['result']['plan']:
19     plan.append(act['name'])
20
21 with open(sys.argv[3], 'w') as f:
22     f.write('\n'.join(plan))
```

Usage: ./planner.py domain.pddl problem.pddl plan.ipc

The deployed planner is a variant of the LAPKT project [12] that is tailored to be extremely fast (and not necessarily optimal). The planner is restricted to 10 s and 500 Mb, but the project is open source and free for anyone to deploy on their own using different resource limits. There is also a cool-off mechanism in place that restricts users from calling the service too often if multiple users are currently requesting plans.

With these limits and functionality combined, the single-threaded deployment of the solver has had a tremendous adoption. At the time of writing, the service has been called approximately 585,000 times, and has been the basis for many courses in AI and planning, as well as a prototyping component for planning-related research (e.g., robotic way-point navigation).

[6]It is not recommended to submit this to an IPC, as it violates many of the rules for entered planners, and the spirit of the contest itself.

The open source element of the server infrastructure has also served as a useful basis for several other planning-like services. The common thread among these services is the functionality to take a domain and problem PDDL as input in order to produce a text output (e.g., problem analysis or reformulation).

2.1 *Solver Future*

The future of the solver pillar of PD is to do a more thorough evaluation of the planners out there that are capable of solving problems quickly. The Agile track of the regularly held IPC is an excellent source for this. Ultimately, we want to use the planner that solves the majority of problems as fast as possible, and the techniques that go into making this a reality are not the same as those that lead to a successful IPC planner (e.g., parsing and grounding time becomes a major issue when only 10 s is provided).

We also hope to expand on the capability of the service to provide planning-related information. Currently, the set of ground actions in the produced plan is provided. Further options could include reasons a plan cannot be found, syntax errors on the PDDL sent directly, metrics on the plan itself, etc.

3 Editor Planning Domains

Perhaps the largest of the PD initiatives is an online editor for PDDL. The initiative is similar to previous efforts such as the model acquisition tool itSIMPLE [14], the PDDL Studio software [11], and the online basic editor myPDDL [13]. The PD Editor builds on the myPDDL editor in order to tie together the other PD initiatives, and includes features found in PDDL Studio and itSIMPLE. Among the standard features of an editor (e.g., syntax highlighting, bracket matching, and code folding), the following custom features have been developed:

- The PD Solver is integrated so that solutions can be computed and displayed during editing (see Fig. 5.3). Custom deployed solvers may also be used instead of the default.
- PDDL-specific auto-completion can be used.
- Domain and problem files can be imported from remote servers by browsing through the PD API by Collection, Domain, and Problem.
- Problem analysis can be conducted using an online version of TorchLight [6]; deployed using the same infrastructure as the open source PD Solver.

Two larger features were added subsequent to the introduction of the PD initiative: sessions and plugins. We discuss each in deeper detail here.

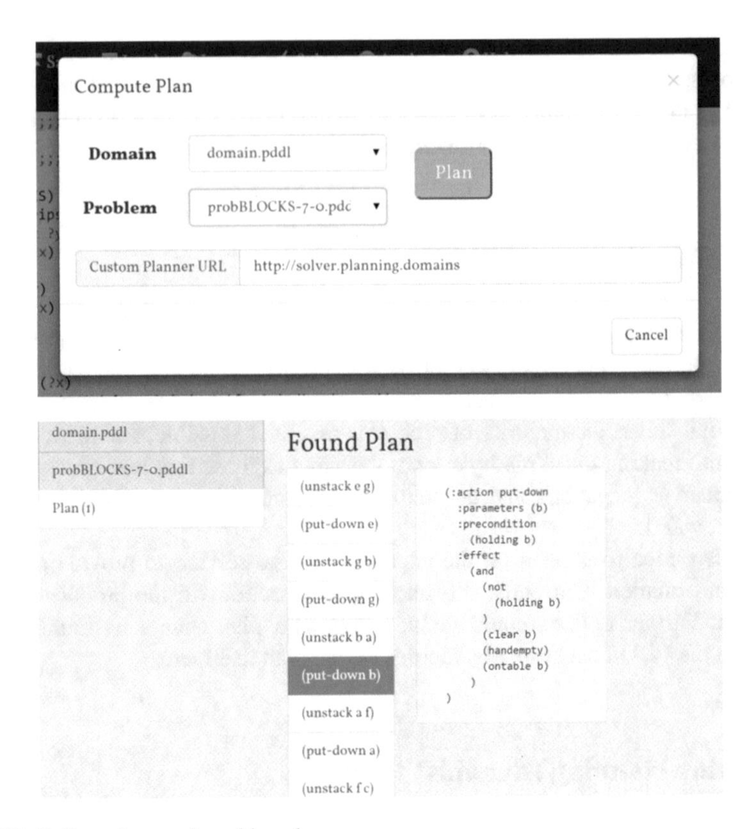

Fig. 5.3 Online solver, and resulting plan

3.1 Plugin Framework

In 2016, the editor went through a major revision to enable a full plugin functionality, and much of the features moved to this modular architecture (e.g., loading problems, invoking the remote solver, etc.). This opened the door to democratizing the development of the online editor, and allows for a rich configuration of the editor for specific use-cases.

One central use-case is in the classroom setting. We observed several examples of custom plugin configurations that captured dedicated solver URL's, initial PDDL tabs, etc. These configurations were shared with the student body for work during the course, and customized by the instructor.

Currently, the list of publicly shared plugins includes:

1. **Save Tabs**: Provides the ability to save the content of all PDDL tabs.
2. **Solver**: Connects the editor to the online solver service.
3. **Timeline Viewer**: Allows the display of temporal plans via an interactive Gantt chart.

4. **SDAC Translator**: Translates problems with state-dependent action costs into ones without.
5. **Torchlight**: Provides in-depth problem analysis on reachability and other features.
6. **Misc PDDL Generators**: Collection of utilities for common modeling patterns (e.g., dealing with numbers, grids, connected networks, etc.).

3.2 Session Functionality

Building on the plugin architecture, and its capability to export the state and configuration of all plugins currently loaded, session functionality was introduced to store this data remotely. Combined with the (enabled-by-default) *Save Tabs* plugin, this provides a rich mechanism for continuing work on a collection of PDDL files at a later time.

The implementation provides both read-only and read/write links for sharing. While the latter provides a natural ability to continue work at a later time, the former has opened the door to sharing exemplary PDDL files or plugin configurations. Links to pre-configured editor views appear in the recently published book on PDDL [5], as primary links for courses in AI, and generally as a means for debugging modeling errors; a session is saved, shared as a read-only version, and modifications re-shared as a new session.

3.3 Editor Future

There are many features planned for the PD Editor, and many surround the plugin framework that is now in place. The modeling process is where the vast majority of PDDL errors occur as models are incorrectly specified, and there is arguably very little tooling available to help in this process. Some of the ideas targeted for future plugin development include:

- An interactive tutorial to introduce newcomers to PDDL.
- General syntax checking for valid PDDL.
- Computation of symmetry analysis and mutex information.
- Reachability analysis on fluents and actions (indicating those that can never be seen in a plan).

Two larger-scale initiatives are underway to improve on the editor capabilities (and that of Planning.Domains in general), and we detail them in the following section. Finally, beyond extending the editor with new features to facilitate modeling, we also intend to enhance the collaborative nature of the tool by adding support for multiple simultaneous users.

4 Education Planning Domains

The PD Editor has been embraced as a teaching tool in several AI and automated planning courses across the world, highlighting the need of resources targeted for the purpose of education. Hence, the fourth pillar of PD aims to collect available information for learning (teaching) automated planning and modeling techniques. The initial distribution of contents targets:

1. Course materials such as slides, hands-on exercises, and workshops.
2. Online resources such as video lectures, demos, and reference manuals.
3. Project ideas, such as course or honors projects, and general activities that could be used for education.

The education pillar is in its infancy, but it stands to become a vital resource for the education of planning techniques and technologies.

5 What Is Next for Planning.Domains

The Planning.Domains initiative will continue to be a community driven collection of initiatives both for and by the planning community. Already, we have given a sense of what is on the horizon for each of the pillars. However, there are two major projects undergoing rapid development that are worth noting.

5.1 Planimation

Planimation is a modular and extensible open source framework to visualize sequential solutions of planning problems specified in PDDL. Such visualizations make solutions more amenable to humans interacting with planners, and assist in the modeling process as well as in the education of AI planning. Planimation introduces a preliminary declarative PDDL-like animation profile specification, expressive enough to synthesize animations of arbitrary initial states and goals of a benchmark with just a single profile.

PDDL is used to specify a model of a planning problem, and a planner synthesizes a solution which can take the form of a sequence, policy, or a tree, depending on the model being solved [4]. Solutions for classical planning take the form of a sequential plan expressed as text-based keywords identifying the *grounded* actions mapping the initial state to a goal state. Special tools exist to validate the soundness of a plan given a problem specification in PDDL, and assist on the detection of errors in existing solvers [8]. Other tools have been developed to assist the PDDL modeling process [1, 14]. For instance, itSIMPLE assists the analysis of PDDL by translating state chart diagrams encoded by a modeler into Petri-Nets

[14]. Other tools provide insights about *solvers* through search tree visualizations of a given search algorithm [10], or visualizations of the internal decision-making process of a solver [2].

If the PDDL specification is syntactically correct but fails to model the classical planning problem the modeler had in mind, then there is no tool to give feedback and facilitate detecting the source of failure. For example, missing preconditions are prevalent and hard to detect if the only feedback is the name of the actions in a valid plan. *Planimation*, a plan visualizer, intends to close the feedback loop and animate a plan given a PDDL specification, relying on visual cues to help modelers find the sources of mental and model misalignment. Furthermore, the tool diminishes the effort needed to understand and explain the dynamics encoded by PDDL problems, as plans explain themselves visually.

PDDL encodes a transition system declaratively by providing typically a single *domain* file specifying actions and predicates, and a *problem* file defining objects, the initial state, and goal states. Actions changing the valuation of predicates induce the state transition we aim to animate visually. In order to do so, we provide a third PDDL-like *animation* file that specifies a sprite for each object, and the animation behavior triggered *when a predicate becomes true in a state*. The declarative visual animation language decouples the visualization engine in the same way PDDL decouples models from solvers. PDDL modelers can extend their problems with a single animation profile and visualize the plans returned by existing solvers [3]. For example, the animation profile to visualize any Blocksworld problem can be viewed at http://editor.planning.domains/#read_session=yiCWKZREGv.

The animation language has been developed taking into account the diversity represented within the IPC domains. New special purpose functions can easily be added to the visual solver. Indeed, more functions may be needed in the future, but the current language has been sufficient to animate a variety of domains such as *Blocksworld, Grid, Logistics*, and *Towers of Hanoi*. All animations result in translations in space, scaling, or appear/disappear effects on objects. For more information we refer the reader to the existing repositories and documentation at http://github.com/planimation.

User Interface Functionality The Planimation UI is shown in Fig. 5.4. It consists of the following panels:

- a *steps panel*, which shows the plan and action currently being executed. Any action a is clickable, and sets the visualization to the resulting state $s' = f(s, a)$.
- The *Step Info* panel shows the preconditions and effects of the current action;
- the *Animation* panel displays the animation canvas and *darkens* objects if they are in their goal position;
- the *Control* panel allows animation speed changes and playback controls;
- the *Subgoal* panel shows all the goal predicates and changes their color when they are satisfied. Clicking on a subgoal opens a dropdown list with all the steps in which the subgoal is satisfied, jumping to such step if selected.
- The last panel is the *header*, which contains help information, a button to show the goal state, and a button to download the visualization in a JSON VFG format. The file can be used to load Planimation without invoking a planner.

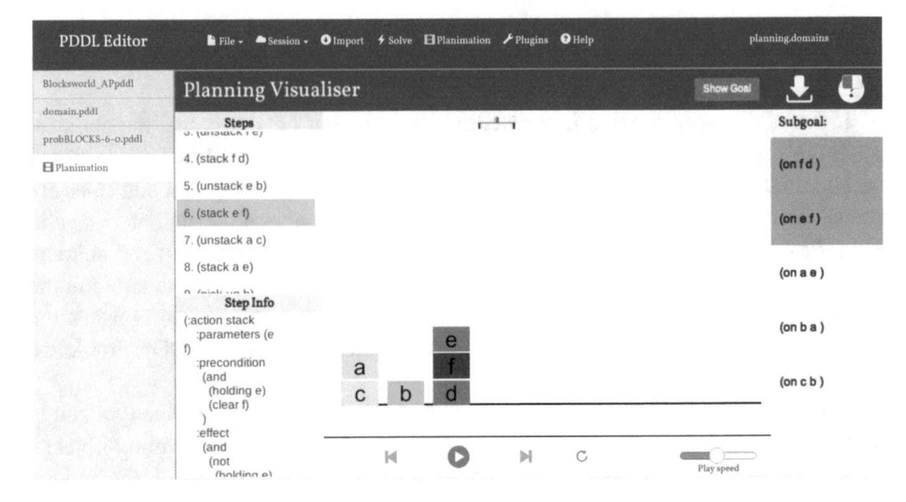

Fig. 5.4 Planimation animating Blocksworld

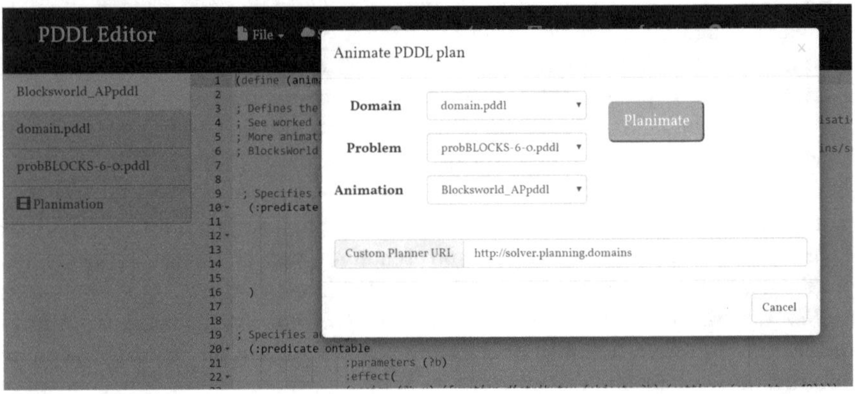

Fig. 5.5 Planimation plugin for editor.planning.domains

How to Use It In order to visualize a PDDL domain, the modeler has to write an animation profile following the syntax accepted by Planimation. Once the animation profile is ready, an animation can be synthesized by uploading the problem, domain, and animation PDDL files into https://planimation.planining.domains. Any planner in the cloud using the PD Solver RESTful API can be used to solve the problem.

Alternatively, the modeler can load the Planimation plugin within the PD Editor, and generate animations as new tabs (see Fig. 5.5). Multiple animations can be generated using different tabs and animation profiles.

5.2 VSCode Integration

In 2018, a plugin was released for Microsoft's Visual Studio Code (VSCode) editor to assist in the creation of PDDL models.[7] This effort shared many of the same motivations and functionality as the editor.planning.domains initiative, with a greater emphasis on the advanced PDDL modeler using common software engineering practices (e.g., unit tests, compilation analysis, etc.). As of 2019, there is a concerted effort underway to unify both the online editor and offline VSCode plugin.

The ultimate vision for the integration is to allow for universal plugins to be written for both editors. An early example of this is the solver functionality; both editors using the same online service and API to retrieve classical plans. Another common integration is access to all of the benchmark problems at api.planning.domains. A screenshot of this functionality is shown in Fig. 5.6.

Aside from the shared solver and API access, early integration efforts have focused on a seamless integration of the session functionality. Sessions can be loaded and modified in both editors, allowing for users to move between their choice

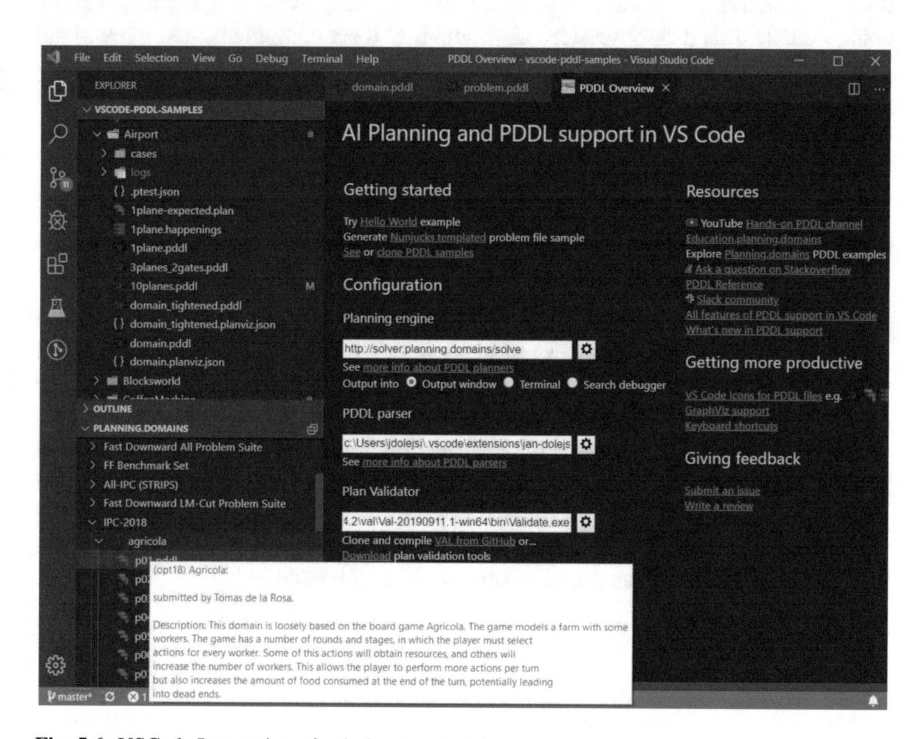

Fig. 5.6 VSCode Integration of api.planning.domains

[7] https://marketplace.visualstudio.com/items?itemName=jan-dolejsi.pddl.

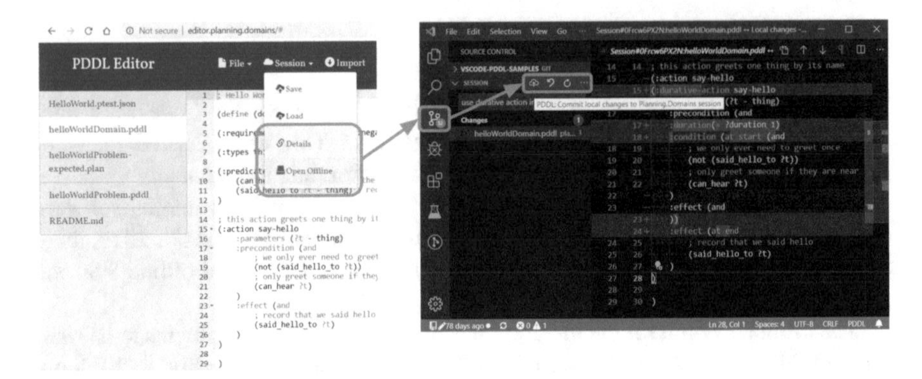

Fig. 5.7 VSCode Integration of editor.planning.domains sessions

of editor and continue the modeling process. Figure 5.7 shows an early prototype of this functionality in action.

The two-way channel of sessions between the online and offline editors opens the door to a much richer experience in the education setting as well. Instructors are now able to create a read-only session, distribute it to a set of students, and allow them to maintain their own individual sessions shared with the instructor. Analogous to version controlled repositories that are forked, shared, and monitored, this method of instruction brings modern teaching pedagogy to the setting of PDDL modeling.

6 Conclusion

In this chapter we have described the various aspects of the Planning.Domains initiative: (1) an online API for programmatic access to planning problems; (2) a remote service for invoking a quick satisfiable planner; (3) an online editor for the modeling of planning problems; and (4) a collection of educational resources.

The various efforts are community driven, and represent a shifting landscape in the current capability of modeling tools and techniques for authoring PDDL. We invite you to first and foremost try the services for yourself and let us know what you think, and ultimately would ask for anyone interested to join us in contributing to the variety of services so that the community may benefit.

References

1. Barreiro, J., Boyce, M., Do, M., Frank, J., Iatauro, M., Kichkaylo, T., Morris, P., Ong, J., Remolina, E., Smith, T., et al.: Europa: A platform for AI planning, scheduling, constraint programming, and optimization. 4th International Competition on Knowledge Engineering for Planning and Scheduling (ICKEPS) (2012)

2. Chakraborti, T., Fadnis, K.P., Talamadupula, K., Dholakia, M., Srivastava, B., Kephart, J.O., Bellamy, R.K.: Visualizations for an explainable planning agent. arXiv preprint arXiv:1709.04517 (2017)
3. Chen, G., Ding, Y., Edwards, H., Hin Chau, C., Hou, S., Johnson, G., Sharukh Syed, M., Tang, H., Wu, Y., Yan, Y., Tidhar, G., Lipovetzky, N.: Planimation. ICAPS system demonstration (2019)
4. Geffner, H., Bonet, B.: A concise introduction to models and methods for automated planning. Morgan & Claypool Publishers (2013)
5. Haslum, P., Lipovetzky, N., Magazzeni, D., Muise, C.: An Introduction to the Planning Domain Definition Language. Morgan & Claypool Publishers (2019)
6. Hoffmann, J.: The TorchLight Tool: Analyzing Search Topology Without Running Any Search. In: Proceedings of the System Demonstrations, in the 21th International Conference on Automated Planning and Scheduling. pp. 37–41 (2011)
7. Howey, R., Long, D., Fox, M.: Val: Automatic plan validation, continuous effects and mixed initiative planning using PDDL. In: 16th IEEE International Conference on Tools with Artificial Intelligence. pp. 294–301. IEEE (2004)
8. Howey, R., Long, D., Fox, M.: VAL: Automatic plan validation, continuous effects and mixed initiative planning using PDDL. In: 16th IEEE International Conference on Tools with Artificial Intelligence. pp. 294–301. IEEE (2004)
9. Lipovetzky, N., Geffner, H.: Width and Serialization of Classical Planning Problems. In: ECAI. pp. 540–545 (2012)
10. Magnaguagno, M.C., Pereira, R.F., Móre, M.D., Meneguzzi, F.: Web planner: A tool to develop classical planning domains and visualize heuristic state-space search. In: Proceedings of the Workshop on User Interfaces and Scheduling and Planning, UISP. pp. 32–38 (2017)
11. Plch, T., Chomut, M., Brom, C., Barták, R.: Inspect, edit and debug PDDL documents: Simply and efficiently with PDDL studio. System Demonstrations and Exhibits at ICAPS pp. 15–18 (2012)
12. Ramirez, M., Lipovetzky, N., Muise, C.: Lightweight Automated Planning ToolKiT. http://lapkt.org/ (2015), accessed: 2019-06-7
13. Strobel, V.: myPDDL—Knowledge Engineering for PDDL. http://pold87.github.io/myPDDL/ (2015), accessed: 2016-03-18
14. Vaquero, T.S., Silva, J.R., Ferreira, M., Tonidandel, F., Beck, J.C.: From Requirements and Analysis to PDDL in itSIMPLE3.0. Proceedings of the Third International Competition on Knowledge Engineering for Planning and Scheduling, ICAPS 2009 pp. 54–61 (2009)

Chapter 6
Modeling Planning Tasks: Representation Matters

Lukáš Chrpa

1 Introduction

Domain-independent planning decouples planning task description, specified in a description language (e.g., PDDL), and planning engines that accept the task description as an input and generate plans (if they exist). A planning domain model gives general description of the environment and actions of a given domain while a planning problem specifies concrete objects, an initial state, and a goal. Planning domain model together with planning problem description form a planning task. Hence it is typical that one domain model can be used for a class of planning tasks.

Whereas the planning community, driven by International Planning Competitions (IPC),[1] focuses on developing advanced planning techniques for solving planning tasks, the engineering process of developing models of planning tasks is also important, although as shown in the last International Competition on Knowledge Engineering for Planning and Scheduling (ICKEPS) it does not receive much attention from the community [16]. Efficiency of a model of a planning task, or a domain model is however a crucial factor determining performance of planning engines and hence the usability of the model in a real application.

This chapter focuses on existing concepts that can be leveraged for developing efficient domain models while emphasizing common weaknesses of the state-of-the-art planning engines and discussing options how the weaknesses can be overcome by efficient modeling. Specifically, we will discuss macro-operators that, roughly

[1]http://ipc.icaps-conference.org.

L. Chrpa (✉)
Faculty of Electrical Engineering, Czech Technical University in Prague, Prague, Czech Republic

Faculty of Mathematics and Physics, Charles University in Prague, Prague, Czech Republic
e-mail: chrpaluk@fel.cvut.cz

M. Vallati, D. Kitchin (eds.), *Knowledge Engineering Tools and Techniques for AI Planning*, https://doi.org/10.1007/978-3-030-38561-3_6

speaking, represent "shortcuts" in the state space, entanglements that reduce size of the search space, "bagged" representation that alleviates unwanted symmetries between objects of the same type, and, finally, Domain Control Knowledge that, roughly speaking, provides guidance for planning engines through the search space. Each type of extra knowledge has certain benefits and limitations (e.g., macro-operators might help to find a plan in fewer steps but might considerably increase branching factor) that will be discussed in order to provide useful tips and trick for domain modelers.

2 Outer Entanglements

As state-of-the-art planning engines perform grounding in a preprocessing step, i.e., they instantiate all predicates and operators, the size of the representation might grow considerably that, in consequence, causes higher CPU time overheads and memory requirements. Reducing the size of grounded representation can therefore have a positive impact on performance of planning engines.

Outer Entanglements [14, 18] aim at reducing the size of planning task representation by eliminating possibly useless instances of planning operators. Specifically, Outer Entanglements are relations between planning operators and predicates whose instances are present in the initial state or the goal. These relations capture a useful knowledge that some planning operators are needed only to modify initial situations (e.g., picking up a package at its initial location) or achieve goal situations (e.g., delivering a package to its goal location). Consequently, a limited number of instances of "entangled" operators have to be considered in the planning process. Eliminating some operators' instances often makes some atoms unreachable (e.g., a package cannot be in other than initial or goal location). Hence, the size of task representation is smaller too and planning engines have to make less effort to find solution plans.

Formally, Outer Entanglements are defined as follows [18]:

Definition 1 Let Π be a planning task, where I_{Π} is the initial state and G_{Π} is the goal. Let o be a planning operator and p be a predicate defined in the domain model of Π. We say that operator o is **entangled by init (resp. goal)** with a predicate p in Π if and only if $p \in pre(o)$ (resp. $p \in add(o)$) and there exists π, a solution plan of Π, such that for every action $a \in \pi$ being an instance of o and for every atom p_{gnd} being an instance of p, it holds: $p_{gnd} \in pre(a) \Rightarrow p_{gnd} \in I_{\Pi}$ (resp. $p_{gnd} \in add(a) \Rightarrow p_{gnd} \in G_{\Pi}$). We also say that π **satisfies** the entanglement (by init or goal) conditions.
Henceforth, entanglements by init and goal are denoted as **outer entanglements**.

To illustrate the meaning of Definition 1, we consider the well-known BlocksWorld domain [35]. An *entanglement by init* between the unstack(?x ?y) operator and the (on ?x ?y) predicate captures that if an instance of on (e.g., (on a b) is required by a corresponding instance of unstack (e.g., unstack(a b)), then that instance of on (e.g., (on a b) is present in the initial state. Similarly, an *entanglement by goal* captures the stack(?x ?y) operator and the (on ?x ?y) predicate captures that if an instance of on (e.g., (on a b) is achieved by a corresponding instance of stack (e.g., stack(a b)), then that instance of on (e.g., (on a b) is present in the goal.

In order to leverage Outer Entanglements in a planner-independent fashion, they can be encoded into the planning task representation. This can be done by introducing static predicates into the preconditions of "entangled" operators such that only instances "complying" with the entanglements are considered in the planning process. Let Π be a planning task, I be its initial state, and G be its goal. Let an operator o be entangled by init (resp. goal) with a predicate p in Π (o and p are defined in the domain model of Π). Then the task Π is reformulated as follows [18]:

1. Create a predicate p' (not defined in the domain model of Π) having the same arguments as p and add p' into the domain model of Π.
2. Modify the operator o by adding p' into its precondition. p' has the same arguments as p which is in precondition (resp. positive effects) of o.
3. Create all possible instances of p' which correspond to instances of p listed in I (resp. G) and add the instances of p' to I.

Adding a static predicate p' into the precondition of o causes that only instances of o "complying" with the entanglement are considered in the planning process. Figure 6.1 depicts the encoding of an entanglement by init between the unstack operator and the predicate on. In our terminology, unstack(?x ?y) refers to o, on(?x ?y) to p, and entinit_on(?x ?y) to p'.

The aim of Outer Entanglements is to decrease the number of operator instances that planning engines have to reason with for solving the given planning task. In the BlocksWorld example, we can observe that the original model considers quadratic number of the unstack and stack operators (with respect to the number of blocks), while the "entangled" model considers linear number of these operators. Consequently, in can be observed that each block can be at most in four states: stacked on the initial block, held by the robotic hand, put on the table, and stacked on the goal block. Moreover, unless the initial and goal blocks are the same, once the block is unstacked from the initial position it cannot be returned to it and after the block is stacked on the goal position it can no longer be moved.

```
(:action unstack
:parameters (?x ?y - block)
:precondition (and (clear ?x)(handempty)(on ?x ?y)(entinit_on ?x ?y))
:effect (and (holding ?x)(clear ?y)
             (not (clear ?x))(not (handempty))(not (on ?x ?y)))
)
```

Fig. 6.1 An example of the encoding of an entanglement by init between the unstack operator and the on predicate

Deciding Outer Entanglements is PSPACE-complete (as well as classical planning) and in literature, they are learnt from training plans, solutions of simple planning tasks [18]. The learning method, however, might not be accurate, i.e., learn entanglements that do not generalize (i.e., they do not hold for the whole class of tasks), or miss some entanglements (e.g., training plans are too suboptimal). On the other hand, Outer Entanglements might be specified by domain engineers as they have a good knowledge of the domain.

The success of Outer Entanglements depends mostly on how they can reduce the size of the state space. This applies, for instance, in the BlocksWorld domain and its variants (e.g., Depots, Matching-BlocksWorld) or in logistic types of domains (e.g., TPP, Gripper). For deeper insights about Outer Entanglements see [14, 18].

3 Macro-Operators

Macro-operators (macros) are encoded in the same way as ordinary planning operators, but encapsulate sequences of planning operators [28]. Technically, an instance of a macro is applicable in a state if and only if a corresponding sequence of operators' instances is applicable in that state and the result of the application of the macro's instance is the same as the result of application in the corresponding sequence of operators' instances. Macros can be added to the original domain models, which gives the technique the potential of being *planner independent*. Informally speaking, macros can be understood as shortcuts in the search space allowing planning engines to generate plans in fewer steps.

Formally, a macro $o_{i,j}$ is constructed by assembling planning operators o_i and o_j (in that order) as follows. Let Φ and Ψ be mappings between variable symbols (we need to appropriately rename variable symbols of o_i and o_j to construct $o_{i,j}$).

- $pre(o_{i,j}) = pre(\Phi(o_i)) \cup (pre(\Psi(o_j)) \setminus add(\Phi(o_i)))$
- $del(o_{i,j}) = (del(\Phi(o_i)) \setminus add(\Psi(o_j))) \cup del(\Psi(o_j))$
- $add(o_{i,j}) = (add(\Phi(o_i)) \setminus del(\Psi(o_j))) \cup add(\Psi(o_j))$

For a macro to be *sound*, no instance of $\Phi(o_i)$ can delete an atom required by a corresponding instance of $\Psi(o_j)$, otherwise they cannot be applied consecutively. Whereas it is obvious that if a predicate deleted by $\Phi(o_i)$ (and not added back) is the same (both name and variable symbols) as a predicate in the precondition of $\Psi(o_j)$, i.e., $(del(\Phi(o_i)) \setminus add(\Phi(o_i)) \cap pre(\Psi(o_j)) \neq \emptyset)$, then the macro $o_{i,j}$ is unsound, another source of macro unsoundness is often not being even considered in literature. Instantiating two or more macro's variables to a same object might make the macro unsound. For example, in the BlocksWorld domain, a macro pickup-stack(?x ?y) that has (clear ?x)(ontable ?x)(clear ?y)(handempty) in its precondition can be instantiated into pickup-stack(A A) that is applicable if (clear A)(ontable A)(handempty) is true in a current state. However, actions (pickup A) and stack(A A) cannot be applied consecutively because (pickup A) deletes (clear A) which is required by stack(A A) and hence pickup-stack(?x

```
(:action pick-up-stack
 :parameters ( ?x - block ?y - block)
 :precondition (and (clear ?x)(ontable ?x)(handempty)
                    (clear ?y)(not (= ?x ?y)))
 :effect (and (clear ?x)(handempty)(on ?x ?y)(not (ontable ?x))
              (not (holding ?x))(not (clear ?y)))
)
```

Fig. 6.2 The pickup-stack macro in PDDL

?y) is unsound. These situations can be easily identified by checking whether the same object substitution leads to the situation in which the first operator deletes a precondition for a second operator. To make the affected macro sound, *inequality constraints* have to be added to macro's precondition (e.g., (not (= ?x ?y)) is added into pickup-stack(?x ?y)'s precondition) [13]. For illustration, see Fig. 6.2 in which the pickup-stack macro is depicted.

Longer macros, i.e., those encapsulating longer sequences of original planning operators can be constructed iteratively by the above approach.

The use of macros dates back to 1970s when the REFLECT system was developed [20]. In 1980s, that MORRIS system was developed [31]. MORRIS learns macros from parts of plans appearing frequently (S-macros) or being potentially useful despite having low priority (T-macros). Noteworthy, S-macros have a planner-independent aim as they tend to be used frequently in the planning process, while T-macros aim at specific weaknesses of a given planning engine. Macro Problem Solver [28] learns macros for particular non-serializable sub-goals (e.g., in Rubik's cube). Korf [28] also shows that use of macros can reduce computational complexity.

Besides a clear advantage, that is, providing shortcuts in the search space, macros often have many instances that in consequence increase branching factor as well as memory requirements. Therefore, there exist works that aim at opposite, either eliminate "redundant" actions whose effects can be achieved by sequences of other actions [24], or split operators into simpler ones [2]. That said, it is important that benefits of macros outweigh their drawbacks, which is also known as the *utility problem* [31].

As planner-dependent macro generation systems, we can mention the SOL-EP version of Macro-FF [7], or Marvin [19] that generate macros that help to escape local heuristic minima of the well-known FF planner [25]. Specifically, FF uses enforced hill climbing search that if stuck in local heuristic minima performs breadth-first search to find a state with better (lower) heuristic value. The FF heuristics represents an estimation of the (nearly optimal) cost of delete-relaxed plans (action delete effects are not considered). That said, macros aim to make FF heuristic to be monotonically decreasing, therefore, mitigating (and ideally eliminating) the breadth-first search episodes.

Wizard [32] that is a planner-independent macro generating system that learns macros from training plans by leveraging genetic programming. The learning process incorporates a cross-validation phase in which generated macros are evaluated on a set of test problems being solved by a given planner. Hence, the aim is to learn macros that maximize performance of a given planning engine in a given domain. Such macros can be considered as planner-specific (despite being generated by a planner-independent technique).

In planner-independent settings, a frequent occurrence of sequences of actions in training plans usually plays a pivotal role in macro generation systems. Examples of such systems include the work of Chrpa [13] in which macros are learnt by analyzing dependencies between actions in training plans. Dulac et al. [21] exploit n-gram algorithm to analyze training plans to learn macros. DBMP/S [27] applies Map Reduce for learning macros from a larger set of training plans. The CA-ED version of MacroFF [7] generates macros according to several pre-defined rules (e.g., the "locality rule" stemming from Component Abstraction) that apply on adjacent actions in training plans. Noteworthy, CAP [4] leverages Component Abstraction for generating sub-goal specific macros.

BloMa [17] is a macro learning approach that exploits block deordering [34] such that training plans are deordered into "blocks," i.e., sequences of actions that have to be applied in a given order, that are further combined into "macroblocks" according to several rules describing relationships between blocks. Macroblocks that are frequent in training plans are assembled into macros. The advantage of BloMa is that it can learn useful longer macros, for example, a macro capturing shaking a cocktail and cleaning the shaker afterwards.

MUM [11] aims at learning "instance-wise" macros. In other words, macros learnt by MUM should have a comparable number of instances to the original operators. To do so, MUM exploits Outer Entanglements by applying them on macros reducing the number of their instances. Outer Entanglements also serve as a heuristic for macro generation such that operators with the entanglement by init relation go first while operators with the entanglement by goal relation go last. In the ideal case, macros directly connect initial state of an object with its goal state (e.g., pick-move-drop). Frequency of macro occurrence in training plans is a secondary criterion.

Critical Section Macros [10], inspired by Critical Sections in parallel computing in which processes deal with shared resources, capture activities that use limited resources (e.g., a robotic hand). In planning, resource availability and use is represented by mutex predicates, for example, (handempty) and (holding ?x), respectively. Then we can identify *locker* and *releaser* operators that locks and releases the resource respectively. For example, unstack is a locker as it deletes (handempty) and achieves (holding ?x), while putdown is a releaser as it deletes (holding ?x) and achieves (handempty). Straightforwardly, a Critical Section Macro starts with a locker and ends up with a releaser. In between, the macro might contain *users* that contain the resource use predicate in their preconditions (e.g., a paint operator that paints the block held by the robotic hand), or *gluing operators* that have to be present in the macro for some reason (e.g., a move operator that moves the robotic hand between tables). Critical Section Macros hence represent the whole activities from locking to releasing resources and, consequently, encapsulate the whole period of resource use. As it is often the case that the resource can be locked by multiple objects, delete-relaxation heuristics that are widely used in planning tend to (heavily) underestimate the plan cost as they assume the resource can be used by multiple objects at the same time. Critical Section Macros since they "bypass" the resource use part, therefore, mitigate the discrepancy between delete-relaxed heuristic estimation and the real cost of the plan. On top of that, the aggressive version of the Critical Section Macros generation technique removes the original operators that are replaced by the generated macros. This approach has shown to be very efficient in a couple of domains [10].

Besides systems that learn macros by training on a set of small problems, there are several systems such as DHG [3] or OMA [12] that extracts planner-independent macros online, i.e., without the training phase. These systems, however, usually underperform the learning ones [12].

From the Knowledge Engineering perspective, macros can be designed and encoded manually by an engineer. Alternatively, an engineer can manually choose macros generated by any of the existing macro generation systems. Helpful macros usually appear frequently in plans, connect an initial state of an object with its goal state, replace original operators, or address a weakness of a class of planning engines (e.g., escaping from local minima of a heuristic function). Unhelpful macros often have too many instances, or appear less frequently in plans. Although there are some observations determining macro utility in general, it might largely depend on a specific planning engine as well as a specific planning task (in a given domain). Hence, an engineer should test his/her macros on a set of non-trivial planning tasks and a selected planning engine to see how the performance improves (or degrades).

4 Bagged Representation

Representing particular objects explicitly is typical for the STRIPS representation on which the PDDL language is based. However, in some cases we require to know only quantities of some types of objects rather than distinguishing between each individual one. For example, in the IPC 1998 version of the Gripper domain, the task is to move balls from one room to another with a robot having two grippers. Typically, each ball is represented as an individual object. That, however, introduces unnecessary symmetries as the planner might reason about which ball has to be moved first (and which ball next and so on). As all the balls have to be moved to the other room anyway, it is straightforward that it does not matter in which order the balls are moved.

Generally speaking, instead of representing such objects individually, it is better to represent the numbers of these objects satisfying particular conditions (e.g., how many balls are in each room). The natural way to represent the numbers of objects is by numeric fluents [22]. Such a representation, however, goes beyond classical planning and would require planners supporting numeric fluents (e.g., Metric-FF [26] or LPG [23]).

To keep in the classical planning representation, we can leverage *bagged representation* [33]. In a nutshell, instead of numeric fluents we use *predicate counters* that represent numbers of objects satisfying given conditions. That involves a class of objects where each object represents specific number (integer), starting at zero up to a given upper bound. The upper bound has to be specified up front as the STRIPS representation does not allow to create objects during planning. As we already know the number of objects we want to count, the upper bound for the numeric objects can be determined from that.

Technically speaking, predicates having the object o we want to count as a parameter, i.e., $p(o, x_1, \ldots, x_n)$, are replaced with predicate counters, i.e., $p_o(n, x_1, \ldots, x_n)$, where n represents the number of p's instances true in the given state under the same instantiations of $x_1, \ldots, x_n$. For example, let (at room1 ball1), (at room1 ball2), (at room2 ball3) be true in some state, then the corresponding predicate counters are (at-ball room1 n2), (at-ball room2 n1), where n1, n2 represent numbers 1 and 2, respectively.

Operators that delete k instances of the p predicate with the same object o decrement the corresponding predicate counter p_o by k. Analogously, operators that add l instances of the p predicate with the same object o increment the corresponding predicate counter p_o by l. For this purpose, additional predicates specifying arithmetics between numeric objects have to be defined. For example, we can define (sum ?n1 ?n2 ?n3) that represents $n3 = n1 + n2$, or if we increment/decrement predicate counters only by one, we can define (next ?n1 ?n2) that represents $n2 = n1 + 1$.

An example of bagged representation of the pick and drop operators is shown in Fig. 6.3.

```
(:action pick
:parameters (?n1 ?n0 ?obj ?room ?gripper)
:precondition (and (ball ?obj)(room ?room)(gripper ?gripper)
                   (at-robby ?room)(free ?gripper)
                   (next ?n0 ?n1)(count ?obj ?room ?n1))
:effect (and (carry ?obj ?gripper)(not (count ?obj ?room ?n1))
             (count ?obj ?room ?n0)(not (free ?gripper))))

(:action drop
:parameters (?n1 ?n0 ?obj ?room ?gripper)
:precondition (and (ball ?obj)(room ?room)(gripper ?gripper)
                   (carry ?obj ?gripper)(next ?n0 ?n1)
                   (at-robby ?room)(count ?obj ?room ?n0))
:effect (and (not (count ?obj ?room ?n0))(count ?obj ?room ?n1)
             (free ?gripper)(not (carry ?obj ?gripper))))
```

Fig. 6.3 Bagged representation of the pick and drop operators

5 Procedural Domain Control Knowledge

Domain Control Knowledge (DCK) captures useful information that guides search for solution plans. One idea how to express DCK is via *search-control rules* with PRODIGY [8] as one of the earliest planning systems that incorporated such rules. PRODIGY rules were in form of expert-system rules guiding search decisions. More recent and more efficient systems used versions of Linear Temporal Logic to express control rules. The representatives of such systems are TALplanner [29] and TLPlan [5].

Control rules can be understood as *state-centric* DCK, i.e., they express what state and/or sequences of states are permissible. In contrast, *action-centric* DCK focuses on representing how actions should be ordered in solution plans. A good example of action-centric DCK is Procedural DCK that leverages a procedural programming paradigm such that the DCK is in the form of sequence of instructions determining which actions can be applied in a given step [6].

Procedural DCK can be encoded in Golog-like programs alongside with a planning task. Let A be a set of actions and ψ a Boolean formula over the atoms defined in the planning task. Then atomic programs are specified as follows [6]:

1. nil—an empty program
2. a—a single action ($a \in A$)
3. any—any action
4. $\psi?$—a test action

If $\sigma, \sigma_1, \sigma_2$ are programs, then the following are also programs [6]:

1. $(\sigma_1; \sigma_2)$—a sequence of programs
2. **if** ψ **then** σ_1 **else** σ_2—a conditional statement
3. **while** ψ **do** σ—a while loop
4. $\sigma*$—a non-deterministic iteration
5. $(\sigma_1|\sigma_2)$—a non-deterministic choice of programs
6. $\pi(x-t)\sigma$—a non-deterministic choice of a variable x of a type t

The atomic programs, or instructions, in other words, refer to actions that can be applied in a given step. An exception is the test action that verifies whether a given formula is true in the current state. The programs define three types of non-deterministic choices: performing the program zero or more times $\sigma*$, a simple choice between two programs $(\sigma_1|\sigma_2)$, and a variable instantiation choice $\pi(x-t)\sigma$.

The programs can be translated into a Finite State Automata that are then compiled into the planning task specification [6]. Hence the programs can be exploited in a planner-independent way, although the compilation introduces some ADL features (e.g., quantified preconditions) that might not be supported by all classical planning engines. Noteworthy, Finite State Automata can be adapted to represent DCK directly [15] as shown in the following section.

Besides Golog-like programs, *planning programs* that can be understood as Procedural DCK have recently been introduced [1]. Similarly to Golog-like programs, planning programs are sequences of instructions, where an instruction is an action or a conditional jump. The latter differs from Golog-like programs as it allows to jump to a specified instruction if the condition is satisfied. The if and while statements can be easily represented by conditional jumps. On top of that, planning programs support defining and calling procedures [1].

6 Transition-Based Domain Control Knowledge

Besides Procedural DCK, an engineer can exploit *Transition-Based Domain Control Knowledge* (TB-DCK) that is inspired by Finite State Automata [9, 15]. TB-DCK represents a "grammar" of solution plans which is, roughly speaking, knowledge about ordering of planning operators in plans. On top of that, TB-DCK allows to define extra preconditions that can be used to restrict applicability of some instances of planning operators that are not useful.

In principle, TB-DCK consists of a set of DCK states and transitions that refer to which planning operators can be applied under specified conditions in a given planning state. The formal definition of TB-DCK follows [15].

Definition 2 *Transition-based Domain Control Knowledge (TB-DCK)* is a quadruple $\mathcal{K} = (S, O, T, s_0)$, where S is a set of DCK states, $s_0 \in S$ is the initial DCK state, O is a set of planning operators, and T is a set of transitions in the form (s, o, C, s'), where $s, s' \in S$, $o \in O$, and C is a set of constraints, where each constraint is in the form:

- $p, \neg p$—a predicate p must or must not be present in the current planning state
- g:p—a predicate p is an open goal in the current planning state

To illustrate the concept of TB-DCK, we consider a simple logistic domain, where packages have to be delivered from their initial locations to their goal locations by a truck that can carry at most one package. All locations are connected. We define four predicates: (at-truck ?l)—the truck is at location ?l; (at ?p ?l)—a package ?p is at location ?l; (in ?p)—a package ?p is in the truck; (empty-truck)—a truck is empty (no package is in it). Then, we define three planning operators: Drive(?from ?to)—the truck moves from the location ?from to a location ?to; Load(?p ?l)—a package ?p is loaded into the truck at the location ?l; and Unload(?p ?l)—a package ?p is unloaded from the truck at the location ?l. We may observe that (1) an empty-truck has to be moved only to locations where some package is waiting for being delivered, and (2) if a package is loaded to the truck (in its initial location), then the truck has to move to package's goal location, where the package is then unloaded. Such an observation can be encoded as a TB-DCK as depicted in Fig. 6.4. In particular, the DCK state s_0 represents that the truck is empty, s_1 represents that the truck has just been loaded with a package, and s_2 represents that the package is ready to be unloaded in its goal location.

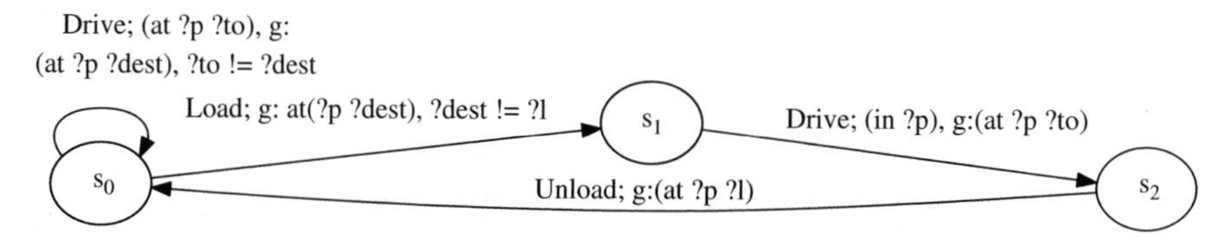

Fig. 6.4 Transition-based DCK for our simple logistic domain

Conceptually, a generic planning algorithm iterates by non-deterministically selecting an applicable action in the current state and updating the current state by applying the action, until a goal state is reached (or no action is applicable). With TB-DCK, the generic planning algorithm is extended such that it non-deterministically selects an outgoing transition from the current DCK state and then it non-deterministically selects an action that is an instance of the operator associated with the transition and the transition constraints are met. Besides updating the state of the environment, the current DCK state is updated as well (the incoming DCK state of the selected transition).

To illustrate the benefits of planning with TB-DCK, we can observe in our running example that the action selection is done only while being in the DCK state s_0—the truck can load a package that has not yet been delivered, or move to a location in which there is a package that has not yet been delivered. In the DCK states s_1 and s_2, there is only one option—move to the goal location of the loaded package, or unload the package, respectively.

One of the advantages of TB-DCK is that it can be encoded into a planning task and thus exploited by standard planning engines. The environment description of a TB-DCK enhanced planning task is extended by "DCK state" predicate, i.e., (DCK-state ?s), and open goal predicates that are "twins" of the goal predicates (i.e., they have the same arguments but different names). In the initial state of the enhanced planning task, we have an instance of the DCK state predicate corresponding to the initial DCK state (e.g., (DCK-state s0)), and instances of the open goal predicates corresponding to their "twins" present in the goal.

Transitions are encoded as planning operators in a TB-DCK enhanced domain model. In particular, let (s_a, o, C, s_b) be a transition, then the schema of the associated planning operator o is extended by:

- Adding the DCK state predicate (DCK-state sa) into o's (positive) precondition as well as o's delete effects, and the DCK state predicate (DCK-state sb) into o's add effects (noteworthy if $sa = sb$, then we need to only add (DCK-state sa) into o's precondition),
- Adding constraints in form $p, \neg p$ into o's positive and negative precondition, respectively,
- For a constraint in form g:p, adding the open goal predicate "twin" of p into o's (positive) precondition,
- Adding p's open goal "twin" into o's delete effects if o has p in its add effects.

```
(:action Drive-empty
:parameters (?from ?to ?dest - location ?p - package)
:precondition (and (at-truck ?from)(at ?p ?to)(DCK-state s0)
    (open-goal-at ?p ?dest)(not (= ?to ?dest)))
:effect (and (not (at-truck ?from))(at-truck ?to))
)
(:action Drive-full
:parameters (?from ?to - location ?p - package)
:precondition (and (at-truck ?from)(DCK-state s1)
         (in ?p)(open-goal-at ?p ?to))
:effect (and (not (at-truck ?from))(at-truck ?to)
         (not (DCK-state s1))(DCK-state s2))
)
(:action Load
:parameters (?p - package ?l ?dest - location)
:precondition (and (at-truck ?l)(at ?p ?l)(empty-truck)(DCK-state s0)
        (open-goal-at ?p ?dest)(not (= ?to ?dest)))
:effect (and (not (at ?p ?l))(not (empty-truck))(in ?p)
          (not (DCK-state s0))(DCK-state s1))
)
(:action Unload
:parameters (?p - package ?l - location)
:precondition (and (at-truck ?l)(in ?p)(DCK-state s2)
              (open-goal-at ?p ?l))
:effect (and (not (in ?p))(empty-truck)(at ?p ?l)
          (not (open-goal-at ?p ?l))
          (not (DCK-state s2))(DCK-state s0))
)
```

Fig. 6.5 Modified planning operators (in PDDL) of our simple logistics domain with respect to the transition-based DCK as in Fig. 6.4

Noteworthy, more transitions can be associated with the same planning operators. Hence, operators encoding these transitions have to have unique names. In our running example, we have two transitions referring to the Drive operator. So, we create two operators, for instance, Drive-empty and Drive-full to reflect two different transitions in the given TB-DCK (see Fig. 6.4). The TB-DCK enhanced domain model of our running example is depicted in Fig. 6.5.

TB-DCK has the strongest impact in domains where goals can be achieved one by one, or where it can help to avoid dead-ends. Of course, if the domain is particularly challenging, i.e., NP- or PSPACE-hard, then TB-DCK cannot bridge the complexity gap. For details, see [9].

7 A Case Study: The Spanner Domain

The Spanner domain has been introduced in the learning track of IPC 2011. It is about a worker who can collect spanners on the way to the gate, where the worker has to tighten nuts. However, after tightening a nut, the spanner will become unusable. Also, the worker cannot go back, hence the worker has to collect all the spanners before arriving to the gate. An illustrative example is depicted in Fig. 6.6 and the original PDDL representation as in the IPC 2011 is shown in Fig. 6.7.

Although the original domain model is encoded naturally, i.e., accurately reflecting the domain requirements, most state-of-the-art planning engines struggle with such a model despite it is easy to solve for a domain-dependent algorithm (e.g., pick-up all the spanners at a given location before moving on). There are two main reasons: (1) unnecessary symmetries as we do not have to distinguish between individual spanners (as well as nuts), and (2) deep dead-ends that are undetectable by delete-relaxed heuristics as it wrongly assumes that one spanner can be used to tighten all the nuts.

The first issue can be addressed by exploiting bagged representation, so the model would consider the number of spanners at each location as well as the

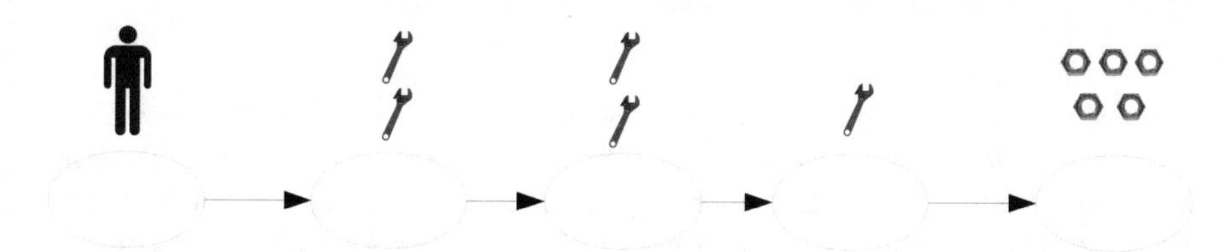

Fig. 6.6 An illustrative example of the Spanner domain

```
(:action walk
        :parameters (?start - location ?end - location ?m - man)
        :precondition (and (at ?m ?start)(link ?start ?end))
        :effect (and (not (at ?m ?start)) (at ?m ?end)))

(:action pickup_spanner
        :parameters (?l - location ?s - spanner ?m - man)
        :precondition (and (at ?m ?l)(at ?s ?l))
        :effect (and (not (at ?s ?l))(carrying ?m ?s)))

(:action tighten_nut
        :parameters (?l - location ?s - spanner ?m - man ?n - nut)
        :precondition (and (at ?m ?l)(at ?n ?l)(carrying ?m ?s)
                          (useable ?s)(loose ?n))
        :effect (and (not (loose ?n))(not (useable ?s)) (tightened ?n)))
)
```

Fig. 6.7 The PDDL representation of the Spanner domain as in the IPC 2011

```
(:action walk
        :parameters (?start - location ?end - location ?m - man)
        :precondition (and (at ?m ?start)(link ?start ?end)
                           (at-count ?start c0))
        :effect (and (not (at ?m ?start)) (at ?m ?end)))

(:action pickup_spanner
        :parameters (?l - location ?m  man ?n0 ?n1 ?n2 ?n3  counter)
        :precondition (and (at ?m ?l)(next ?n0 ?n1)(count ?l ?n1)
                        (next ?n2 ?n3)(carry-count ?m ?n2))
        :effect (and (not (at-count ?l ?n1))(at-count ?l ?n0)
                     (not (carry-count ?l ?n2))(carry-count ?m ?n3)))

(:action tighten_nut
        :parameters (?l - location ?m - man ?n  nut ?n2 ?n3  counter)
        :precondition (and (at ?m ?l)(at ?n ?l)(next ?n2 ?n3)
                           (carry-count ?m ?n3)(loose ?n))
        :effect (and (not (loose ?n))(tightened ?n)
                     (not (carry-count ?l ?n3))(carry-count ?m ?n2)))
)
```

Fig. 6.8 The optimized PDDL representation of the Spanner domain

number of spanners carried by the worker. Noteworthy, bagged representation can be also used to represent the number of untightened nuts (for the sake of clarity we refrain from representing nuts in bagged representation). The second issue can be addressed by adding an extra precondition to the walk operator restricting its applicability to situations where no spanner is present at a current worker location.[2] Adding such a precondition is easy if the spanners are represented by bagged representation, otherwise (the individual spanner representation) an engineer would have to exploit quantifiers (supported by ADL). The optimized representation of the Spanner domain addressing both the above issues is shown in Fig. 6.8.

8 Conclusion

Designing and developing good quality planning domain models is a process aiming at completeness and accuracy of the models, so they reflect the requirements of the real-world domain, as well as their operationality, so the planning engines can efficiently reason with them [30]. This chapter provided an overview of existing techniques that can help engineers to develop models that are more efficient for

[2] Adding extra preconditions can be understood as a part of specifying TB-DCK with only one DCK state where each operator is associated with one transition.

planning engines. Although some knowledge can be extracted from training plans (e.g., macros, entanglements), more sophisticated Domain Control Knowledge has to be specified by domain engineers (e.g., Procedural or Transition-based DCK).

Another important aspect regarding clarity of the Knowledge Engineering process is to separate a "raw" domain model, which captures the physics of the environment, and additional knowledge (e.g., DCK) making the model more efficient. Although for generic planning engines they have to be put together into an "enhanced" domain model, which can be done automatically, keeping them separate make them easier to maintain. Also, a small change in requirements might not only propagate to the "raw" domain model but also, and sometimes considerably, into the shape and possible usefulness of additional knowledge.

Acknowledgement This research was funded by the Czech Science Foundation (project no. 18-07252S).

References

1. Aguas, J.S., Celorrio, S.J., Jonsson, A.: Computing programs for generalized planning using a classical planner. Artif. Intell. **272**, 52–85 (2019). https://doi.org/10.1016/j.artint.2018.10.006
2. Areces, C., Bustos, F., Dominguez, M., Hoffmann, J.: Optimizing planning domains by automatic action schema splitting. In: Proceedings of ICAPS (2014)
3. Armano, G., Cherchi, G., Vargiu, E.: Automatic generation of macro-operators from static domain analysis. In: Proceedings of the 16th European Conference on Artificial Intelligence, ECAI'2004, including Prestigious Applicants of Intelligent Systems, PAIS 2004, Valencia, Spain, August 22–27, 2004. pp. 955–956 (2004)
4. Asai, M., Fukunaga, A.: Solving large-scale planning problems by decomposition and macro generation. In: ICAPS. pp. 16–24 (2015)
5. Bacchus, F., Kabanza, F.: Using temporal logics to express search control knowledge for planning. Artificial Intelligence **116**(1–2), 123–191 (2000). https://doi.org/http://dx.doi.org/10.1016/S0004-3702(99)00071-5, http://www.sciencedirect.com/science/article/pii/S0004370299000715
6. Baier, J.A., Fritz, C., McIlraith, S.A.: Exploiting procedural domain control knowledge in state-of-the-art planners. In: Proceedings of ICAPS. pp. 26–33. Providence, Rhode Island (September 22–26 2007), http://www.cs.toronto.edu/~jabaier/publications/bai-fri-mci-icaps07.pdf
7. Botea, A., Enzenberger, M., Müller, M., Schaeffer, J.: Macro-FF: improving AI planning with automatically learned macro-operators. Journal of Artificial Intelligence Research (JAIR) **24**, 581–621 (2005)
8. Carbonell, J., Etzioni, O., Gil, Y., Joseph, R., Knoblock, C., Minton, S., Veloso, M.: Prodigy: An integrated architecture for planning and learning. SIGART Bull. **2**(4), 51–55 (Jul 1991). https://doi.org/10.1145/122344.122353, http://doi.acm.org/10.1145/122344.122353
9. Chrpa, L., Barták, R.: Enhancing domain-independent planning by transition-based domain control knowledge. In: The 33rd Workshop of the UK Planning and Scheduling Special Interest Group (PlanSIG) (2015)
10. Chrpa, L., Vallati, M.: Improving domain-independent planning via critical section macro-operators. In: Proceedings of Thirty-Third AAAI Conference on Artificial Intelligence (2019)
11. Chrpa, L., Vallati, M., McCluskey, T.L.: Mum: A technique for maximising the utility of macro-operators by constrained generation and use. In: Proceedings of the International Conference on Automated Planning and Scheduling, ICAPS. pp. 65–73 (2014)

12. Chrpa, L., Vallati, M., McCluskey, T.L.: On the online generation of effective macro-operators. In: Proceedings of IJCAI. pp. 1544–1550 (2015)
13. Chrpa, L.: Generation of macro-operators via investigation of action dependencies in plans. Knowledge Eng. Review **25**(3), 281–297 (2010). https://doi.org/10.1017/S0269888910000159
14. Chrpa, L., Barták, R.: Reformulating planning problems by eliminating unpromising actions. In: Eighth Symposium on Abstraction, Reformulation, and Approximation, SARA 2009, Lake Arrowhead, California, USA, 8–10 August 2009 (2009)
15. Chrpa, L., Barták, R.: Guiding planning engines by transition-based domain control knowledge. In: Principles of Knowledge Representation and Reasoning: Proceedings of the Fifteenth International Conference, KR 2016, Cape Town, South Africa, April 25–29, 2016. pp. 545–548 (2016), http://www.aaai.org/ocs/index.php/KR/KR16/paper/view/12806
16. Chrpa, L., McCluskey, T.L., Vallati, M., Vaquero, T.: The fifth international competition on knowledge engineering for planning and scheduling: Summary and trends. AI Magazine **38**(1), 104–106 (2017), http://www.aaai.org/ojs/index.php/aimagazine/article/view/2719
17. Chrpa, L., Siddiqui, F.H.: Exploiting block deordering for improving planners efficiency. In: Proceedings of the Twenty-Fourth International Joint Conference on Artificial Intelligence, IJCAI. pp. 1537–1543 (2015)
18. Chrpa, L., Vallati, M., McCluskey, T.L.: Outer entanglements: a general heuristic technique for improving the efficiency of planning algorithms. J. Exp. Theor. Artif. Intell. **30**(6), 831–856 (2018). https://doi.org/10.1080/0952813X.2018.1509377
19. Coles, A., Fox, M., Smith, A.: Online identification of useful macro-actions for planning. In: Proceedings of ICAPS. pp. 97–104 (2007)
20. Dawson, C., Siklóssy, L.: The role of preprocessing in problem solving systems. In: Proceedings of IJCAI. pp. 465–471 (1977)
21. Dulac, A., Pellier, D., Fiorino, H., Janiszek, D.: Learning useful macro-actions for planning with n-grams. In: 25th IEEE International Conference on Tools with Artificial Intelligence, ICTAI 2013, Herndon, VA, USA, November 4–6, 2013. pp. 803–810 (2013)
22. Fuentetaja, R., de la Rosa, T.: Compiling irrelevant objects to counters. special case of creation planning. AI Commun. **29**(3), 435–467 (2016). https://doi.org/10.3233/AIC-150692
23. Gerevini, A., Saetti, A., Serina, I.: Planning with numerical expressions in LPG. In: Proceedings of the 16th European Conference on Artificial Intelligence, ECAI'2004, including Prestigious Applicants of Intelligent Systems, PAIS 2004, Valencia, Spain, August 22–27, 2004. pp. 667–671 (2004)
24. Haslum, P., Jonsson, P.: Planning with reduced operator sets. In: Proceedings of AIPS. pp. 150–158 (2000)
25. Hoffmann, J.: FF: the fast-forward planning system. AI Magazine **22**(3), 57–62 (2001), http://www.aaai.org/ojs/index.php/aimagazine/article/view/1572
26. Hoffmann, J.: The metric-FF planning system: Translating "ignoring delete lists" to numeric state variables. J. Artif. Intell. Res. **20**, 291–341 (2003). https://doi.org/10.1613/jair.1144
27. Hofmann, T., Niemueller, T., Lakemeyer, G.: Initial results on generating macro actions from a plan database for planning on autonomous mobile robots. In: ICAPS. pp. 498–503 (2017)
28. Korf, R.: Macro-operators: A weak method for learning. Artificial Intelligence **26**(1), 35–77 (1985)
29. Kvarnström, J., Doherty, P.: TALplanner: a temporal logic based forward chaining planner. Annals of Mathematics and Artificial Intelligence **30**(1–4), 119–169 (2000)
30. McCluskey, T.L., Vaquero, T.S., Vallati, M.: Engineering knowledge for automated planning: Towards a notion of quality. In: Proceedings of the Knowledge Capture Conference, K-CAP 2017, Austin, TX, USA, December 4–6, 2017. pp. 14:1–14:8 (2017). https://doi.org/10.1145/3148011.3148012
31. Minton, S.: Quantitative results concerning the utility of explanation-based learning. In: Proceedings of AAAI. pp. 564–569 (1988)
32. Newton, M.A.H., Levine, J., Fox, M., Long, D.: Learning macro-actions for arbitrary planners and domains. In: Proceedings of the International Conference on Automated Planning and Scheduling, ICAPS. pp. 256–263 (2007)

33. Riddle, P.J., Barley, M.W., Franco, S., Douglas, J.: Automated transformation of PDDL representations. In: Proceedings of the Eighth Annual Symposium on Combinatorial Search, SOCS 2015, 11–13 June 2015, Ein Gedi, the Dead Sea, Israel. pp. 214–215 (2015)
34. Siddiqui, F.H., Haslum, P.: Block-structured plan deordering. In: AI 2012: Advances in Artificial Intelligence—25th Australasian Joint Conference, Sydney, Australia, December 4–7, 2012. Proceedings. pp. 803–814 (2012). https://doi.org/10.1007/978-3-642-35101-3_68
35. Slaney, J., Thiébaux, S.: Blocks world revisited. Artificial Intelligence **125**(1–2), 119–153 (2001)

Part II
Interaction, Visualisation, and Explanation

Chapter 7
An Interactive Tool for Plan Generation, Inspection, and Visualization

Alfonso E. Gerevini and Alessandro Saetti

Abstract In mixed-initiative planning systems, humans and AI planners work together for generating satisfactory solution plans or making easier solving hard planning problems, which otherwise would require much greater human planning efforts or much more computational resources. In this approach to plan generation, it is important to have effective plan visualization capabilities, as well to support the user with some interactive capabilities for the human intervention in the planning process. This paper presents an implemented interactive tool for the visualization, generation, and revision of plans. The tool provides an environment through which the user can interact with a state-of-the-art domain-independent planner, and obtain an effective visualization of a rich variety of information during planning, including the reasons why an action is being planned or why its execution in the current plan is expected to fail, the trend of the resource consumption in the plan, and the temporal scheduling of the planned actions. Moreover, the proposed tool supports some ways of human intervention during the planning process to guide the planner towards a solution plan, or to modify the plan under construction and the problem goals.

Keywords Interactive planning · Mixed-initiative planning · Plan visualization and inspection · Graphical user interfaces for planning

1 Introduction

In the AI planning literature, many approaches to plan generation or revision combining automated techniques with human-driven decisions have been proposed (e.g., [1, 4, 6, 7, 10, 11, 31, 38–40]). The rationale of these interactive, mixed-initiative approaches is that the collaborative joint work of a human and an AI planner can be much more effective than either human planning or fully automated

---

A. E. Gerevini · A. Saetti (✉)
Dipartimento di Ingegneria dell'Informazione, Università degli Studi di Brescia, Brescia, Italy
e-mail: alfonso.gerevini@unibs.it; alessandro.saetti@unibs.it

M. Vallati, D. Kitchin (eds.), *Knowledge Engineering Tools and Techniques for AI Planning*, https://doi.org/10.1007/978-3-030-38561-3_7

planning alone, in terms of problem solvability, planning speed, or user satisfaction about the quality of the generated solutions.

For example, a mixed-initiative planning system has been successfully applied to the Mars Exploration Rovers Mission project, which involved two NASA rovers for the ground exploration of Mars [3]. As argued in [3], the complexity of this project and the aggressive operations plan made using an automated tool for generating the daily activity plans necessary. However, also the human involvement during the planning process was needed. The activity plan needed to be presented, critiqued and, hopefully, accepted. If the plan had been constructed fully automatically, it would have been too difficult to analyze for humans. Another concern in this application was the infeasibility of formally encoding and effectively utilizing during automated plan generation all the knowledge that characterizes plan quality.

As argued by Ferguson and Allen [11], in mixed-initiative planning the description of a plan that the system provides to the user should be richer than just a list of action names with the associated temporal information (e.g., for each action, its start time and expected duration [12, 20]). In particular, for an interactive planning tool it is essential to have specialized user interface capabilities explaining the reasons why an action has been planned, or why, in the context of the plan under consideration, it is expected that its execution will fail. Moreover, it is desirable that the system supports an effective visualization and inspection of the plan, which helps the user to understand the ongoing planning process, the decisions taken by the planner, and the feasibility and quality of the solution plan proposed by the system to the user.

An adequate description of the current plan and the planning process that led to its generation is very useful to the user for deciding the possible human interventions in order to (a) guide the planning process for a faster generation of a solution, (b) constrain the plan under construction so that, e.g., it contains certain actions or crosses some particular intermediate states specified by the user, or (c) modify the problem goals during planning. However, plan visualization is a scarcely investigated area in AI planning, and only very few planning systems currently incorporate a user interface with effective plan visualization capabilities (e.g., [25, 26]).

In the last years, automated domain-independent planning has dramatically improved in terms of planning performance and especially speed [34]. However, to the best of our knowledge, all modern domain-independent planners (e.g., [14, 21, 22, 27, 35]) have been developed for a fully automated planning context, and are not equipped with an interactive tool supporting their use in a mixed-initiative framework. This limits their applicability to real-world applications where, often, the domain experts want to analyze the plan generated by the planner and possibly refine some portion(s) of it, before committing to its execution. The output information of all recent efficient domain-independent planners is given to the user only in a simple high-level textual form, indicating some very general information about the planning process and describing the generated plan as a simple list of actions with the relative scheduled start times and durations. Moreover, there is no possibility of human intervention to guide the planning process, to inspect the generated plan, and to possibly revise it.

In this work, we concentrate on a successful approach to fully automated domain-independent planning with the aim of making it more suitable for mixed-initiative planning, and provide a general tool for plan visualization, inspection, and generation. This approach, which is implemented in the well-known LPG planner, is based on some particular graphs called *action graphs* for representing the plans during search, and on a stochastic local search procedure for searching a solution plan [14, 16, 17]. LPG is a modern planner supporting the standard language PDDL2.2 [12, 20, 23], and belonging to the so-called satisficing style of planning, which in the last 15 years has received significant interest in the AI planning community (e.g., [13, 23, 28, 34]).[1] Other efficient satisficing planners have been recently proposed, but LPG remains competitive for many existing benchmark domains, especially for metric-temporal domains (see, e.g., [9, 17]).

The main contribution of the work presented in this paper is a new interactive environment for the visualization, inspection, generation, and revision of plans. The tool implementing this environment is called InLPG, and uses LPG as the underlying automated planning system. The user can interact through InLPG with the underlying planner about

- the plan under construction or revision: e.g., the user requests a temporal scheduling for some planned action that is different from the one decided by the planner, or imposes that some particular action must be in the solution plan;
- the planning problem under consideration: e.g., the user communicates to the system that some particular goal can be ignored, or adds some new goals;
- the automated planning process: e.g., when the search process of the planner is trapped into a local minimum the user pauses the search of the planner and modifies the planner decisions about the next search states to explore, so that the planner learns how to escape from similar local minima visited during the search.

Moreover, the proposed tool supports plan visualization through various (dynamic) views of the plan, such as a Gantt chart of the planned actions, a constraint graph for the temporal constraints in the plan, a resource graph for monitoring and describing the trend of the resource consumption in the plan, a graphical representation of the main data structure representing the search states and partial plans explored by the automated planner, several plots describing the trend of the search process.

Although built on top of a specific planner, the proposed tool is very general, because LPG is domain-independent and most of the various plan visualization techniques of InLPG can be also applied to an input plan taken from a plan library, generated by another planner or completely created by humans.

---

[1]The word satisficing was coined by Simon to mean "rational enough" [36], and subsequently it was adopted by the optimization community to mean "good enough." This term has also been adopted by the planning community to indicate planners aimed at computing plans of good quality, but with no guarantee of their optimality w.r.t. a specified plan metric [23].

The paper is organized as follows. Section 2 introduces the necessary background and preliminaries about automated AI planning and the LPG planner. Section 3 describes the main components of the proposed interactive planning tools. Section 4 gives a walk-through example about possible interactions and plan visualization during a plan generation session through the proposed tool. Section 5 presents some experimental results about using the proposed tool for solving some planning problems more effectively than with a fully automated planner alone. Finally, the last section is devoted to the conclusions and possible future work.

2 Preliminaries

In this section, we introduce the necessary background and the preliminaries about automated AI planning and the LPG planner.

2.1 The Planning Problem

Informally, the plan generation is a task consisting in determining and organizing a set of actions (a plan) whose execution transforms a given initial state of the world into a new state satisfying some desired goals. More formally, a (propositional) *planning problem* Π is a tuple $\langle L, A, s_0, g \rangle$, where

- L is a set of positive literals called the *Facts* of Π;
- A is the set of *Actions* of Π; each action a is a triple $\langle P(a), E(a)^+, E(a)^- \rangle$, where
  - $P(a) \subseteq L$ is a set of positive literals called the *preconditions* of a,
  - $E(a)^+ \subseteq L$ is a set of positive literals called the *positive effects* of a,
  - $E(a)^- \subseteq L$ is a set of negative literals called the *negative effects* of a;
- $s_0 \in 2^L$ is the *initial world state* of Π;
- $g \subseteq L$ is a set of literals called the *goals* of Π.

For instance, consider a simple problem of delivering some packages by trucks from some depots to some customers. Figure 7.1 shows the facts and actions of such a problem using the standard Planning Domain Definition Language (PDDL) [12, 20, 23]. Predicates `(package ?p)`, `(location ?l)`, and `(truck ?t)` specify that `?p`, `?l`, and `?t` are a package, a location, and a truck, respectively; predicate `(at ?x ?l)` specifies that either truck or package `?x` is at location `?l`; predicate `(in ?p ?t)` specifies that package `?p` is inside truck `?t`; finally, predicate `(delivered ?p ?l)` specifies that package `?p` has been delivered to the customer at location `?l`. Action `(load ?p ?t ?l)` represents the movement of package `?p` from location `?l` into truck `?t`; action `(unload ?p ?t ?l)` represents the opposite movement; action `(drive ?t ?from ?to)` represents the movement of truck `?t` from location

```
(define (domain SimpleLogistic)
(:predicates (package ?p) (location ?l) (truck ?t)
             (at ?x ?l) (in ?p ?t) (delivered ?p ?l))

(:action load
 :parameters (?p ?t ?l)
 :precondition (and (package ?p) (truck ?t) (location ?l) (at ?p ?l) (at ?t ?l))
 :effect (and (not (at ?p ?l)) (in ?p ?t)))

(:action unload
 :parameters (?p ?t ?l)
 :precondition (and (package ?p) (truck ?t) (location ?l) (in ?p ?t) (at ?t ?l))
 :effect (and (not (in ?p ?t)) (at ?p ?l)))

(:action drive
 :parameters (?t ?from ?to)
 :precondition (and (truck ?t) (location ?from) (location ?to) (at ?t ?from))
 :effect (and (not (at ?t ?from)) (at ?t ?to)))

(:action deliver
 :parameters (?p ?l)
 :precondition (and (package ?p) (location ?l) (at ?p ?l) (at ?p ?l))
 :effect (and (not (at ?p ?l)) (delivered ?p ?l))))
```

Fig. 7.1 PDDL specification of the facts and actions for a simple package delivery problem. Each action (predicate) specification has a set of parameters making it a schema of ground actions (predicates) where parameters are replaced with objects of the planning problem

```
(define (problem truck-1)
(:domain SimpleLogistic)
(:objects Truck1 NY Wa Bo package1 package2 package3)
(:init (truck Truck1) (location NY) (location Wa) (location Bo)
       (package package1) (package package2) (package package3)
       (at truck1 Wa) (at package1 Bo) (at package2 Bo) (at package3 NY))
(:goal (and (delivered package1 NY)(delivered package2 Wa)(delivered package3 Wa))))
```

Fig. 7.2 An example of PDDL specification of the initial world state and the set of goals of a simple instance of the package delivery problem

`?from` to location `?to`; finally, action `(delivery ?p ?l)` represents delivering package `?p` to the customer at location `?l`.

The tokens starting with "?" indicate parameters, which have to be substituted with problem objects (airplanes, trucks, and locations) specified in the description of the problem (for an example, see Fig. 7.2). Hence, each PDDL action (predicate) description corresponds to a schema of ground actions (predicates) where each parameter is instantiated.

Figure 7.2 shows the PDDL specification of an initial world state and a set of goals for the package delivery problem with three packages (`package1`, `package2`, and `package3`), three locations New York, Washington, and Boston (`NY`, `Wa`, and `Bo`, respectively), and one truck (`truck1`). In the initial state, `package1` and `package2` are at Boston, `package3` is at New York, and, finally, `truck1` is at Washington. In our example, the goals of the problem are that `package1` is delivered at New York, and `package2` and `package3` at Washington.

An action a is executable in a world state s if the preconditions of a are satisfied in s, i.e., $P(a) \in s$. The state s' resulting from the execution of a in s is obtained by

Fig. 7.3 An example of a plan for the package delivery problem in Fig. 7.2

```
0:  (DRIVE TRUCK1 WA BO)
1:  (LOAD PACKAGE1 TRUCK1 BO)
2:  (LOAD PACKAGE2 TRUCK1 BO)
3:  (DRIVE TRUCK1 BO NY)
4:  (LOAD PACKAGE3 TRUCK1 NY)
5:  (UNLOAD PACKAGE1 TRUCK1 NY)
6:  (DRIVE TRUCK1 NY WA)
7:  (UNLOAD PACKAGE2 TRUCK1 WA)
8:  (UNLOAD PACKAGE3 TRUCK1 WA)
9:  (DELIVER PACKAGE1 NY)
10: (DELIVER PACKAGE2 WA)
11: (DELIVER PACKAGE3 WA)
```

```
(:durative-action drive
 :parameters (?t ?from ?to)
 :duration (= ?duration (drive-time ?from ?to))
 :condition (and (at start (at ?t ?from)) (movable ?t))
                 (at start (>= (fuel ?t) (required-fuel ?from ?to))))
 :effect (and (at start (not (at ?t ?from))) (at end (at ?t ?to))
              (at end (increase (total-fuel-used) (required-fuel ?from ?to)))
              (at end (decrease (fuel ?t) (required-fuel ?from ?to)))))
```

Fig. 7.4 An example of an action with duration, scheduling constraints, and consuming resources

adding the positive effects of a to s and removing the negative effects of a from s, i.e., $s' = s \cup E(a)^+ \setminus E(a)^-$.

A solution of a planning problem is a plan, that is a partially ordered set of (executable) actions, whose execution from the initial world state achieves the problem goals. If two actions in a plan are not ordered, they can be executed with any relative ordering. Figure 7.3 shows an example of solution plan for our running example.

The previous definition of the planning problem has some strong simplifying assumptions. In real-world problems, actions take time, consume resources, and in some cases their execution can only occur during some predefined time windows where one or more necessary conditions hold. Moreover, usually the quality of a solution plan takes resource consumption and makespan (overall plan duration) into account.

Figure 7.4 shows an example of a more complex model of the action moving truck `?t` from a location `?from` to a location `?to` that can be formalized through version 2.1 of PDDL [12]. Constraint `(= ?duration (drive-time ?from ?to))` imposes that the duration of action `(move ?t ?from ?to)` is equal to the value of numerical fluent `(drive-time ?from ?to)`.[2] Condition `(movable ?t)` is a special fact that holds only during some predefined time windows specified in the description of the problem initial state, and that imposes some scheduling constraints to action `(move ?t ?from ?to)`. Numerical condition `(>= (fuel ?t) (required-fuel ?from ?to)` constraints action `(move ?t ?from ?to)` to start its execution only if the amount of fuel of truck `?t` is greater than the value of numerical fluent `(required-fuel ?from ?to)`. At the end of the action, effect

[2]In PDDL, numerical fluents are functions over real values.

```
(define (problem truck-1)
(:domain SimpleLogistic)
(:objects Truck1 NY Wa Bo package1 package2 package3)
(:init (truck Truck1) (location NY) (location Wa) (location Bo) (package package1)
 (package package2) (package package3) (at truck1 Wa) (at package1 Bo) (at package2 Bo)
 (at package3 NY) (at 8.00 (movable truck1)) (at 20.00 (movable truck1))
 (at 30.00 (movable truck1)) (at 42.00 (not (movable truck1)))
 (= (drive-time NY Wa) 3.0) (= (drive-time NY Bo) 4.0) (= (drive-time Wa NY) 3.0)
 (= (drive-time Wa Bo) 8.0) (= (drive-time Bo NY) 4.0) (= (drive-time Bo Wa) 8.0)
 (= (capacity truck1) 50) (= (required-fuel NY Wa) 40.6) (= (required-fuel NY Bo) 7.31)
 (= (total-fuel-used) 0) (= (required-fuel Wa NY) 40.6) (= (required-fuel Wa Bo) 35.6)
 (= (fuel truck1) 0) (= (required-fuel Bo NY) 7.31) (= (required-fuel Bo Wa) 35.6))
(:goal (and (delivered package1 NY)(delivered package2 Wa)(delivered package3 Wa)))
(:metric minimize (total-time)))
```

Fig. 7.5 An example of PDDL specification of the initial world state and the set of goals of a simple package delivery problem with action durations, scheduling constraints, and numerical fluents

`(decrease (fuel ?t) (required-fuel ?from ?to))` decreases the value of numerical fluent `(fuel ?t)` by the value of numerical fluent `(required-fuel ?from ?to)`, and, similarly, `(increase (total-fuel-used) (required-fuel ?from ?to)` increases the value of numerical fluents `(total-fuel-used)`.

Figure 7.5 shows the PDDL description of the initial world state and the goal state of a package delivery problem in which each package must be delivered to a location by a certain deadline. The truck movements take time, can happen only during some time windows (when the `movable` conditions hold), and consume some amounts of fuel, depending on the distance traveled between locations. Finally, the "metric" field specifies that the quality of plans is measured in terms of makespan. For the objects of our running example, the (initial) value of fluents `(drive-time ?from ?to)`, `(required-fuel ?from ?to)`, `(total-fuel-used)` and the time windows when the `(movable ?t)` conditions hold are defined in the initial state.

2.2 *The LPG Planner*

In the rest of this section, we give a brief description of the planner, LPG [14, 17], used by our interactive tool. While LPG supports the complex action models and planning problems that can be specified by PDDL2.2, for simplicity, we will focus the presentation on simple propositional planning problems. For a more comprehensive description of LPG, the interested reader can refer to the previously cited papers.

Like in partial-order causal-link planning, (e.g., [29, 32, 33]), LPG searches a solution plan in a search space of partial plans that are represented by *linear action graphs*. A linear action graph is a variant of the well-known planning graph representation [2]. Starting from an initial LA-graph, LPG uses a stochastic local search process that transforms it into a LA-graph representing a valid plan through the iterative application of some search steps modifying the graph. We describe first the linear action graph representation, and then the general search process of LPG in the space formed by these graphs.

2.2.1 Plan Representation Through LA-Graphs

A linear action graph (LA-graph) $\mathcal{A}$ for a planning problem Π is a directed acyclic leveled graph alternating a *fact level* and an *action level*. Each (action) level of the graph corresponds to a time step. Fact levels contain *fact nodes*, each of which is labeled by a literal representing a fact that is purposeful for Π. Each fact node f at a (fact) level l is associated with a *no-op* action node at (action) level l representing a dummy action having the literal labeling f as its only precondition and positive effect. The purpose of the no-op nodes is propagating the possible truth of a fact from one (action) level to the next one. Each action level contains one action node labeled by the name of a problem action that it represents and the no-op nodes corresponding to that level.

An action node labeled a at an (action) level l is connected by incoming edges from the fact nodes at (fact) level l representing the preconditions of a (*precondition nodes*) and by outgoing edges to the fact nodes at (fact) level $l + 1$ representing the effects of a (*effect nodes*). The initial level contains the special action node a_{start} and the last level the special action node a_{end}. The effect nodes of a_{start} represent the facts of the problem initial state, and the precondition nodes of a_{end} the problem goals. Figure 7.6 gives an example of a simple LA-graph containing five action nodes (a_{start}, a_1, a_2, a_3, a_{end}) and several fact nodes representing ten facts.

A pair of actions (possibly no-ops) can be constrained by a *persistent mutex relation* [12], i.e., a binary constraint imposing that the involved actions can never occur in parallel in a valid plan. An LA-graph $\mathcal{A}$ also contains a set Ω of *ordering constraints* between actions in the (partial) plan represented by the graph. These constraints are (1) constraints imposed during search to deal with *mutually exclusive* actions: if an action node a at level l of $\mathcal{A}$ interferes with an action node b at a level

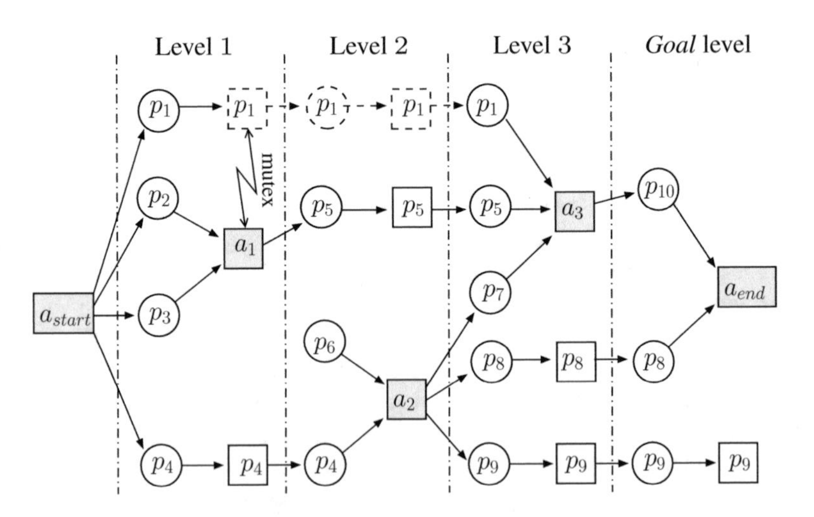

Fig. 7.6 An example of LA-graph. Square nodes are action nodes; circle nodes are fact nodes. Dashed edges form chains of no-ops blocked by interfering (mutex) actions

after l, then a is constrained to finish before the start of b and (2) constraints between actions implied by the causal structure of the plan: if an action a is used to achieve a precondition of an action b, then a is constrained to finish before the start of b. The set Ω of ordering constraints in the LA-graph of Fig. 7.6 is

$$\{a_{start} \prec a_1,\ a_{start} \prec a_2,\ a_{start} \prec a_3,\ a_1 \prec a_3,\ a_2 \prec a_3,$$
$$a_1 \prec a_{end},\ a_2 \prec a_{end},\ a_3 \prec a_{end}\}.$$

For instance, $a_1 \prec a_3 \in \Omega$, because the pair of actions involved in the constraint is mutex (a_1 deletes p_1 that is precondition of a_3), and the level of a_3 is greater than the level of a_1. $a_2 \prec a_3 \in \Omega$, because a_2 achieves precondition p_7 of a_3.

An LA-graph $\mathcal{A}$ represents the (partial) plan formed by the actions labeling the action nodes of $\mathcal{A}$ scheduled at the earliest possible time steps according to the ordering constraints in Ω. Concerning the LA-graph of Fig. 7.6, the earliest execution time step for actions a_1 and a_2 is 1, while for action a_3 it is 2. Hence, the plan represented by the graph of Fig. 7.6 is $\pi = \langle\{a_1, a_2\}, \{a_3\}\rangle$.

An action graph can represent a plan that is not valid for the problem under consideration. A plan is not valid when it contains at least one *flaw*, i.e., an action with precondition nodes that are not *supported*. A precondition node q at a level i of an LA-graph $\mathcal{A}$ is supported if in $\mathcal{A}$ there is an action node at level $i-1$ representing an action with positive effect q. An LA-graph without flaws represents a valid plan, and is called a *Solution Graph*. For example, the plan represented by the LA-graph of Fig. 7.6 is not a valid plan, because it contains action node a_2 having precondition node p_6 that is not supported at level 2.

2.2.2 Local Search in the Space of LA-Graphs

Given the PDDL specification of a planning problem, LPG uses a stochastic local search process for computing a solution graph from an LA-graph that initially contains only a_{start} and a_{end}. Figure 7.7 gives a high-level description of the procedure for searching a solution graph (valid plan) in LPG. Step 1 initializes the current LA-graph $\mathcal{A}$ to the empty plan. At each iteration of the loop 2–6, step 3 selects an unsupported precondition in $\mathcal{A}$. In order to resolve the selected flaw, LPG

1. **Set** $\mathcal{A}$ to an empty plan (containing only a_{start} and a_{end});
2. **While** the current LA-graph $\mathcal{A}$ contains a flaw **do**
3. Select a flaw σ in $\mathcal{A}$;
4. Identify the search **neighborhood** $N(\mathcal{A}, \sigma)$;
5. Weight the elements of $N(\mathcal{A}, \sigma)$ using a **heuristic function** E;
6. **Choose** an LA-graph $\mathcal{A}'$ in $N(\mathcal{A}, \sigma)$ according to E and a noise parameter; **set** $\mathcal{A}$ to $\mathcal{A}'$;
7. **Return** $\mathcal{A}$.

Fig. 7.7 The general scheme of the search procedure of LPG

uses two basic graph modifications consisting in an extension of $\mathcal{A}$ to include a new action node, or a reduction of $\mathcal{A}$ to remove an action node (and the relevant edges).

When an action node is added to a level l, the LA-graph is extended by one level, all action nodes from l are shifted forward by one level, and the new action is inserted at level l. Similarly, when an action node is removed, the LA-graph is "shrunk" by one level (a more detailed description is given in [14]).

Step 4 of Fig. 7.7 identifies the search neighborhood of $\mathcal{A}$ for the selected flaw σ. The *neighborhood* $N(\sigma, \mathcal{A})$ of $\mathcal{A}$ for σ is the set of LA-graphs obtained from $\mathcal{A}$ by applying a graph modification that resolves σ. Step 5 evaluates each element $\mathcal{A}'$ in $N(\sigma, \mathcal{A})$ using a *heuristic evaluation function* estimating the number of additional search steps required to find a solution graph from $\mathcal{A}'$. Then, step 6 selects an element with the lowest estimated search cost from $N(\sigma, \mathcal{A})$ using a "noise parameter" [14]. This parameter introduces some randomization in the choice of $\mathcal{A}'$, which is useful to "escape" from search states corresponding to local minima when the local search process is trapped. This procedure is repeated until either the current LA-graph contains no flaw or a maximum number of iterations is exceeded (step 2). Finally, step 7 returns the computed solution LA-graph.

3 Architecture of InLPG

In this section, we describe the architecture of the proposed interactive tool for the visualization, generation, and revision of plans, along with its main components.

3.1 Architecture Overview

The architecture of InLPG is sketched in Fig. 7.8. It consists of five main components that are integrated in the LPG planner:

- *The Input Module*, which inputs the files containing the description of the planning problem under consideration;
- *The Search Process Monitor*, which monitors the search process, and at each step of the search displays the information about the current search state;
- *The Search State Monitor*, which provides different views of the current state during the search process;
- *The Search State Editor*, which provides some tools for human-driven changes to the plan under construction;
- *Search Process Editor*, which provides some tools for human-driven changes to the search process and to the current planning problem.

Our environment includes an *open-controllable* version of LPG, i.e., all the decision points of the search procedure sketched in Fig. 7.7 can be controlled by an external process that, in our context, is under the control of a human user. In

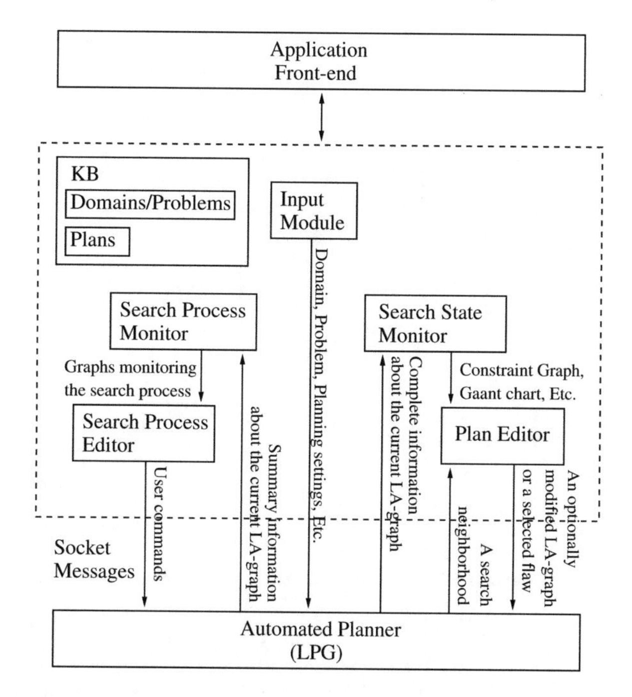

Fig. 7.8 A sketch of the main components of InLPG and their interactions

particular, at each search step, the user can select a plan flaw to repair (step 3 of the procedure in Fig. 7.7), modify the definition of the search neighborhood (step 4), and select a graph modification among those that generate the elements in the search neighborhood (step 5). Specifically, LPG runs as a separate process, and it communicates with the rest of the environment through socket messages. The decisions about the search process taken by the user through InLPG overwrite the decisions taken by the heuristics of LPG.

Figure 7.9 shows two screenshots of the user interface. The left frames show the Gantt chart of the plan computed at the 368th search step (upper screenshot) and the trend of some resources during the execution of the plan (bottom screenshot). The plan is flawed, because it contains actions that cannot be executed (the darker boxes in the Gantt chart, which in the actual screen are red). The information in this frame can be moved to the secondary frame (right frame), as displayed by the bottom screenshot, or into different windows. This latter option is particularly useful if the user wants to compare different plans.

The quality of the displayed plan is 1350.[3] The right-hand side of the upper screenshot contains four plots. The first three plots (starting from the top) show, for each search step, the number of flaws (1st plot), the number of actions (2nd plot), and the makespan of the plan constructed at the current search step (3rd plot). The

[3]The quality of the plan is automatically measured according to the metric expression specified in the problem formulation. In this example, the quality is expressed by the duration of the plan.

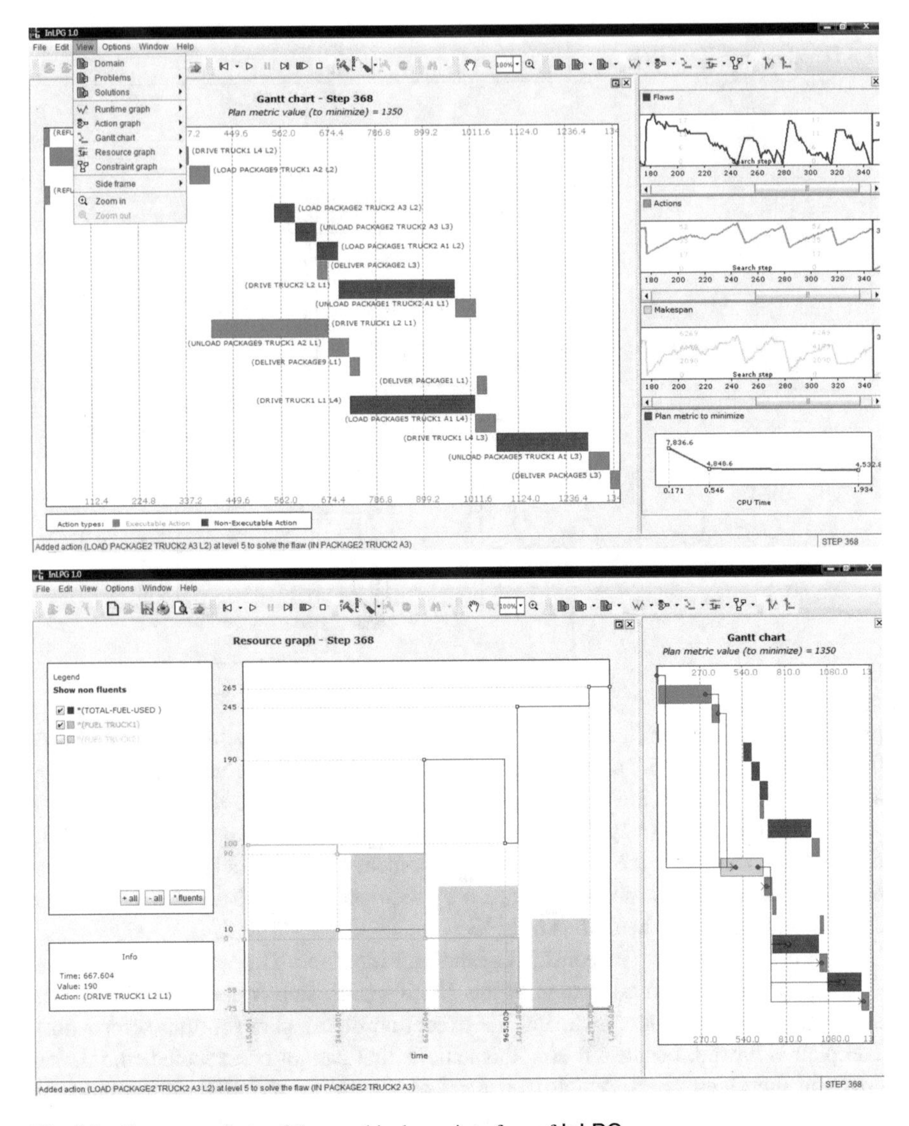

Fig. 7.9 Two screenshots of the graphical user interface of InLPG

bottom plot informs the user about the trend of the quality of the solutions computed so far: for the example of Fig. 7.9, LPG first found one solution with quality 7836.6 using 0.171 CPU-seconds; subsequently, a better solution with quality 4868.6 was found using about 0.4 CPU-second, and finally another slightly better solution with quality 4522.8 was generated using about 1.5 CPU-second.

In the next subsections, we will give a detailed description of the components integrated in our environment, and in the next section we will illustrate an example of possible interactions with InLPG using its components.

3.2 Input Module

By using standard acquisition tools the user inputs the basic planning information to our environment: a *domain file* containing the action schemata and the predicates of the planning problem to solve (e.g., the PDDL description in Fig. 7.1), a *problem file* describing the problem initial state and goals (e.g., the PDDL descriptions in Figs. 7.2 and 7.5), and, optionally, a plan file containing the description of a plan taken from a plan library (for an example, see the textual description in Fig. 7.3). The language for encoding the planning information in these files is version 2.2 of PDDL [12, 20, 23]. The possibility of loading a plan is particularly useful for solving plan adaptation problems [18, 19], in which the input plan is modified to become a solution of the planning problem. In addition, the user can change the default values of some technical parameters of the search algorithm implemented in LPG. A complete list of such parameters is described in [15].

After the necessary information has been acquired, the input module verifies the syntax of the PDDL files, and, if they are syntactically correct, it sends a message to LPG in order to start a planning process for solving the given planning problem.

3.3 Search Process Monitor

At each search step, LPG sends a message to the search process monitor containing the following basic information about the current LA-graph: the number of flaws in the LA-graph, the number of actions in the represented plan, and the makespan of this plan. The search process monitor processes this information and plots the corresponding graphs in order to visualize a variety of information about the ongoing search process. For example, if the user sees that the number of actions or the plan makespan is much higher than the desired value, then she is informed that the search process is most likely visiting a portion of the search space that is faraway from the portion where the desired solution is located. Moreover, if the number of flaws does not decrease with the search steps, then the search might be trapped in a local minimum or plateau. Figures 7.9 (right frame of the upper screenshot) and 7.10 show two examples of these plots.

Identifying which are the search steps where the planner makes wrong decisions that are crucial for the success of the search process can be difficult. The plots of the search process monitor help the user to identify them. The intervention of a human to revise the wrong decisions made by the planner for these steps could be very important in order to effectively guide the process towards a solution plan. In our context, when the search process visits a portion of the space which contains no solution LA-graph, the number of flaws does not significantly decrease. The plot of the number of flaws can indicate this problematic behavior in which the planner is continuously making wrong decisions, and hence it can suggest that a human-driven choice could improve the search.

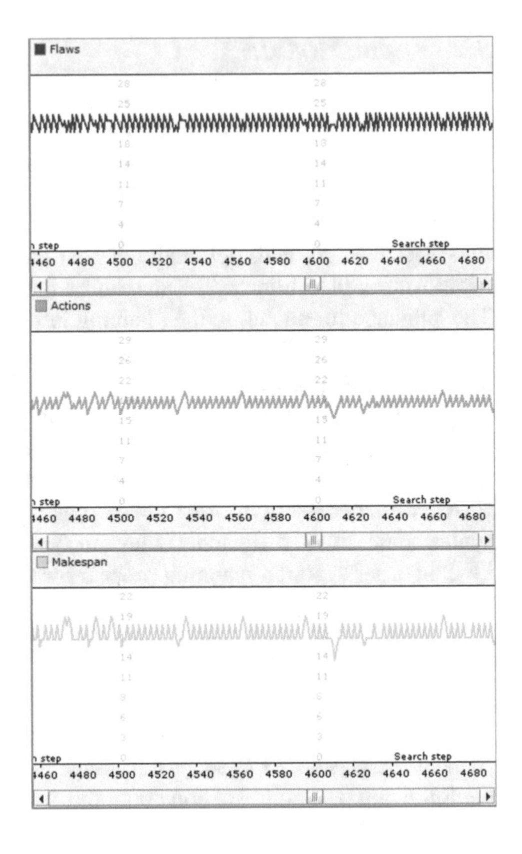

Fig. 7.10 A portion of the screen of the user interface showing an example of the plots displayed by the search process monitor: the number of flaws (upper plot), the number of actions in the current LA-graph (middle plot), and duration of the plan under construction (bottom plot). On the x-axis we have the corresponding sequence of performed search steps

For example, according to the upper plot in the right frame of Fig. 7.9, the planning system is working well: at several search steps the number of flaws in the current LA-graph is zero.[4] On the contrary, according to the plots about the number of actions, flaws and duration of the plan in Fig. 7.10, LPG is not approaching a solution LA-graph.

3.4 Search State Monitor

When the user intervenes in the search process of LPG, e.g., by pausing the search process, LPG sends a detailed description of the current search state (LA-graph) to the search state monitor. The search state monitor processes such information by computing the following information:

[4]When a search step reaches an LA-graph with no flaw, the planner has found a valid plan. However, this plan is given in output only if its quality improves the quality of the previous output plan. In the example of Fig. 7.9, some valid plans are computed, but only the one found at about the 350th step is given as the third output plan.

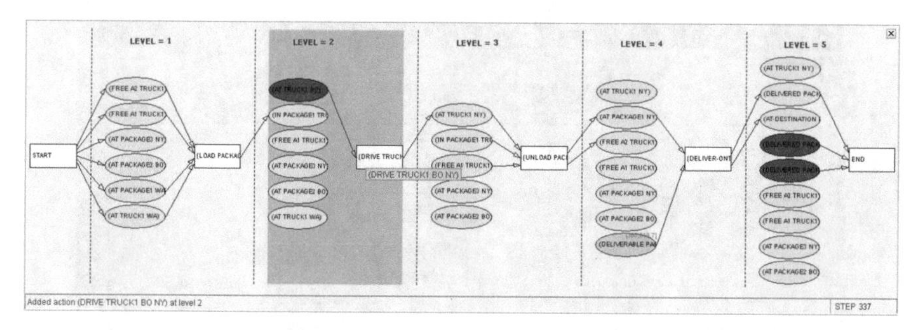

Fig. 7.11 A portion of the screen of the user interface showing the current LA-graph. Square nodes are action nodes; elliptical nodes are fact nodes. Dark elliptical nodes (red nodes in the actual screen) are plan flaws (unsatisfied action preconditions). For lack of space, the label of some nodes is abbreviated. By moving the mouse on a node, a tooltip displays the corresponding full label. The darkened level (blue area in the actual screen) is the level of the last change performed by the planning process

- a complete graphical representation and some compact representations of the current LA-graph;
- a graphical representation of the temporal constraints involving the actions in the plan represented by the current LA-graph;
- a textual description of the plan represented by the current LA-graph;
- a Gantt chart of the actions in the current plan; and
- a graph showing the trend of the involved resources during the plan.

All these graphs are dynamic. At each search step, they are automatically recomputed and, in order to guarantee their readability, they are automatically scaled and appropriately displayed. For example, at each search step, the temporal constraints involving the actions in the plan change because either an action is removed or inserted, and thus the search state monitor recomputes an appropriate graphical organization of the nodes in the revised constraint graph, in order to avoid edge crossing and to reduce the edge length. The new constraint graph is computed by GRAPHVIZ (http://www.graphviz.org/), an automated tool for layered drawing of directed graphs. Moreover, the nodes of the graphs can be clicked to obtain information on the represented action or to change some property of the action (e.g., a new start time for the represented action). Figures 7.11, 7.12, 7.14, and 7.15 give examples of the LA-graph, constraint graph, Gantt chart, and graph of the resources, respectively.

By looking at the graphical representations of the computed plan, the user can evaluate the current search state and realize (or at least hypothesize) how the current plan has to be modified. In particular, the user interface highlights the graphical objects corresponding to plan actions that are not executable, and provides information exploiting why they are not executable (for instance, because of a precondition is not satisfied, a scheduling constraint is violated, or not enough resources are available in the state where the action is executed).

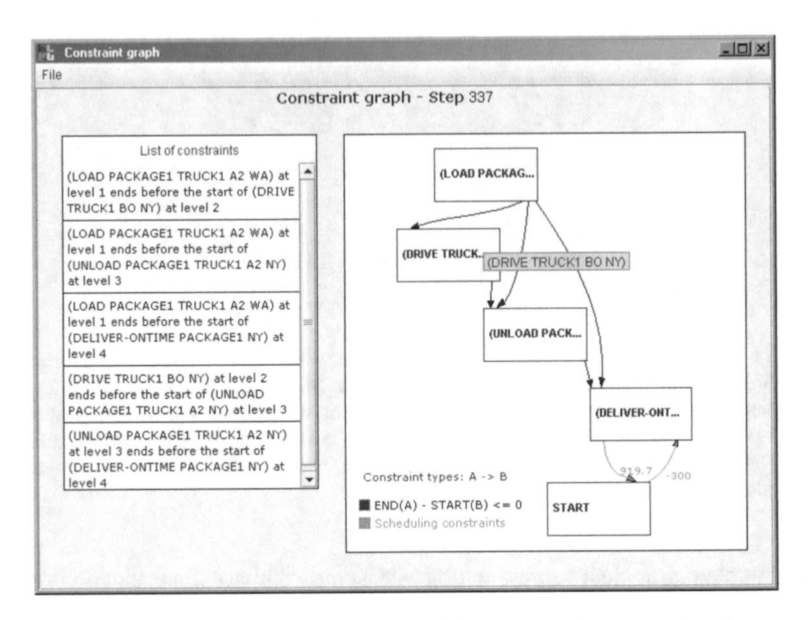

Fig. 7.12 A window of the user interface containing the graph representing the temporal constraints in the current LA-graph. The label of the nodes is abbreviated. By moving the mouse on a node, a tooltip displays the corresponding full label

3.5 *Plan Editor*

The plan editor is activated when the user revises the plan under construction. In a mixed-initiative planning context, the possibility for the user to inspect a plan and revise it through a plan editor is useful both during the process of constructing high quality plans and during the process of adapting an existing plan to satisfy the requirements of a new planning problem.

The plan editor allows the user to remove undesired actions, to add new actions (possibly satisfying some preconditions that currently are unsatisfied), and to reschedule an action in the plan. In the following, we describe the tools provided by the plan editor for supporting these plan revisions. If the user adds (removes) an action from the current plan, the plan editor sends a special message to LPG containing the desired (undesired) action selected by the user. Then, the planner computes a new LA-graph obtained from the previous one by adding (removing) the corresponding action node. Similarly, the plan editor allows the user to constrain the start time of an action in the plan to be after a desired time, and the end time of an action to be before a desired time. Moreover, the plan editor allows the user to restart the search from an empty plan, instead of continuing it from the current plan, which can be a good option when the current plan contains too many undesired actions or too many actions violating the desired scheduling constraints.

When the plan under construction is not valid, the plan editor allows the user to repair "by hand" a flaw in the current partial plan. This interaction requires the usage of some handshaking messages. The plan editor sends a message to the planner containing the flaw selected by the user; then, LPG sends a new message to the component containing the elements in the search neighborhood for the selected flaw. The plan editor displays such a neighborhood (each element is compactly represented by the corresponding action addition/removal that eliminates the flaw under consideration), while the user selects an element from the neighborhood (for an example, see Fig. 7.16). Finally, the plan editor sends a message describing the graph modification selected by the user to the planner, which then computes a new LA-graph obtained by applying the selected graph modification.

3.6 Search Process Editor

The search process editor allows the user to control the progress of the search. The user can inspect and run the search by different modalities: with "no interruption," "step by step" and by "multi steps." For instance, the user can "pause" the search at any time, inspect the current plan and LA-graph and then continue the search step-by-step, i.e., the search progresses only one step and then waits that the user clicks the command button to proceed for the next step. If the user observes that the heuristics of LPG make an incorrect choice when repairing the selected flaw, the search process editor allows her to move the search one step *backward*, so that she can intervenes and forces the planner to make an alternative decision among a set of alternatives provided by the system. The multi-step modality is very useful to obtain a graphical animation of the search progress. Under this modality, for each search step all graphs provided by the environment are automatically updated and re-displayed after k-milliseconds, where k is the speed of the animation that can be set by the user.

Moreover, the search process editor gives the user a tool for affecting the future search steps of the planning process by modifying the definition of the search neighborhood for every flaw in the current plan. For example, by inspecting the (partial) plan computed so far, the user realizes that, in order to achieve a desired solution plan, some actions should never be removed from the plan under consideration. The search process editor allows her to specify this constraint to the search, and, in this case, the search process editor sends a special message to the planner imposing that in the rest of the search process the removal of these actions will not be part of any search neighborhood.

The search process editor also allows the user to associate a "breakpoint" with a flaw in the plan under construction. When this happens, the editor sends a special message to the planner containing the selected flaw, which modifies the standard behavior of the planner in the following way. Whenever the planning process of LPG selects such a flaw to repair, the process is interrupted; the system presents all possible options for repairing the flaw to the user; and the user choices one of these options repairing the flaw "by hand."

For each flaw σ repaired by hand and search neighborhood N, the search process editor memorizes the successor action graph (a graph modification) selected by the user. In the successive search steps, if the planning process of LPG attempts to repair flaw σ again, evaluating a search neighborhood *similar* to N, then the successor action graph selected by the planner is the graph obtained by performing the graph modification previously selected by the user, which could be different from the action graph that the planner would select from the neighborhood according to its heuristic evaluation. Let N_{curr} be the neighborhood for solving the flaw σ under consideration, the similarity between N_{curr} and N is measured by $\frac{|N_{curr} \cap N|}{\max\{|N_{curr}|, |N|\}}$. A similarity threshold t can be customized as an input setting of the graphical user interface. Thus, the search process repairs flaw σ using the graph modification previously chosen by the user for σ with neighborhood N, if the similarity measure between N_{curr} and N is greater than or equal to t; it repairs flaw σ using the graph modification selected according to the default criteria of LPG, otherwise.

Finally, the search process editor allows the user to impose *intermediate goals*, i.e., facts that must be true at some point in the plan. In Sect. 5, we will show that the use of appropriate intermediate goals can be helpful to speed up the planning process.

4 Walk-through Example of a User Interaction

This section illustrates through a simple example how a user can interact with our environment in order to inspect and improve the plan generation in LPG. The figures in this section are screenshot portions of the user interface.

The considered planning problem is a very simple instance of the `Trucks` problem developed for the fifth International Planning Competition [8, 13]. This problem instance concerns moving three packages (`package1`, `package2`, and `package3`) between New York, Washington, and Boston (`NY`, `Wa` and `Bo`, respectively) by one truck (`truck1`) under certain constraints (the general `Trucks` problem involves an arbitrary number of objects, cities, and trucks) [8, 13]. The `Trucks` planning problem is similar to the package delivery problem with action durations, scheduling constraints, and numerical fluents, previously defined in Sect. 2. The main differences are the time windows defined in the initial state of the planning problem which impose scheduling constraints to the actions delivering packages instead of to the actions moving trucks, and that the loading space in each truck is decomposed into a collection of areas, which are organized by a spatial map imposing an order to their access and usage. In our example, the loading space inside the truck is decomposed in only two areas (`a1` and `a2`).

It should be noted that, although the running example is based on a logistic planning problem, our tool is completely domain independent, and hence, without changes to the implementation of the system, it can be used to support planning in any domain specified by PDDL2.2.

Figure 7.10 gives a screenshot of a portion of the screen showing the information about the planning process solving the running problem. The upper plot shows the number of flaws in the LA-graph at each search step; the middle plot shows the number of action nodes in the LA-graph; finally, the bottom plot shows the makespan of the plan represented by the LA-graph. At each search step, the search process monitor processes the basic information about the current LA-graph that it has received from LPG, and it updates these plots. By inspecting these plots, the user notices that from search step 220 to step 340, the number of flaws, the number of actions, and the plan duration do not significantly change. This information indicates that the local search procedure used by the planner is trapped into a local minima, and suggests the user intervention in the involved search steps, in order to better guide the choices of the planner. Hence, the user decides to pause the search process by clicking the corresponding command button in the tool bar of the graphical user interface. As a consequence of this, the search process editor sends a command to the planner interrupting the search, the planner sends a complete description of the current computed LA-graph to the search state monitor, and the search state monitor displays a graphical representation of such a graph, which can be analyzed by the user. Figure 7.11 gives a screenshot of a portion of the screen containing the graphical representation of the current LA-graph at the search step where the user has interrupted the search process. The LA-graph contains at level 1 action `(load package1 truck1 a2 Wa)` (abbreviated with "`(load packag..`"), at level 2 action `(drive truck1 Bo NY)` (abbreviated with "`(drive truck..`"), at level 3 action `(unload package1 truck1 a2 NY)` (abbreviated with "`(unload pack..`"), at level 4 action `(deliver-ontime package1 NY)` (abbreviated with "`(delivered-ont..`"), and at level 5 the special action a_{end} ("END").[5]

The effect nodes of an action that are also precondition nodes of other actions at the successive levels of the graph indicate the reasons why such an action has been planned. For example, action `(drive truck1 Bo NY)` is in the plan because `truck1` must be at New York before `package1` is unloaded from `truck1` to New York city (this is the activity represented by the action node at level 3). The unsupported precondition nodes of an action are plan flaws making this action non-executable. Action node `(drive truck1 Bo NY)` is not executable because node `(at truck1 Bo)` is not supported in the LA-graph.

Then, the user opens the interface window visualizing the temporal constraints for the actions in the LA-graph of Fig. 7.11, which is shown in Fig. 7.12. The nodes in Fig. 7.12 correspond to actions in the plan represented by the LA-graph, while the edges represent the temporal constraints between them. For example, the edge from action `(load package1 truck1 a2 Wa)` (abbreviated with "`(load packag..`") to action `(drive truck1 Bo NY)` (abbreviated with "`(drive truck..`") imposes that in the current plan `package1` is loaded from Washington on area `a2` of `truck1` before `truck1` is moved from Boston to New York. The edges connecting the special action node a_{start} ("START") and action `(deliver-ontime package1 NY)` (abbreviated with "`(deliver-ont..`") impose that `package1` must be delivered to

[5] Action `(load ?p ?t ?a ?l)` represents the movement of package `?p` from location `?l` onto area `?a` in truck `?t`, while action `(unload ?p ?t ?a ?l)` represents the opposite movement.

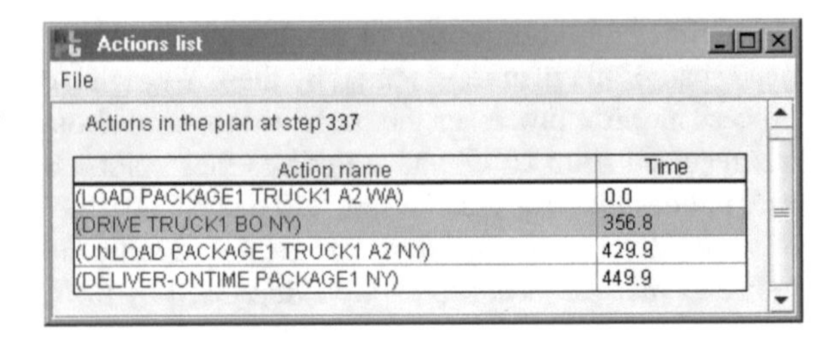

Action name	Time
(LOAD PACKAGE1 TRUCK1 A2 WA)	0.0
(DRIVE TRUCK1 BO NY)	356.8
(UNLOAD PACKAGE1 TRUCK1 A2 NY)	429.9
(DELIVER-ONTIME PACKAGE1 NY)	449.9

Fig. 7.13 A window of the user interface containing the textual description of the plan represented by the computed LA-graph

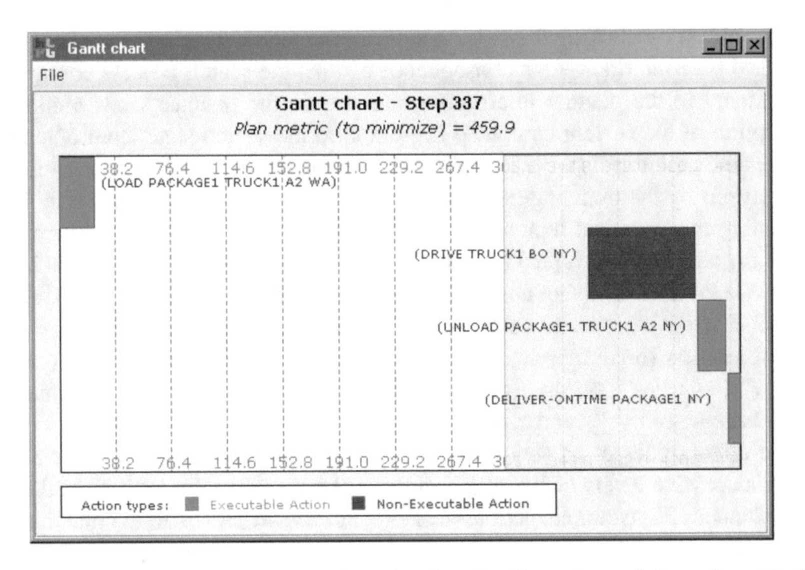

Fig. 7.14 A window of the user interface showing the Gantt chart of the actions labeling the nodes of the LA-graph in Fig. 7.11. The gray area represents the time window during which action `(deliver-ontime package1 NY)` must be scheduled in order to satisfy the input delivery constraints of the example. The chart indicates that action `(drive truck1 Bo NY)` is not executable

New York at a time between 300 and 919.7 time units since the beginning of the plan (this constraint is defined in the input problem specification).

Figure 7.13 shows the textual description of the plan under construction represented by the LA-graph of Fig. 7.11. At time 0.0, `package1` is loaded from Washington on area `a2` of `truck1`; at time 356.8, `truck1` is moved from Boston to New York; at time 429.9, `package1` is unloaded from area `a2` of `truck1` to New York; finally, at time 449.9, `package1` is delivered to the customer in New York. Figure 7.14 shows the Gantt chart of the actions in the plan. The darker boxes (the red boxes in the actual screen) in a Gantt chart represent actions in the LA-graph that are not executable. In our example, action `(drive truck1 Bo NY)` is not executable. By inspecting the frame of the graphical user interface containing the LA-graph (Fig. 7.11), the user can immediately see that this action is not executable because of the unsupported precondition `(at truck1 Bo)`. This action needs that,

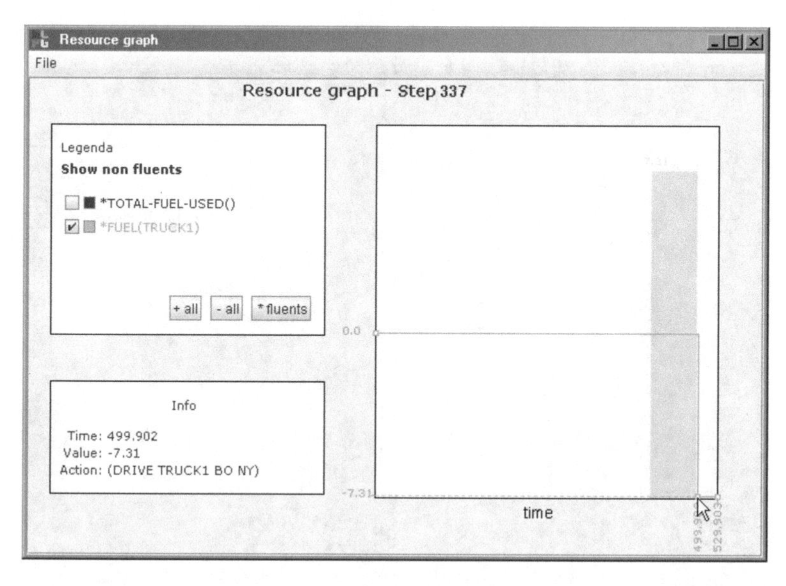

Fig. 7.15 A simple example of the resource consumption chart of the user interface for resource `(fuel truck1)` in the running example. On the x-axis we have the plan execution time. The gray area in the graph represents the fuel level required by action `(drive truck1 Bo NY)` over all its execution

at the time of its execution, `truck1` is at Boston, while at that time `truck1` is at Washington (`(at truck1 Wa)` is true in the world state identified with level 2 of the LA-graph).

Let us assume that action `(drive truck1 Bo NY)` consumes 7.31 fuel units, and hence that an amount of fuel greater than or equal to 7.31 is required over all its execution. Figure 7.15 shows an example of the user interface resource graph showing the trend of the fuel consumption for `truck1` during the execution of the plan under consideration. In the initial state, the fuel level of `truck1` is 0; at the end of action `(drive truck1 Bo NY)` the fuel level decreases from 0 to -7.31, and remains this quantity until the end of the plan. When in the resource consumption chart the line representing a certain resource crosses the gray area (green area in the actual screen) corresponding to the usage of the resource for a certain action, such an action is not executable in the context of the current plan. In particular, in Fig. 7.15 action `(drive truck1 Bo NY)` is not executable, because, at the time when it is expected to be executed, there is not enough fuel.

By looking at the graphs of Figs. 7.11, 7.12, 7.14, and 7.15, which are computed and appropriately displayed by the search state monitor, the user can realize that, in order to deliver `package1` to New York on time, driving `truck1` to New York from Washington is better than driving from Boston. Hence, the user decides to intervene by using the tools of the search state editor to revise the current (partial plan): she clicks the right mouse button on the box representing action `(drive truck1 Bo NY)` in the Gantt chart and, by using the context menu that is activated, she

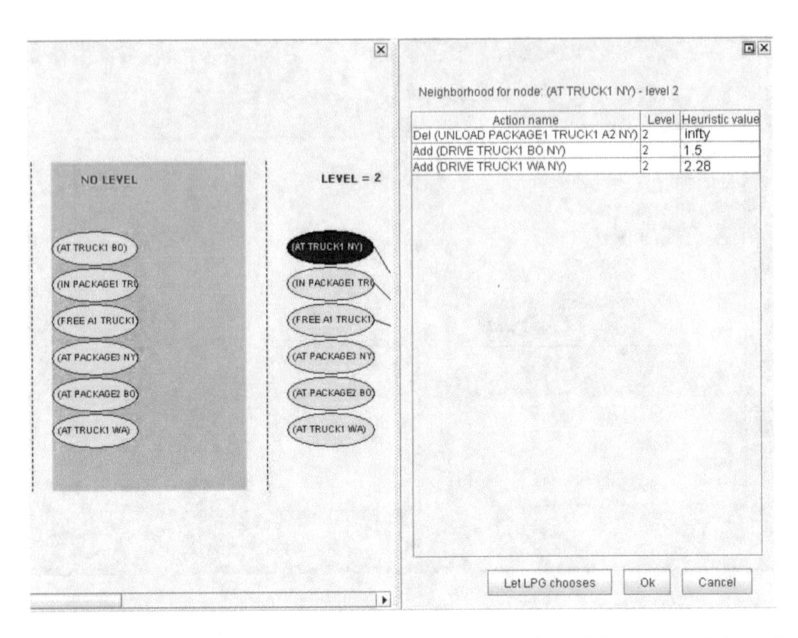

Fig. 7.16 Two frames of the user interface containing a portion of the current LA-graph (left frame) and the search neighborhood of such a graph for repairing the flaw `(at truck1 NY)` at level 2 of the LA-graph (right frame)

selects the option of removing this action from the current plan. Then, she clicks the right mouse button on the box representing the flawed node `(at truck1 NY)` in the modified LA-graph $\mathcal{A}'$, and, by using the context menu that is activated, she selects the option for repairing such a flaw. The right frame of Fig. 7.16 is then automatically displayed to her. This window indicates the possible graph modifications for repairing the selected flaw, i.e., it shows the search neighborhood of $\mathcal{A}'$ for repairing `(at truck1 NY)`. The search neighborhood is formed by three graphs obtained from $\mathcal{A}'$ by (1) removing action node `(unload package1 truck1 a2 NY)` at level 2; (2) adding action node `(drive truck1 Wa NY)` at level 2; and (3) adding action node `(drive truck1 Bo NY)` at level 2. The flaw repair frame of the user interface also shows the heuristic values of each graph modification, i.e., the values of the heuristic function used by planner LPG to evaluate each element in the search neighborhood (see third column of the table in the right frame of Fig. 7.16). Lower heuristic values indicate better possible graph modifications (neighborhood elements). Note that, since the *only* action supporting goal `(at truck1 NY)` is action `(unload package1 truck1 a2 NY)`, the heuristic value corresponding to removing this action is infinity (approximated by using very large number).[6] The

[6]In LPG, the evaluation of the successor LA-graph obtained by removing an action a supporting a precondition g is the estimated number of search steps required to support g by planning actions different from a.

heuristic value of adding action (drive truck1 Bo NY) is 1.5, and the heuristic value for adding action (drive truck1 Wa NY) is 2.28. This indicates that, if the user did not intervene to determine the choice of the graph modification to make for repairing flaw (at truck1 NY), the planner would re-insert action (drive truck1 Bo NY) into $\mathcal{A}'$. On the contrary, in our example, we assume that the user selects the element in the search neighborhood corresponding to the insertion of action (drive truck1 Wa NY) at level 2 of $\mathcal{A}'$. Therefore, in the resulting plan truck1 is moved to New York from Washington instead of from Boston.

Since the user also wants that in the *final* plan truck1 arrives at New York from Washington instead of from Boston, by using the tools of the search process editor, she modifies the rest of the search process as follows. She clicks the right mouse button on the box corresponding to action (drive truck1 Wa NY) in the new generated graph and, by the context menu that is activated, she selects the option imposing that the action is never removed during the search process. Finally, the user clicks on the "play" button in the tool bar of the user interface in order to resume the automated search process of LPG.

In general, when a local search procedure is in proximity to a local minimum, it is likely that it moves towards such a minimum, and then it gets stuck in that portion of the search space. As we will see in the next section, if this happens, few human-driven search steps could significantly improve the search process, because they would take the search away from the local minimum. The interactive graphical user interface of InLPG provides significant assistance to perform such a form of mixed-initiative planning, which in our simple running example is illustrated by the user-driven substitution of action (drive truck1 Bo NY) with action (drive truck1 Wa NY).

5 Experiments

In this section, we present the results of an experimental study aimed at testing the effectiveness of our mixed-initiative approach to improving plan generation using planner LPG.[7]

The experimental tests were run using an Intel Xeon(tm) 3 GHz machine, with 1 Gbytes of RAM. As test domains we used five instances of the Philosophers planning problem [24], and four instances of the Storage problem [8, 13], which are standard benchmarks originally designed for the international planning competition, for which LPG does not perform very well.

In the Philosophers problem, the state of some processes of a concurrent system has to be changed in order to cause a system deadlock. The communication between these processes is performed via message queues or shared access to global

[7] LPG is written in C and is available from http://lpg.ing.unibs.it, while the user interface is written in Java and will soon be made publicly available.

Table 7.1 Performance of LPG, InLPG guided by a human user who repairs some flaws "by hand" during the search process (InLPG "flaw repair"), and InLPG guided by a human user who defines some intermediate goals at the beginning of the search process (InLPG "intermediate goals") for nine instances of the planning problems `Storage` and `Philosophers`

Problem	LPG		InLPG "flaw repair"			InLPG "intermediate goals"		
	CPU-Time	TSS	CPU-Time	TSS	HSS	CPU-Time	TSS	HSS
`Storage-15`	2.09	1412	0.46	26	3	0.55	30	1
`Storage-20`	7.75	901	1.05	43	7	1.39	55	1
`Storage-25`	227.9	8130	2.41	86	11	2.98	107	1
`Storage-30`	421.1	6047	7.86	115	15	8.53	121	1
`Philosopher-1`	1.16	4292	0.08	37	3	0.04	18	1
`Philosopher-2`	–	–	0.11	62	5	0.05	27	1
`Philosopher-3`	–	–	0.10	81	8	0.05	40	1
`Philosopher-4`	–	–	0.13	106	12	0.05	47	1
`Philosopher-5`	–	–	0.18	183	15	0.05	56	1

CPU-seconds (2nd, 4th, and 7th columns), total number of search steps indicated with TSS (3rd, 5th, and 8th columns), and number of human-driven search steps indicated with HSS (6th and 9th columns). "–" means no solution found by LPG within 10 CPU-minutes

variables. The `Storage` problem involves spatial reasoning, and concerns moving a certain number of crates from some containers to some depots by hoists. Inside a depot, each hoist can move according to a specified spatial map connecting different areas of the depot.

The experiment was performed by two users having some general background knowledge on artificial intelligent (undergraduate students who had attended an introductory AI course). First, the users familiarized with the planning domains and the interactive framework of InLPG. Then, they were asked to solve the test problems using InLPG, i.e., interacting with LPG to solve the problem (when LPG failed) or to reduce the number of search steps performed by the planner. For each experimental test, the maximum elapsed time limit was 10 min, after which the problem under consideration was considered unsolved.

Table 7.1 shows the results of two experiments. In the first experiment (4th–6th columns), the users have interrupted the planning process of LPG, appropriately repairing some crucial flaws by hand, and then restarted the planning process.[8] For the problems of `Philosophers`, the crucial planning decisions concern the choice of the process states where the deadlock may happen. For the problems of `Storage`, the crucial planning decisions concern the choice of the spatial location inside a depot where a crate is stored. Basically, for the considered instances of both these problems, these crucial decisions are determined by the choice of the final

[8]In this experiment, the similarity threshold is set to 1, i.e., the memorized human decisions are reused only if the current neighborhood is the same as the neighborhood previously evaluated by the user.

actions achieving the problem goals, and hence they can be suggested to the planner by few human-driven search steps.

In the second experiment (7th–9th columns), the human user intervention is entirely at the beginning of the planning process of LPG: the user specified few significative intermediate goals suggesting the particular way in which the problem goals can be achieved in the plan.

LPG solved all the considered `Storage` instances, but, in both our experiments, it performed many more search steps, and was almost two orders of magnitude slower than InLPG. For the `Philosopher` instances, the performance gap was even more dramatic, because, for these problem instances, the search process of LPG is often trapped into local minimum. The only problem that was completely automatically solved by LPG is the simplest one, while, in both our experiments, InLPG efficiently solved all the considered problems.

6 Related Work

Like several approaches proposed in the mixed-initiative planning framework, our approach is based on the idea of human-in-the-loop control of planning visualization [37, 38]. A difference of our approach is the strong use of plan visualization tools.

The graphical representation of a linear action graph in our system is in line with the plan representation proposed by Allen and Ferguson for human-machine collaborative planning [11]: it provides the reasons why an action is being planned, and the reasons why actions are not executable in the computed plan. For complex planning domains, when these reasons concern numerical resources, our environment provides a graph showing the trend of the resources over the plan execution period.

One among the most related work is the PRODIGY system [5, 6, 39]. Like InLPG, a human can break the planning process of PRODIGY, and impose to the planner particular choices based on the current planning scenario. Both PRODIGY and InLPG allow case-based reasoning (i.e., the retrieval of plans in order to compute a new solution plan starting from them, instead of from the empty plan) and plan generation, and they are anytime planners, i.e., they generate a sequence of valid plans with increasing quality. However, PRODIGY uses a state-space non-linear planner and follows a backward-chaining search procedure, while our approach searches in the space of action graphs through a local search procedure, and it uses heuristics based on relaxed plans [17]. Also, we developed tools for supporting human decisions (such as different views of the plan under construction), and our approach supports a more expressive planning language including some complex constructs (including action durations, scheduling constraints, managing of numerical amounts of resources).

Another related work is MAPGEN [3], the interactive environment used by NASA for the Mars Exploration Rover mission. Like LPG, MAPGEN supports planning in complex domains, but it is based on EUROPA, an advanced version of HSTS [30], which uses domain specific input knowledge, while our approach is fully domain-independent.

7 Conclusions

We have presented InLPG, an implemented interactive tool for the visualization, inspection, generation, and revision of plans, which supports a form of "human-in-the-loop" control of planning that is typical in the mixed-initiative approach to plan generation. InLPG includes a graphical interface through which the user can interact with a state-of-the-art domain-independent planner, obtaining an effective visualization of a variety of information about the plan under construction or inspection, as well as about the undergoing planning process. Moreover, the tool provides some capabilities allowing the user to intervene during the planning process to modify the problem goals, the plan under construction or the planner heuristic decisions at search time. InLPG assists the user by a number of instruments, including:

- a dynamic graphical visualization of the plan under construction in terms of (1) the LA-graph data structure used by the underlying planner, (2) the temporal constraint graph of the actions in the plan, (3) a Gantt chart representation of the plan, (4) a graphical visualization of the resource consumption over time of the plan, (5) some plots about the trend of the undergoing search process, and (6) a forward and backward step-by-step execution of the underlying planner;
- a visual explanation of the reason why an action has been planned, or why, in the context of the current plan, its execution is expected to fail;
- a user interface including a set of graphical tools for an effective visualization of a variety of plan information, such as scrollable windows for panning the whole displayed graphs; image zooming for highlighting portions of the graphs; instruments for searching an action, a fact or a flaw in the plan under construction and in the corresponding LA-graph.

The paper includes a detailed example illustrating a possible use of the proposed interactive system, and presents the results of an experiment indicating that the user interaction with system can be very helpful to solve hard planning problems. Few human-driven search steps or intermediate goals specified through the user interface of InLPG can significantly help the underlying planner LPG to reach a solution plan, which otherwise would be much harder to find for the planner (and for the user) alone.

In the planning literature, a number of mixed-initiative planning systems have been proposed (e.g., [1, 4, 6, 7, 10, 11, 31, 38–40]). While many of these systems are based on domain-dependent planners or use domain specific knowledge, our approach builds on a recent domain-independent planner. Like InLPG, PRODIGY allows the user to interact with a domain-independent planner [5, 6, 39]. However, there are many differences between PRODIGY and InLPG. A major difference (which concerns the comparison with all other existing systems supporting some form of mixed-initiative planning) is the planning approach of the underlying planner: PRODIGY and InLPG use very different search procedures, heuristics, and search spaces. Moreover, InLPG supports an expressive standard planning language including some practically useful features that are not supported by version 4.0 of PRODIGY (e.g., action scheduling constraints). Finally, the user interface of InLPG has some plan visualization capabilities that are not included in the PRODIGY system (version 4.0).

Current and future work concerns the extension of InLPG with further innovative techniques of information visualization for improving the readability of large plans. We also intend to augment the options offered by the tool to the human intervention during the plan construction, and to conduct some experiments to test the usability of the tool for novice and expert users of planning technology. In particular, we believe that the proposed tool can be useful also in an educational context to support teaching and learning AI planning. Currently, we are testing the use of InLPG with this purpose for an AI course at the master degree level.

References

1. J. Allen and G. Ferguson, Human-machine collaborative planning, in *Proc. of the 3rd Int. NASA Workshop on Planning and Scheduling for Space* (2002).
2. A. Blum and M. Furst, Fast planning through planning graph analysis, in *Artificial Intelligence*. **90**(1997) 281–300.
3. J. Bresina, A. Jonsson, P. Morris, and R. K. Activity planning for the Mars Exploration Rovers. in *Proc. of the 15th Int. Conf. on Automated Planning and Scheduling*, (Monterey, California, USA, 2005), pp. 40–49.
4. M. Cox and C. Zhang, Planning as mixed-initiative goal manipulation, in *Proc. of the 15th Int. Conf. on Automated Planning and Scheduling*, (Monterey, California, USA, 2005), pp. 282–291.
5. M. T. Cox and M. Veloso, Supporting Combined Human and Machine Planning, in *Proc. of the 2nd Int. Conf. on Case-Based Reasoning*, (Providence, Rhode Island, USA), pp. 531–540.
6. M. T. Cox and M. Veloso, Controlling for unexpected goals when planning in a mixed-initiative setting. in *Proc. of the 8th Portuguese Conf. on Artificial Intelligence*, (Coimbra, Portugal, 1997), pp. 309–318.
7. K. Currie and A. Tate, O-plan: the open planning architecture, in *Artificial Intelligence*. **52**(1991):49–86.
8. Y. Dimopoulos, A. Gerevini, P. Haslum, and A. Saetti, The benchmark domains of the deterministic part of IPC-5, in *Abstract Booklet of the competing planners of ICAPS-06*, (Cumbria, UK, 2006), pp. 14–19.
9. P. Eyerich, R. Mattmüller, and G. Röger, Using Context-Enhanced Additive Heuristics for Temporal Numerical Planning, in *Proc. of the 19th Int. Conf. on Automated Planning and Scheduling*, (Thessaloniki, Greece, 2009), pp. 130–137.
10. G. Ferguson, J. Allen, and B. Miller, TRAINS-95: Towards a mixed-initiative planning assistant, in *Proc. of the 3rd Conf. on Artificial Intelligence Planning Systems*, (Edinburgh, UK, 1996) pp. 70–77.
11. G. Ferguson and J. F. Allen, Arguing about plans: Plan representation and reasoning for mixed-initiative planning, in *Proc. of the 2nd Int. Conf. on AI Planning Systems*, (Chicago, Illinois, 1994), pp. 43–48.

12. M. Fox and D. Long, PDDL2.1: An extension to PDDL for expressing temporal planning domains, in *Journal of Artificial Intelligence Research.* **20**(2003):61–124.
13. A. Gerevini, P. Haslum, D. Long, A. Saetti and Y. Dimopoulos, Deterministic Planning in the Fifth International Planning Competition: PDDL3 and Experimental Evaluation of the Planners, in *Artificial Intelligence.* **173**(2009):619–668.
14. A. Gerevini, A. Saetti, and I. Serina, Planning through stochastic local search and temporal action graphs, in *Journal of Artificial Intelligence Research.* **20**(2003):239–290.
15. A. Gerevini, A. Saetti, and I. Serina, An empirical analysis of some heuristic features for local search in LPG, in *Proc. of the 14th Int. Conf. on Automated Planning and Scheduling*, (Whistler, Canada, 2004), pp. 171–180.
16. A. Gerevini, A. Saetti, and I. Serina, An approach to temporal planning and scheduling in domains with predictable exogenous events, in *Journal of Artificial Intelligence Research.* **25**(2006):187–231.
17. A. Gerevini, A. Saetti, and I. Serina, An Approach to Efficient Planning with Numerical Fluents and Multi-Criteria Plan Quality, in *Artificial Intelligence.* **172**(2009):899–944.
18. A. Gerevini and I. Serina, Fast plan adaptation through planning graphs: Local and systematic search techniques, in *Proc. of the 5th Int. Conf. on Artificial Intelligence Planning and Scheduling*, (Breckenridge, Colorado, USA, 2000), pp. 112–121.
19. A. Gerevini and I. Serina, Efficient Plan Adaptation through Replanning Windows and Heuristic Goals, in *Journal of Algorithms in Cognition, Informatics and Logic.* **102**(2010):287–323.
20. M. Ghallab, A. Howe, C. Knoblock, D. McDermott, A. Ram, M. Veloso, D. Weld, D. Wilkins, PDDL – The Planning Domain Definition Language, CVC TR98-003/DCS TR-1165 (1998), Yale Center for Computational Vision and Control, available at http://cs-www.cs.yale.edu/homes/dvm/,
21. M. Helmert, The Fast Downward Planning System, in *Journal of Artificial Intelligence Research.* **26**(2006) 191–246.
22. J. Hoffmann and B. Nebel, The FF Planning System: Fast Plan Generation Through Heuristic Search, in *Journal of Artificial Intelligence Research.* **14**(2001):253–302.
23. J. Hoffmann and S. Edelkamp, The deterministic part of IPC-4: An overview, in *Journal of Artificial Intelligence Research.* **24**(2005):519–579.
24. J. Hoffmann, S. Edelkamp, S. Thiebaux, R. Englert, F. Liporace and S. Trueg, Engineering Benchmarks for Planning: the Domains Used in the Deterministic Part of IPC-4, in *Journal of Artificial Intelligence Research.* **26**(2006):453–541.
25. N. Lino and A. Tate, A visualisation approach for collaborative planning systems based on ontologies, in *Proc. of the 8th Int. Conference on Information Visualisation*, (London, England, UK, 2004), pp. 807–811.
26. N. Lino, A. Tate, and Y.-H. Chen-Burger. Semantic support for visualisation in collaborative AI planning. In *Proc. of the Workshop on The Role of Ontologies in Planning and Scheduling* (2005).
27. N. Lipovetzky and H. Geffner, Best-First Width Search: Exploration and Exploitation in Classical Planning, in *Proc. of the 31st AAAI Conference on Artificial Intelligence*, (San Francisco, USA, 2017), pp. 3590–3596.
28. D. Long and M. Fox, The 3rd international planning competition: Results and analysis, in *Journal of Artificial Intelligence Research.* **20**(2003):1–59.
29. D. McAllester and D. Rosenblitt, Systematic nonlinear planning, in *Proc. of the 9th National Conf. on Artificial Intelligence*, (Anaheim, California, USA, 1991), pp. 634–639.
30. N. Muscettola, HSTS: Integrating Planning and Scheduling, in *Intelligent Scheduling*, eds. M. Zweben and M.S. Fox (Morgan Kauffmann, San Francisco, USA, 1994), pp. 169–212.
31. K. L. Myers, P. A. Jarvis, W. M. Tyson, and M. J. Wolverton, A mixed-initiative framework for robust plan sketching, in *Proc. of the 13th Int. Conf. on Automated Planning and Scheduling*, (Trento, Italy, 2003), pp. 256–265.
32. X. Nguyen and S. Kambhampati, Reviving partial order planning, in *Proc. of the 17th Int. Joint Conf. on Artificial Intelligence* , (Seattle, Washington, USA, 2001), pp. 459–464.

33. J. Penberthy and D. Weld, UCPOP: A sound, complete, partial order planner for ADL, in *Proc. of the 3rd Int. Conf. on Principles of Knowledge Representation and Reasoning* (Cambridge, Massachusetts, USA, 1992), pp. 103–114.
34. F. Pommerening, A. Torralba, T. Balyo, *The ninth international planning competition* (2018), https://ipc2018.bitbucket.io
35. S. Richter, M. Westphal, The LAMA Planner: Guiding Cost-Based Anytime Planning with Landmarks, in *Journal of Artificial Intelligence Research*, **29**(2010):127–177.
36. H. A. Simon, Models of Man, (John Wiley & Sons Inc., New York, USA, 1957).
37. A. Tate, In *Advanced Planning Technology: Technological Achievements of the ARPA/Rome Laboratory Planning Initiative*, (AAAI Press, Menlo Park, California, USA, 1996).
38. G. Tecuci, *Proc. of the IJCAI Workshop on Mixed-Initiative Intelligent Systems*, (AAAI Press, Menlo Park, California, USA, 2003).
39. M. Veloso, M. Mulvehill, A., and T. Cox, M, Rationale-supported mixed-initiative case-based planning, in *Proc. of the 9th Conf. on Innovative Applications of Artificial Intelligence*, (Providence, Rhode Island, USA, 1997), pp. 1072–1077.
40. C. Zhang, *Cognitive models for mixed-initiative planning*, (PhD thesis, Wright State University, Computer Science and Engineering Department, Dayton, Ohio, USA, 2002).

Chapter 8
Interactive Visualization in Planning and Scheduling

Roman Barták

Abstract Planning and scheduling are two closely related areas that deal with organizing activities to achieve a particular goal (planning) and allocating these activities to limited time and resources for execution (scheduling). However, regarding the tools supporting the planning and scheduling processes, these two areas are still far from each other. Progress in scheduling has been driven by industry and many techniques and tools to support the scheduling process have been designed. On the other hand, planning is still more an academic topic and, until recently, engineering support of the planning process has been limited. The focus of planning community was mainly on design of efficient planners, but this started to change in recent years and several tools supporting the planning process have been designed. This chapter focuses on interactive visualization of plans and schedules, that is, on the way how plans and schedules can be presented visually to users, and on tools that can work with these visualizations.

Keywords Planning · Scheduling · Visualization · Interactivity

1 Introduction

For both plan and schedule visualization, there is a common way of exploiting intuitive Gantt charts that visualize activities as rectangles organized on a timeline. The concepts of activity and relations between activities are richer in planning. Causal relations are related to precedence relations but have stronger semantics of preserving some property between the activities. Activities themselves have preconditions for their execution and effects modifying the world state and hence it is important to include information about world state in the visualization. The visualization tools should support plans and schedules with flaws and should be

R. Barták (✉)
Charles University, Prague, Czech Republic
e-mail: bartak@ktiml.mff.cuni.cz

M. Vallati, D. Kitchin (eds.), *Knowledge Engineering Tools and Techniques for AI Planning*, https://doi.org/10.1007/978-3-030-38561-3_8

able to highlight these flaws for easier identification by a human user. Last but not least, the visualization tools should support interactivity, where the users are allowed to modify plans and schedules and the tools help users in such modifications, for example, by repairing possible flaws introduced there.

This chapter presents three systems as examples of tools for interactive visualization of plans and schedules. The first system, iGantt, is a pure schedule visualizer that brings the concept of interactive schedule modification and automated repair of flaws [2]. The follow-up system, FlowOpt, extends iGantt to support the whole scheduling processes starting with model design and finishing with analysis of schedules [4]. It gives an example of a complete integrated environment for scheduling. In planning, there existed several pioneering systems providing graphical user interface supporting the planning process: itSimple [10], GIPO [9], and VLEPPO [11]. They are effective tools for modeling and updating planning domains, however, their plan analysis lacks features mentioned above. We will present system VisPlan [8] that is a planning counterpart of iGantt. iGantt and VisPlan are freely available tools written in Java, FlowOpt is a part of commercial system MAK€ [3] and it is a demonstration prototype showing possible capabilities of complex tools supporting the whole scheduling process.

2 Interactive Gantt Chart (iGantt)

iGantt is software to visualize schedules, to interactively modify the schedules, and, finally, to automatically correct violated constraints in the schedules [2]. It is written in Java and hence runs under various operating systems without installation. It is available to download at http://ktiml.mff.cuni.cz/~bartak/CLP/iGantt.html.

There exist many classes of scheduling problems [7], so one needs to describe first, what type of scheduling task is supported by a given tool.

2.1 Problem Specification

Basically, the scheduling problem is defined by resource constraints, by temporal constraints among the activities, and by the objective function [7]. For activities, iGantt assumes classical *non-interruptible activities,* which are probably the most widespread form of activity type. It means that each activity, more precisely, its temporal location in the schedule, is fully determined by its start time and its duration. The end (completion) time of the activity is simply its start time plus its duration as the activity runs from its start till its end. *Interruptible activities* might be temporarily stopped during their execution and hence their active duration could be smaller than the distance between the start and end times as there might be some interruptions, when the activity is not being processed. There exist other types of activities, for example, *elastic (energetic) activities*, where some energy is assigned to the activity and duration of activity is determined by this energy and number of resources allocated to this activity (for example, energy 4 implies duration 4, if the activity runs on a single resource, and duration 2, if the activity runs on two resources). iGantt supports only *unary resources*, where each activity is pre-allocated to some resources, and occupies those resources from its start time till its completion. Note that a single activity can be allocated to several unary resources and hence occupies all of them during execution. The choice of supported resources is closely related to supported type of activities. Again, there exist other types of resources, for example, a cumulative resource allows processing more activities in parallel while respecting resources capacity. Resources might also be more complex by assuming set-up times between activities, etc. The only direct relation between activities, that iGantt supports, is a *precedence constraint*. If activity A precedes activity B then A must finish no later than B starts. Precedence constraints are the most widely used temporal constraint among the activities. They can be specified more precisely in the form of temporal constraints, for example, by defining minimal and maximal distances between the activities. The choice of simple precedence constraints in iGantt was intentional as together with other supported constraints, it guarantees existence of a valid schedule under some easy-to-verify conditions (see later). Regarding the objective function, no specific function is assumed as visualization just displays an existing schedule. Nevertheless, during schedule repairs, iGantt is trying to make the schedule as compact as possible, which corresponds to using makespan as the objective function. Makespan is defined as the distance between the start of the first activity in the schedule and the end of the last activity, and this distance is being minimized during scheduling. Again, it is probably the most widely used objective function in scheduling. In summary, iGantt works with unary resources, non-interruptible activities, and precedence constraints between the activities. This is a quite general specification of a scheduling problem. The reader may see easily that widely used types of scheduling problems, such as job-shop scheduling problems, fit this problem specification.

Notice that in scheduling problems, the constraints are given explicitly and directly. Schedule is defined by a set of activities, each having its start time, duration, and resource(s) to which it is allocated and by a set of precedence relations between the activities. This is also the input to the iGantt software.

2.2 *Visualization of Schedules*

There is not much to be invented about the form of visualization of schedules and, as the name of software indicates, the classical Gantt chart is used to visualize schedules in iGantt. Basically, there are two possible views of the schedule, one is focusing on the precedence constraints between the activities and one is focusing on the resource constraints. In the first (activity) view, each activity has its own row in the visualization, the position of activity in this row is defined by its start time, and precedence constraints are visualized as arcs between the activities. In the

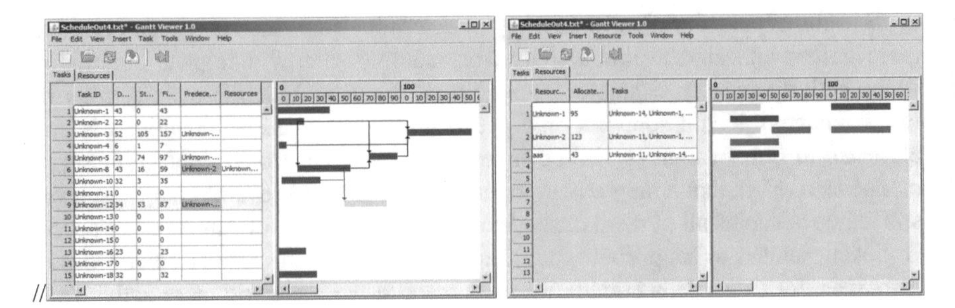

Fig. 8.1 Gantt charts visualization of violated precedence (left) and resource (right) constraint

second (resource) view, each resource has its own row, and activities allocated to that resource are displayed in this row based on their allocation in time. With these views, it is easy to highlight the violation of precedence constraints (activity view) and the violation of resource constraints (resource view). Figure 8.1 shows both views and violation of precedence (left) and resource (right) constraints is identified there. Detecting these violations in a given schedule is straightforward.

2.3 *Interactive Schedule Modifications*

Fully automated scheduling seems like the Holy Grail of scheduling community, but most practitioners frequently require the freedom of manually altering the generated schedules, for example, to introduce some aspects of the particular area that were hard to formalize and hence are not reflected in the automatically generated schedule. iGantt allows modification of any aspect of the scheduling problem. By intuitive drag-and-drop operations, the user can modify allocation of activity in time as well as to resources and change activity duration. It is possible to add and remove activities, resources, and precedence constraints. In fact, the user may start with an empty schedule and design the schedule completely manually. It is a very general concept as the user can modify anything in the schedule. iGantt always visualizes the current schedule and highlights violation of precedence and resource constraints.

2.4 *Automated Schedule Repair*

Despite the high experience of human schedulers, there is a high probability that after a manual modification of a schedule some flaws are introduced to the schedule. This probability is higher if the density of scheduling constraints is large and the constraints are highly coupled. For example, delaying one activity may delay other dependent activities due to precedence constraints between them or due to limited capacity of resources. It might be enough just to detect such violations and report them to the user who will be responsible for manual correction. Nevertheless, such manual corrections may be boring and sometimes very hard because of interconnectivity of the constraints (correction of one flaw introduces other flaws, etc.). iGantt supports fully automated schedule repair, namely we address the problem of correcting precedence constraints and unary resource constraints by shifting locally the affected activities in time.

Let us assume that some initial allocation of all activities to time (an initial schedule) is known. This time allocation may violate some precedence constraints (activity starts before some of its predecessors finishes) or some resource constraints (two or more activities are processed at the same time by the same resource). The goal is to correct the schedule (re-schedule) by shifting the activities in time, that is, to find a feasible schedule that does not violate any constraint. We do not assume changing durations of activities and reallocation to other resources during schedule repair. Moreover, the new schedule should not differ a lot from the initial time allocation of activities. Note that finding a feasible schedule is always possible unless there is a loop in the precedence constraints—activities can always be shifted to later times as there are no deadlines. To minimize the number of changes between the initial and final schedule we apply a local approach, where particular flaws are repaired by local changes of affected activities rather than by generating a completely new schedule from scratch. A local repair may introduce other flaws in the neighborhood which spread like a wave until all flaws are resolved.

We use a three-step approach to repair a schedule. In the first step, loops of precedence constraints are detected, and the user is asked to break each loop by removing at least one precedence constraint from it. This is the only step, where user intervention is used (it is possible to randomly remove some precedence constraint from each loop or even to minimize the number of removed precedence constraints to break all loops, but in our opinion, the human decision is more appropriate in this step). In the second step, we repair all precedence constraints; two methods are used for this repair. Finally, in the third step we repair violation of resource capacity constraints while keeping the precedence constraints valid. Each repair is realized by shifting affected activities locally in time and it is done fully automatically. Let us now describe these repairs in more details; the full technical details including proofs of soundness are available in a separate paper [2].

The precedence constraints are repaired from left to right starting with the earliest precedence constraints. Topological ordering of precedence constraints is used to determine this order (see Fig. 8.2).

Let us now assume that precedence (A→B) is violated, where A and B are activities, and it is the first violated precedence constraint in the topological ordering (all earlier precedence constraints are satisfied). One can shift activity B to some later time to repair the constraint, but this may eventually extend the makespan. Hence, we suggest to, first, shift activity A to earlier time without violating any preceding precedence constraint and only then to shift activity B to a later time; minimal shift necessary to repair the constraint is always used. This is a straightforward way of repairing precedence constraints. Unfortunately, it can shift

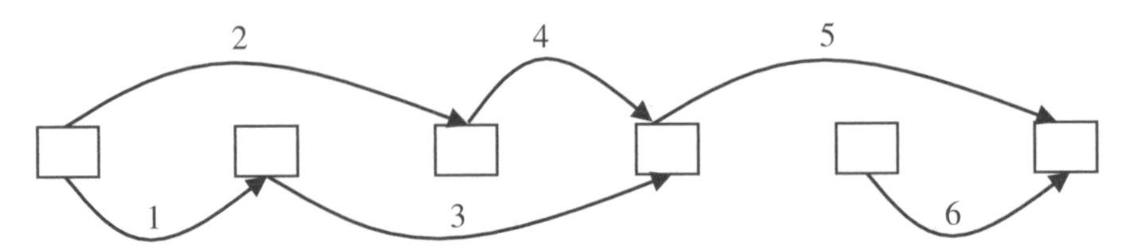

Fig. 8.2 Possible topological ordering of precedence constraints

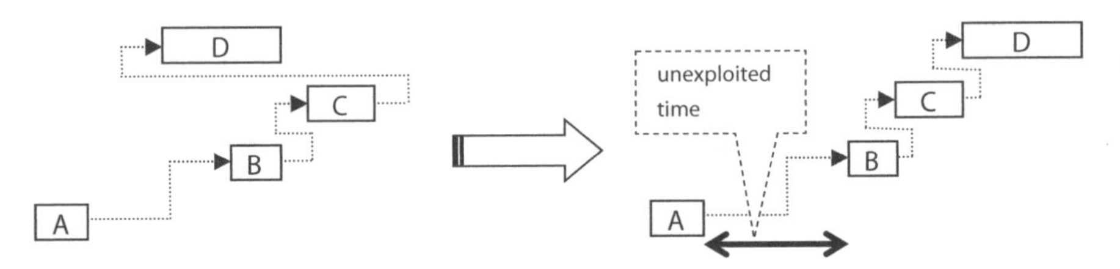

Fig. 8.3 Simple repair of violated precedence does not exploit fully available time on left of D

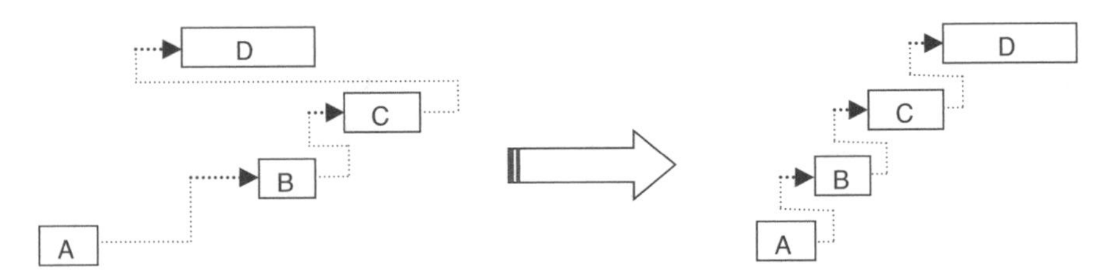

Fig. 8.4 More aggressive repair method exploits better available time on left of D

activities forward more than necessary and hence it can increase makespan more than necessary and make the schedule less compact. Figure 8.3 gives an example of schedule, where repair of violated precedence constraint (C→D) shifts D to later time, while there is some unexploited time available before.

Therefore, iGantt uses a more aggressive method of repair, where the first activity in the violated precedence constraint is shifted slightly more to the left to exploit possible free time there. This may violate some earlier precedence constraints, so the algorithm repairs these violations before continuing with further precedence constraints in the topological order (see Fig. 8.4).

After all precedence constraints are corrected (repaired), the possible resource conflicts are repaired. Recall that activities require for their processing unary resources; it is possible that an activity requires more than one resource (for example, machine, tool, and worker). There is a resource conflict if two (or more) activities require the same resource at the same time. We repair the resources conflicts again from left to right in the following way. If two activities overlap in time and share the same resource, the later activity is shifted to the right to repair the resource conflict. All precedence conflicts are repaired before continuing to the next resource conflict. By sweeping the schedule from past to future we remove

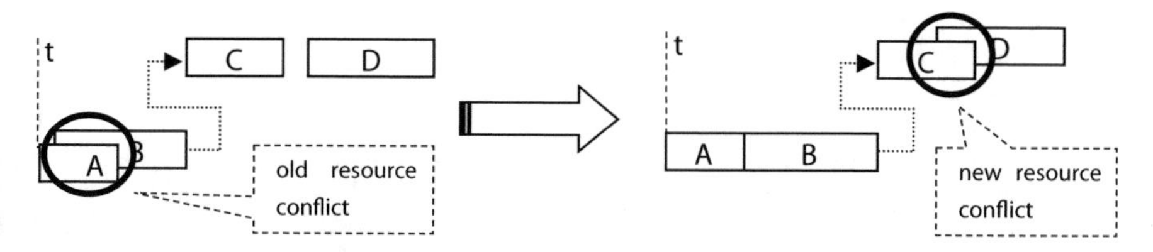

Fig. 8.5 New resource conflicts may appear after repairing another resource, but these conflicts appear only in the future part of the schedule

all violated constraints (recall that there are no deadlines so any activity can be shifted forward). As we shift activities to right only, new conflicts (precedence and resource) may only appear at the not-yet corrected part of the schedule so they will be eventually repaired later (see Fig. 8.5).

3 Interactive Workflow Optimization (FlowOpt)

iGantt was a pioneering system that supports interactive modification of schedules with fully automated schedule repair. Our follow-up project FlowOpt extended its capabilities to a complete system for production workflow optimization [4]. FlowOpt allows users to describe visually and interactively the process of producing any item in the form of a nested workflow with alternatives [1]. After specifying what and how many items should be produced, the system generates a production plan taking in account the limited resources in the factory. The plan is visualized in the form of a Gantt view that uses information about workflows and allows users to arbitrarily modify the plan by selecting alternative processes or allocating activities to different times or resources. Finally, the schedule can be analyzed, the bottleneck parts are highlighted, and some improvements are suggested to the user. Let us now summarize the major functionality of individual modules.

Workflow Editor allows users to create and modify workflows in a visual way. We use the concept of nested workflows [1] that are built by decomposing the top task until the primitive tasks are obtained. This is similar to hierarchical task networks, but without recursion. Three types of decompositions are supported: either the task is decomposed into a sequence of sub-tasks which forms a *serial decomposition* or the task is decomposed into a set of sub-tasks that can run in parallel—a *parallel decomposition*—or finally, the task is decomposed into a set of alternative sub-tasks such that exactly one sub-task will be processed to realize the top task—an *alternative decomposition* (Fig. 8.6). The final primitive tasks are then filled with activities; each activity has a given duration and a set of resources necessary for its processing. The workflow can be built in the top-down way by decomposing the tasks or in the bottom-up way by composing the tasks; both approaches can be used together as the user prefers. In addition to the core nested structure, the user can also specify extra binary constraints between the tasks such as precedence relations,

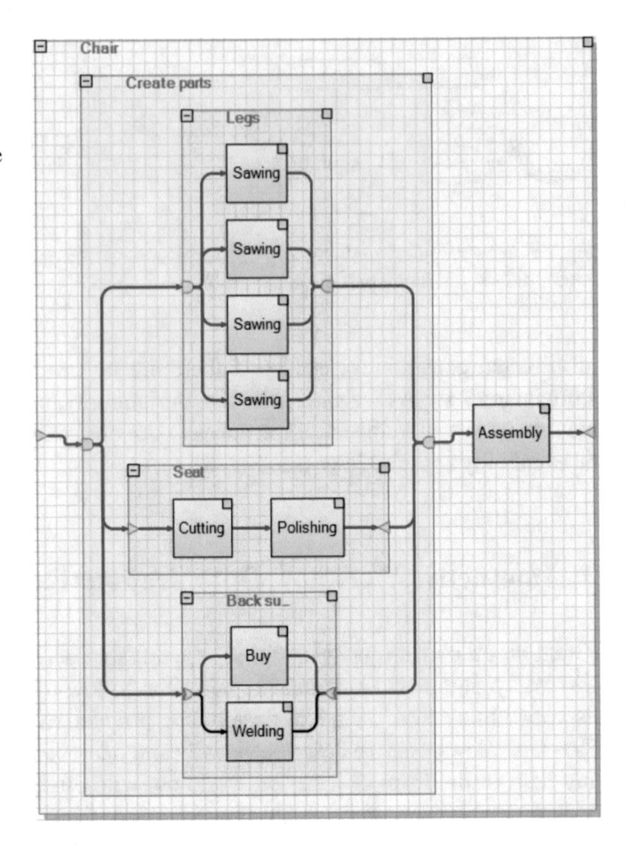

Fig. 8.6 Visualization of nested workflow in the FlowOpt Workflow Editor (from top to down there are parallel, serial, and alternative decompositions)

temporal synchronizations, or causal relations. Everything is done using an intuitive drag-and-drop approach. The system also supports import of foreign workflows and it has the function of fully automated verification of workflows [5].

When the workflows for all items are defined, this is the modeling stage, it is possible to generate production plans. This is as easy as selecting the required items (workflows) in the *Order Manager*, specifying their quantities and required delivery date and starting the *Optimizer* by pressing a single button in GUI. The data about workflows, activities, and resources are automatically converted to the scheduling model and the system produces a schedule that is a selection of activities (tasks) from the workflows (if there are alternatives) and their allocation to time and resources. The Optimizer attempts to optimize both earliness and lateness costs that are derived from the delivery dates.

The generated schedule (production plan) can by visualized in the *Gantt Viewer* that generalizes the ideas of iGantt. This module provides both traditional views of the schedule, namely the task-oriented and resource-oriented views. Because the Gantt Viewer has full access to the workflow specification, it can also visualize the alternatives that were not selected by the Optimizer. The Gantt Viewer allows users to modify any aspect of the production plan using the drag-and-drop techniques.

The user can move activities to different times and resources and change their duration. It is even possible to select another alternative than that one suggested by the Optimizer. Because the Gantt Viewer is aware about all the constraints originating from the workflow specification, it can also highlight violation of any of these constraints. Even more, the Gantt Viewer can automatically repair the flaws that were introduced to the schedule by the user's modifications. Flaws are repaired similarly to iGantt by allowing time shift of activities only, but there already exist techniques that allow schedule repair by selecting a different alternative from the nested workflow [6].

The final module is an *Analyzer* that is responsible for suggesting improvements of the production process. The Analyzer first finds bottlenecks in a given schedule, for example, an overloaded resource. For each bottleneck, the analyzer suggests how to resolve it—this could be by buying a new resource or by decreasing the duration of certain activities (for example, by staff training). Each such improvement is evaluated by the Optimizer. Finally, the system selects a set of improvements such that their combination brings the best overall improvement of the production process under the given constraints such as a limited budged to realize the improvements.

4 Interactive Visualization and Verification of Plan (VisPlan)

Plan analysis is an inevitable part of complete planning systems. With the growing number of actions and causal relations in plan, this analysis becomes a more and more complex and time-consuming process. In fact, plans with hundreds of actions are practically unreadable for humans. In order to make even larger plans transparent and human readable, we have developed a program which helps users with the analysis and visualization of plans. The program called VisPlan [8] finds and displays causal relations between actions, it identifies possible flaws in plans (and thus verifies plans' correctness), it highlights the flaws found in the plan and finally, it allows users to interactively modify the plan and hence manually repair the flaws. The software is available to download at http://glinsky.org/visplan/.

VisPlan is a graphical application (Fig. 8.7) written in Java with the ultimate goal to visualize any plan, to find and highlight possible flaws, and to allow the user to repair these flaws by manual plan modification. VisPlan works with three types of files that the user should specify as program input:

- planning domain file in PDDL
- planning problem file in PDDL
- plan file specified in text format

VisPlan supports STRIPS-like plans and temporal plans. The program recognizes the plan type (strips/temporal) automatically and verifies and visualizes it based on its type. The plan type is determined by the planning domain—durative-actions indicate a temporal plan, actions with no duration indicate a STRIPS-like plan. The following PDDL requirements are currently supported in the program: strips,

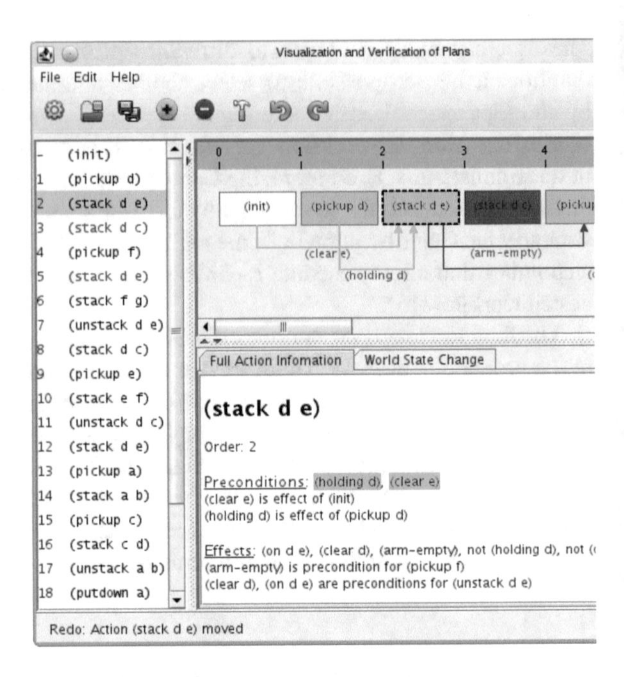

Fig. 8.7 Graphical user interface of VisPlan

typing, negative-preconditions, disjunctive-preconditions, equality, existential-preconditions, universal-preconditions, quantified-preconditions, durative-actions.

Planning domain and problem need to be syntactically correct and mutually consistent (separately parsed planning domain and problem files can be linked with each other). Otherwise, visualization and verification is not performed and errors from the PDDL parser are displayed. Sometimes, PDDL parser encounters errors and issues which are not critical. In these cases, warning and non-critical error messages are displayed and the program continues. Recognized plan actions are given in the following format:

```
start_time: (action_name param1 param2 ...) [duration]
```

In the plan file each action is supposed to be on a separate line. The parser recognizes the lines and creates actions given only in the above-mentioned format. Other lines are ignored. Eventually, a modified plan can be saved either to the original file or to a new text file.

4.1 Plan Verification

Plan verification is automatically executed after the plan is initially loaded and then after each user interaction modifying the plan. The verification process is based on simulation of plan execution and the main idea is to incrementally construct “layers” of facts. Each fact layer is determined by a corresponding set of facts and an action due to which the layer has been created.

At the beginning of the verification, all possible facts (grounded predicates) are instantiated. This domain-specific data remains fixed and is computed only once at the beginning; re-verifications do not change the data. This attitude permits us not to manipulate with the facts during the whole verification process, but to work only with the indexes to the array of grounded facts. Because of that, operations like checking if an action is applicable, application of action's effects, finding missing conditions, etc. are just logical bit-sets operations (where one bit-set has its bits set to true at indexes corresponding to the selected grounded facts). Such operations are very fast.

Unlike facts, only actions present in the plan are grounded (meaning related to an operator with grounded conditions and effects). The operator is found based on matching the planning domain operator and concrete parameters of the action. As mentioned in the previous paragraph, conditions and effects of the grounded operator are represented by bit-sets (pointing to the fix array of grounded facts). The verification process makes sure it has a matching operator available for each examined plan action (otherwise, for instance, when a user adds a new action, the verification process additionally finds and stores the operator). Actions, which do not comply with any operator definition, are marked as invalid and omitted from the verification. Nevertheless, such actions are still displayed (but distinguished from others by a different color and marked as invalid).

There are two special "actions" artificially added into the plan. They are called "init" and "goal" and their aim is to represent the initial state and the goal. A classical plan-space approach is used to define these actions. The init action has empty preconditions and the facts that apply at the initial state are considered as its effects. The goal action has empty effects and the set of facts that need to be satisfied at the final world state are considered as its preconditions. By treating the initial state and the goal as regular plan actions we are able to recognize causal relations also at the margins of the plan without any further work. This way we easily find dependencies on the initial state and, eventually, marking the "goal" action as non-applicable means that the goal conditions are not satisfied.

4.2 Visualization of Sequential and Temporal Plans

As shown in the right-upper frame of Fig. 8.7, plan's actions are visualized as cells (boxes) of fixed size filled by the action name. Each action is colored either green or red (or any other color chosen by the user) depending on whether the action is applicable or non-applicable. Causal relations between the actions are visualized by edges. These edges are annotated by grounded facts that are "passed" between the actions. Only the causal relations for the currently highlighted action are displayed to remove a cluttered view. Display position of the edges is automatically adjusted every time an action is highlighted in order to assure that the edges do not overlap and their labels (describing the causal relations) are fully readable. The edge position adjustment is vertical (with fixed space size between edges), as well as horizontal (source and target points of edges on the same cell have regular space between themselves).

If the process of verification is still going on, actions whose state has not been decided yet are colored grey (or any other color chosen by the user). The state of an action can be one of the following:

- invalid (action does not match any definition in the planning domain file),
- un-decided (action is still being checked by the validation module),
- applicable (action is valid and can be used),
- non-applicable (action cannot be used due to non-satisfied preconditions).

Two special actions, "init" and "goal" are colored differently to distinguish their special meaning. These are the only two actions which cannot be modified in any way.

For the highlighted action, the system displays complete information about the action including the satisfied and violated preconditions and actions giving these preconditions (the right-bottom frame of Fig. 8.7), as well as world change caused by the action. World change illustrates which facts are true prior the action and which after the action. Naturally, world state information is not available for non-applicable actions. Facts that were subject of change (either added or deleted) are marked (by color and/or by strike through their names).

On the left side of the window a list of actions is shown to provide a brief plan summary (the left frame of Fig. 8.7). Actions in the list are sorted by their order/start time and are visually differentiated based on their states. The list gets updated every time a modification is done to the plan. Selecting an action in the list results in adjusting the scrollbar view to comprise the visualized action in the graph and vice versa. If the user needs more space for graphical plan analysis, he/she is free to hide the action summary list completely (as well as informative tab pane at the bottom of the application).

During a plan analysis, the ruler (Fig. 8.7) helps to orientate within a time axis. Its default size of units is one inch (without dependence on user's screen resolution). Size of units can be adjusted by the combo box (upper-right frame of Fig. 8.7) or by dragging any tick of the ruler.

While dragging an action (to change its position), actions providing preconditions and actions using effects of the dragged action are dynamically highlighted, so that the user knows where he/she can drop the action. When actions are swapped it usually changes causal relations between the actions significantly. Due to this fact, highlighting preconditions and effects partially would not provide enough information. Therefore, the plan is re-verified when an action changes its order while dragging. Having such information, the program chooses the correct actions to highlight. Color for highlighting is the same as color for preconditions/effects edges. If actual color of an action is the same as the color for edges when highlighting, another (but similar) color is used then.

Each user has an opportunity to set his/her own user preferences regarding the visual appearance and behavior of software according to the personal needs. The user preferences are saved in the home directory of the user and include various (mostly graphical) settings, for instance:

- colors for actions (each state has its own color), edges (both preconditions and effects), and ruler,
- font size (for different GUI components),
- automatic loading of last successfully loaded files (domain, problem, plan) at start-up,
- default action width in STRIPS-like plans.

4.2.1 Visualization of STRIPS Plans

As the STRIPS plans are sequential, cells representing the actions are displayed in a row. When changing the order of an action by drag & drop, the new order is computed after each movement by checking the horizontal position of the cell being dragged and ruler's units. In the case the new position is different from the current one, a cell placed at that moment on the "new order" position is immediately repositioned to the "current order" position, and thus these two actions swap their position. When the action is finally dropped, it is just placed in the row.

4.2.2 Visualization of Temporal Plans

Ruler units in temporal plans reflect durations of actions. However, as individual durations of actions within a plan can vary a lot, the median duration has been chosen to be the initial ruler unit. Auxiliary ticks are also present on the ruler. All actions (meaning cells) are also guaranteed to have a minimum horizontal size (in order to be visible even if real duration is too small).

Horizontal position of an action is fully determined by its start time and duration. Although actions in temporal plans can overlap with each other, cells representing the actions are positioned in order to be fully visible. This is performed by placing the cells in rows. All cells in the same row have the same vertical position. Cells position adjustment is iterative, and cells are positioned into the first row (from top) where the cell would not overlap with other cells (Fig. 8.8).

When an action is being dragged, in contrast to STRIPS plans, the start time of the action is determined by the horizontal position only (multiplied by the current ruler units). In such a situation re-verification of the plan is done only when the action has changed its position significantly, meaning the relative order of the dragged action margins (start/end) changed with respect to other actions.

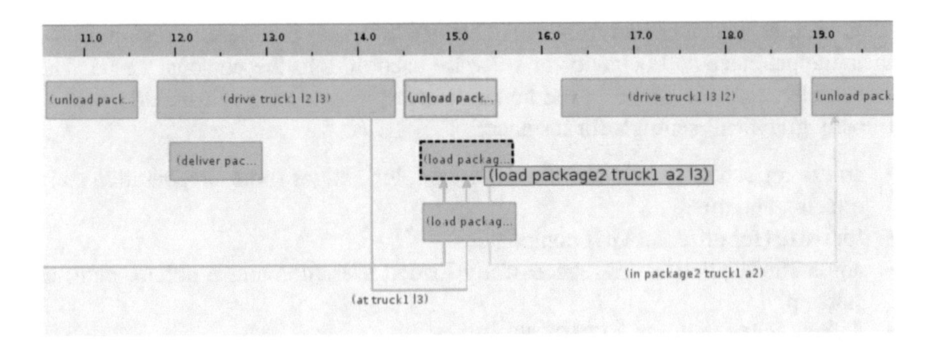

Fig. 8.8 Example of visualization of temporal plans

4.3 *Interactive Plan Modifications*

In addition to visualization of plans the software supports interactive modification of the plan. The following operations with plans are supported:

- inserting new actions (selection of actions and their parameters is automatically restricted to the current planning domain and the problem and offered in the corresponding number of pre-filled combo boxes),
- removing actions,
- modifying actions,
- changing the order of actions in STRIPS plans and start time of action in temporal plans by drag & drop technique.

Modifications are revertible and are under control by undo manager. Undo manager waits for performing an undoable (revertible) modification, which is any of the above. When an undoable change is fired, undo manager clones and saves both the current plan and verificator state (this includes the constructed layers of facts, the causal relations among actions, actions' indexes to layers before and after application, missing conditions). On the one hand, this approach is memory consuming, due to the fact that undo manager stores as many plans and verificator states as the maximal number of possible "undo"s. On the other hand, the approach is time-saving. Re-verification is not needed to be performed after each "undo"/"redo." All the necessary steps include just retrieving previous/next plan and verificator state plus redrawing the graph based on the retrieved plan. In comparison with a memory-saving approach, which would save only modifications' description and would perform opposing action during "undo"/"redo," the chosen approach is easier and more "defect-resistant." That is because it coherently maintains entire plans and states.

Besides the already mentioned plan and verificator state, undo manager saves two more items for user-friendliness and informative purposes. These include id of an action causing an undoable change (in order to select this action and to adjust view to comprise it) and a string describing the change (in order to print informative message onto status panel at the bottom of the application).

Modified plans can be saved in the text format to either the same (initially loaded) file or to a new file (save as).

5 Conclusions

This chapter described three systems for interactive visualization of plans and schedules. Two systems, iGantt and FlowOpt, are used for scheduling problems and they demonstrate advanced prototypes demonstrating capabilities when working interactively with schedules. FlowOpt is in fact a system that supports a complete development cycle from modeling, through solving (scheduling), till schedule visualization and analysis. The system can verify the workflow models and automatically repair manually modified schedules with flaws. It shows possible direction for tools developed for automated planning as the capabilities of tools for planning are behind those for scheduling problems. The last presented system, VisPlan, is a research prototype showing the current capabilities for plan visualization. Though there exist some systems supporting design of planning domain models, as far as we know, there are no tools supporting verification of models and automated correction of plans.

Acknowledgments Roman Barták is supported by the Czech Science Foundation under the project 18-07252S.

References

1. R. Barták, O. Čepek: Nested Precedence Networks with Alternatives: Recognition, Tractability, and Models. *Proceedings of 13th International Conference on Artificial Intelligence: Methodology, Systems, and Applications (AIMSA)*, Varna, Bulgaria, pp. 235–246, 2008.
2. R. Barták, T. Skalický: A local approach to automated correction of violated precedence and resource constraints in manually altered schedules. *Proceedings of Fourth Multidisciplinary International Scheduling Conference: Theory and Applications (MISTA)*, Dublin, Ireland, 2009, pp. 507–517, 2009.
3. R. Barták, C. Sheahan, A. Sheahan: MAK€ – A System for Modelling, Optimising, and Analyzing Production in Small and Medium Enterprises. *Proceedings of 38th Conference on Current Trends in Theory and Practice of Computer Science (SOFSEM),* Špindlerův Mlýn, Czech Republic, pp. 600–611, 2012.
4. R. Barták, M. Jaška, L. Novák, V. Rovenský, T. Skalický, M. Cully, C. Sheahan, T.-T. Dang: FlowOpt: Bridging the Gap Between Optimization Technology and Manufacturing Planners. *Proceedings of 20th European Conference on Artificial Intelligence (ECAI)*, Montpellier, France, pp. 1003–1004, 2012.
5. R. Barták, V. Rovenský: On verification of nested workflows with extra constraints: From theory to practice. *Expert Systems with Applications*, Elsevier, Vol. 41(3), pp. 904–918, 2014.

6. R. Barták, M. Vlk: Hierarchical Task Model for Resource Failure Recovery in Production Scheduling. *Proceedings of 15th Mexican International Conference on Artificial Intelligence (MICAI)*, Cancún, Mexico, pp. 362–378, 2016.
7. P. Brucker: *Scheduling algorithms (4. ed.)*. Springer 2004, pp. I-XII, 1–367.
8. R. Glinský, R. Barták: VisPlan – Interactive Visualisation and Verification of Plans. *Proceedings of the ICAPS Workshop on Knowledge Engineering for Planning and Scheduling (KEPS)*, pp. 134–138, 2011.
9. R.M. Simpson, D.E. Kitchin, T.L. McCluskey: Planning Domain Definition using GIPO. *The Knowledge Engineering Review* 22(2): pp. 117–134, 2007.
10. T. S. Vaquero, J.R. Silva, J.C. Beck: Analyzing Plans and Planners in itSIMPLE3.1. *Proceeding of the ICAPS Workshop on Knowledge Engineering for Planning and Scheduling (KEPS)*. Toronto. Canada, pp. 45–52, 2010.
11. O. Hatzi, D. Vrakas, N. Bassiliades, D. Anagnostopoulos, I. Vlahavas. VLEPPO system, A Visual Programming System for Automated Problem Solving, *Expert Systems with Applications*, Elsevier, Vol. 37 (6), pp. 4611–4625, 2010.

Chapter 9
Argument-Based Plan Explanation

Nir Oren, Kees van Deemter, and Wamberto W. Vasconcelos

Abstract We describe a tool for providing explanation of plans to non-technical users, built on formal argumentation and dialogue theory, and supported by natural language generation and visualisation technologies. We describe how arguments can be generated from domain rules, and how justified arguments can be identified through dialogue, allowing the system to use such a dialogue to explain a plan. We provide information about our prototype system implementation, discussing its current limitations, and identifying potential avenues for future research.

1 Introduction

Automated planners have, together with other technologies, enabled autonomous systems to generate and then execute plans in pursuit of a set of goals with little or no human intervention. While such plans are often better than those a human planner can create, there is a reliance on the correct specifying the initial and goal states, as well as the effects of actions, making such plans brittle in the presence of exceptional (and unexpected) situations.

There is, therefore, a clear need to be able to verify or validate the correctness of the plan specification with regards to the current environmental state. Furthermore, autonomous systems do not operate in isolation, but often form part of a *human-agent* team. In such cases, joint plans dictate both human and autonomous system actions, and mechanisms are required to ensure that humans execute their portion of the plan correctly. If the human actors trust the correctness of a plan, they are

N. Oren (✉) · W. W. Vasconcelos
Computing Science, University of Aberdeen, Aberdeen, UK
e-mail: n.oren@abdn.ac.uk; w.w.vasconcelos@abdn.ac.uk

K. van Deemter
Information & Computing Sciences, University of Utrecht, Utrecht, The Netherlands
e-mail: c.j.vandeemter@uu.nl

M. Vallati, D. Kitchin (eds.), *Knowledge Engineering Tools and Techniques for AI Planning*, https://doi.org/10.1007/978-3-030-38561-3_9

more likely to follow it. One way to engender such trust, which also addresses the validation and verification problem, is to provide an explanation of the generated plan.

We argue that *plan explanation* can serve to improve human trust in a plan. Such plan explanation can take on several forms. *Visual plan explanation* [25] presents the user with a graphical representation of a plan (e.g., with nodes representing actions, edges providing temporal links between actions, and paths representing different plans), and allows for different filters to be applied in order to mitigate cognitive overload. We briefly discuss one instantiation of such techniques in Sect. 4.1.

The second approach for plan explanation that we consider here involves a textual presentation of the plan in natural language, which is created through interaction with the user. This *dialogue based* approach allows a user to ask questions about the plan or about alternative plans, and understand the reasons why specific planning steps were selected. By allowing users to guide the dialogue, the information most relevant to them can be presented, reducing the time needed for them to understand the selected plan, and militating against information overload. An appropriate choice of dialogue will also allow a user to provide new information to the system, allowing re-planning to take place in a natural manner. Our focus in this paper is on how argument and dialogue can be employed to provide plan explanation. This second approach builds on argumentation and formal dialogue theory to select what information to convey. The information is then presented using natural language through the application of Natural Language Generation (NLG), the area of Language Technology where algorithms are developed that can automatically convert "data" into text [11, 23].

The remainder of the chapter is structured as follows. In the next section we provide a brief overview of formal argumentation and dialogue theory, a branch of knowledge representation on which our dialogue based approach is built. Following this, we describe some proof dialogues which can be applied to plan explanation, before describing a plan explanation application we created as part of the "Scrutable Autonomous Systems" project.[1] In Sect. 5 we discuss related work, before identifying current and future avenues of research in Sect. 6, and concluding.

2 Argumentation and Dialogue

The process of explanation can be viewed as the provision of a justification for some conclusion, or equivalently, as advancing some set of arguments which justify the conclusion. Research in formal argumentation theory has described the nature which such justification can take, and we build our textual explanations on this theory. We therefore begin by providing a high-level overview of argumentation, which underpins our approach to plan explanation.

[1] Funded by the Engineering and Physical Sciences Research Council (EPSRC, UK), Grant ref. EP/J012084/1, 2012–2015.

2.1 Abstract Argumentation

Dung's seminal 1995 paper [9] described how, given a set of arguments and attacks between them, one could identify which arguments remain justified. Dung did not consider how arguments were formed, and his approach therefore treats arguments as atomic entities which are part of an abstract *argumentation framework*.

Definition 1 (Argumentation Framework [9]) An argumentation framework is a pair $(\mathcal{A}, \mathcal{D})$ where $\mathcal{A}$ is a set of arguments, and $\mathcal{D} : \mathcal{A} \times \mathcal{A}$ is a binary *defeat* relation over arguments.

An abstract argumentation framework can be represented visually as a graph, with nodes denoting arguments, and edges denoting defeats between them.

An *extension* is a subset of arguments from within $\mathcal{A}$ that is in some sense justified. Perhaps the simplest requirement for a set of arguments to be justified is that they do not contradict, or conflict with each other, as modelled via the defeat relation.

Definition 2 (Conflict Free) Given an argumentation framework $(\mathcal{A}, \mathcal{D})$, a set of arguments $A \subseteq \mathcal{A}$ is *conflict free* if there is no $a, b \in A$ such that $(a, b) \in \mathcal{D}$.

A slightly stronger criteria for an argument to be justified is that no defeat against it should succeed. For this to occur, the argument should either not be defeated, or should be defended from the defeat by some other arguments.

Definition 3 (Defence and Admissibility) Given an argumentation framework $(\mathcal{A}, \mathcal{D})$, an argument $a \in \mathcal{A}$ is defended by a set of arguments $S \subseteq \mathcal{A}$ if, for any defeat $(b, a) \in \mathcal{D}$, it is the case that there is a $s \in S$ such that $(s, b) \in \mathcal{D}$. A set of arguments S is then said to be *admissible* if it is conflict free and if each argument in S is defended by S.

Building on the notion of admissible arguments, we may define *extensions*, which identify discrete groups of arguments that can be considered justified together.

Definition 4 (Extensions) Given an argumentation framework $(\mathcal{A}, \mathcal{D})$, a set of arguments $S \subseteq \mathcal{A}$ is a

- **complete extension** if and only if it is admissible, and every argument which it defends is within S.
- **preferred extension** if and only if it is a maximal (with respect to set inclusion) complete extension.
- **grounded extension** if and only if it is the minimal (with respect to set inclusion) complete extension.
- **stable extension** if it is conflict free and defeats any argument not within it.

While other extensions have been defined (see [2]) for details, these four extensions capture many of the intuitions regarding what it means for a set of arguments to be justified.

It should be noted that for a given argumentation framework, there will be only a single unique grounded extension. However, multiple complete and preferred extensions may exist, as can zero or more stable extensions. An argument is said to be sceptically accepted under a semantics X if it appears in all X extensions; it is credulously preferred if it appears in at least one, but not all such extensions.

2.2 Labellings

Labellings [1] provide another approach to computing extensions. A labelling $\mathcal{L} : \mathcal{A} \rightarrow \{\mathbf{IN}, \mathbf{OUT}, \mathbf{UNDEC}\}$ is a total function mapping each argument to a single label. Informally, **IN** denotes that an argument is justified; **OUT** that it is not, and **UNDEC** that its status is uncertain.

Wu et al. [28] among others demonstrated an equivalence between such labellings and different argumentation semantics. For example, the following constraints are those required for the arguments labelled **IN** to be equivalent to those within a complete extension:

- An argument is labelled **IN** if and only if all its defeaters are labelled **OUT**.
- An argument is labelled **OUT** if and only if at least one of its defeaters is labelled **IN**.
- It is labelled **UNDEC** otherwise.

Maximising the number of **UNDEC** arguments will result in a labelling which is equivalent to the grounded semantics while minimising these arguments will yield a preferred extension. While some argue that labellings are more intuitive (especially to non-technical audiences) than the standard argumentation semantics, the question of how a legal labelling can be identified still remains. Several algorithms for identifying legal labellings have been proposed [17] whose complexity mirrors the complexity of computing the relevant argumentation semantics.

Proof dialogues, which we discuss in Sect. 3 next, are another technique for computing an argumentation semantics. As their name implies, such proof dialogues seek to mirror some form of discussion, building up the elements of an extension as the dialogue progresses. Before considering proof dialogues, we consider how arguments are generated.

2.3 From Knowledge to Arguments

While abstract argumentation allows us to identify which arguments are justified, we must also consider how arguments are generated. In this section we introduce a simple *structured* argumentation framework which allows for the construction of arguments from a knowledge base. The system we consider here is a slight simplification of ASPIC− [4] which in itself is a variant of ASPIC+ which includes several simplifications and enables *unrestricted rebut*, as explained below.

The knowledge base of ASPIC− consists of a set of *strict* and *defeasible* rules. The former encode standard modus ponens, while the latter represent rules whose conclusions hold by default. We write $P \rightarrow c$ where P is a set of literals and c is a literal to denote a strict rule; $P \Rightarrow c$ encodes a defeasible rule. In both cases, P are the rule's premises, and c is the rule's conclusion. We also assume a preference ordering $\prec$ over defeasible rules, and that—given the standard negation operator $\neg$—the set of strict rules is closed under contraposition. In other words, given a strict rule $a \rightarrow b$ in the knowledge base, the rule $\neg b \rightarrow \neg a$ must also be present.

Arguments are constructed by nesting rules. An argument is made up of a set of *sub-arguments* and a single *top rule*. We can formalise this as follows.

Definition 5 Given a set of rules KB and a set of arguments S, we can construct an argument $A = \langle tr, sa \rangle$ where $tr \in KB$ is a rule and $sa \subseteq S$ is a set of arguments such that if tr is of the form $P \rightarrow c$ or $P \Rightarrow c$, then for every $p \in P$ there is an argument $a_p \in sa$ whose top rule has conclusion p, and $sa = \bigcup \{a_p\} \bigcup$ sub-arguments of a_p.

An argument is said to be strict if its top rule is strict, and all of its sub-arguments are strict. The final conclusion of an argument is the conclusion of its top rule, while an argument's conclusions consist of its final conclusion and the final conclusion of all its sub-arguments.

An argument $(A1)$ for a simple strict or defeasible fact can be introduced through a rule with no premises, e.g., $A1 :\rightarrow rw1$. If a second rule $R2 : rw1 \Rightarrow rf$ exists, then a second argument $A2 : A1 \Rightarrow rf$ can be obtained. The top rule of this latter argument is the $R2$, while its sub-argument is $A1$. We illustrate the argument generation process with a running example.

Example 1 Consider a UAV which has two choices regarding where to land, namely runway 1 (*rw1*), or runway 2 (*rw2*). While it believes it has sufficient fuel to reach both runways, *rw1* is further, meaning it will have to utilise its reserve fuel. However, given the runway length and weight of surveillance equipment it is carrying, landing at *rw2* is also considered dangerous. The UAV is programmed to prefer dipping into its fuel reserves over landing on a short runway. This can formally be encoded through the set of rules shown in Table 9.1. In turn, these rules result in the arguments shown in Table 9.2.

As in abstract argumentation, arguments interact with each other via attacks. While ASPIC− considers undercutting attacks (where one rule makes another inapplicable) as well as rebutting attacks (where the conclusions of a rule are in conflict with another rule's premises or conclusions), in this chapter we consider only the latter type of attack.

Definition 6 Given two arguments A and B, argument A attacks B (via *rebut*) if the conclusion of A's top rule is either

Table 9.1 Rules for the UAV example

	Rule	Description
R1	$\Rightarrow \neg ow$	(By default) we are not overweight
R2	$\Rightarrow \neg rf$	(By default) we are not using reserve fuel
R3	$\Rightarrow rw1$	(By default) we will land at rw1
R4	$\Rightarrow rw2$	(By default) we will land at rw2
R5	$rw1 \rightarrow rf$	Landing at rw1 will use reserve fuel
R6	$rw2 \rightarrow ow$	Landing at rw2 will cause us to be overweight

Table 9.2 Arguments obtained from the rules of Table 9.1

A0	$A13 \rightarrow \neg rw1$	A1	$A7 \rightarrow \neg rw2$
A2	$A12 \rightarrow ow$	A3	$A11 \rightarrow rw1$
A4	$A3 \rightarrow rf$	A5	$A13 \rightarrow ow$
A6	$\Rightarrow \neg rf$	A7	$\Rightarrow rw1$
A8	$\Rightarrow \neg ow$	A9	$A6 \rightarrow \neg rw1$
A10	$A7 \rightarrow rf$	A11	$A8 \rightarrow \neg rw2$
A12	$A9 \rightarrow rw2$	A13	$\Rightarrow rw2$

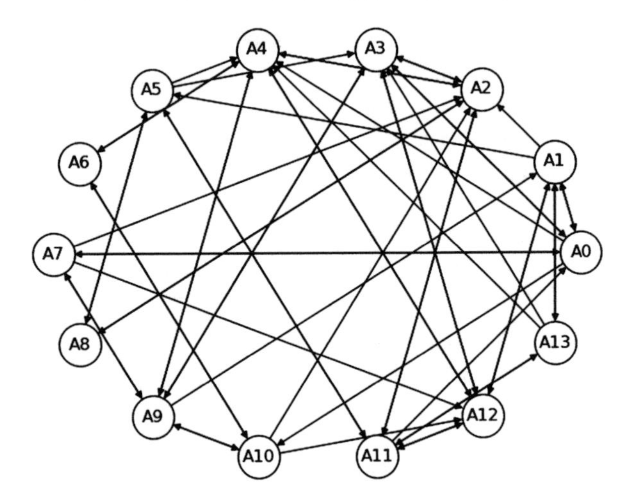

Fig. 9.1 The attacks obtained from the UAV example

– the negation of the conclusion of B's top rule; or
– the negation of the conclusion of B''s top rule where B' is a sub-argument of B.

Example 2 Definition 6 applied to the arguments in Table 9.2 results in the argument graph shown in Fig. 9.1. This argument graph has multiple preferred extensions, but no grounded extension.

Attacks between arguments reflect inter-argument inconsistencies, but do not take preferences or priorities between rules into account. Attacks are transformed into *defeats* when these priorities are considered. For an argument A, if we denote all rules used within it (including the rules used within its sub-arguments) as $Rules(A)$, then such defeats are defined as follows

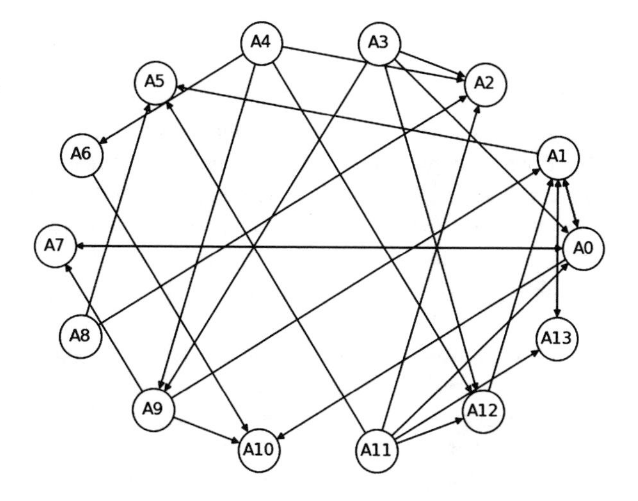

Fig. 9.2 The argument framework obtained when defeats are computed for the UAV example

Definition 7 Argument A defeats an argument B if A attacks B for all $r_A \in Rules(A)$, $r_B \in Rules(B)$, it is the case that $r_A > r_B$.

In the above definition, it is assumed that a strict rule is preferred to all defeasible rules within an argument.

This condition for defeat captures the *weakest link* principle, computing the strength of an argument by considering its weakest rule. Furthermore, it complies with the *democratic* ordering principle, requiring that a single rule within the stronger argument be preferred over all rules within the weaker argument. A discussion of other principles can be found in [18].

Example 3 Figure 9.2 illustrates the abstract argument framework obtained when defeats are computed for the UAV example. The grounded extension for this argument framework is $\{A1, A3, A4, A7, A8, A10, A11\}$. One can therefore conclude that the UAV should land on $rw1$ rather than $rw2$ while making use of reserve fuel and not being overweight.

3 Proof Dialogues

While argumentation can be used to identify appropriate arguments which justify why some plan should be executed, we have not yet considered how such arguments should be presented to a user. In this section, we describe *proof dialogues*, which provide a dialogical approach to justifying arguments. Such a dialogical approach then naturally provides an explanation as to why an argument (and in turn a plan) is justified.

Proof dialogues seek to determine the status of a single argument, i.e., whether it does, or does not appear within an extension according to some semantics (or alternatively what its labelling is). In the remainder of this chapter, we refer to this single argument as the focal argument. In the process of determining the status of the focal argument, the status of other arguments may also become apparent.

Since our focus lies in explaining why a plan was executed, we do not care about what could have been, or could be, but rather what was or is. Within the Scrutable Autonomous Systems project, we therefore concentrated on single extension semantics, namely the grounded and—to a lesser extent—sceptically preferred semantics. Less attention was paid to the latter due to the computational complexity involved in computing the status of arguments under this semantics [2, 24].

In this section we revisit the proof dialogue described in [17] for the grounded semantics, in order to illustrate how such dialogues operate. In the next section, we then consider more advanced proof dialogues which (we argue) are better able to provide explanation than this dialogue. We discuss this point further below.

Proof dialogues are usually represented as a discussion between two players, **Pro**, who wishes to demonstrate that the focal argument appears within the extension under the given semantics, and **Con**, who wishes to demonstrate otherwise.

For an argument to appear within a grounded extension, it must be defended by other arguments within the extension, but cannot (directly or indirectly) defend itself. This suggests the following structure for a proof dialogue where **Pro** and **Con** alternate in advancing arguments.

Opening move: **Pro** introduces the focal argument.

Dialogue moves: If the length of the dialogue is odd (i.e., it is **Con**'s move) then **Con** must introduce an argument that attacks the last argument introduced. Otherwise, if the length of the dialogue is even (i.e., it is **Pro**'s move), then **Pro** must introduce an argument that attacks the last introduced argument, but cannot introduce an argument that they have already introduced. If a player cannot make a move, then the dialogue terminates.

Dialogue termination: The last person to be able to make a move is the winner of the dialogue.

It has been shown that if an argument is in the grounded extension, then there is a sequence of arguments that **Pro** can advance (i.e., a strategy) to win the dialogue.

Perhaps the most significant disadvantage of the dialogue game described above is that they do not describe how a winning strategy may be found. If such a dialogue is used for explanation, and a non-winning strategy is used, than the explanation generated will not be appropriate. We therefore describe an alternative dialogue game for computing whether a focal argument is in the grounded extension, together with an appropriate strategy. This dialogue game was originally introduced in [3], and is referred to as the *Grounded Discussion Game*, abbreviated GDG.

Participants within the game can make four different moves, defined as follows.

- *HTB(A)* stating that "*A* has to be the case". This move, made by **Pro**, claims that *A* has to be labelled **IN** within the legal labelling.

- *CB(B)* stands for "*B* can be the case". This move, made by **Con**, claims that *B* does not necessarily have to be labelled **OUT**.
- *CONCEDE(A)* allows **Con** to agree that *A* has to be the case.
- *RETRACT(B)* allows **Con** to agree that *B* must be labelled **OUT**.

The game starts with the proponent making a HTB statement about the focal argument. In response, **Con** utters one or more CB, CONCEDE, or RETRACT statements. **Pro** makes a further HTB statement in response to a CB move. The precise conditions for each move are as follows:

- HTB(A) is the first move. Alternatively, the previous move was CB(B), and A attacks B.
- CB(A) is moved when A attacks the last HTB(B) move made by **Pro**; A has not been retracted, and no CONCEDE or RETRACT move is applicable.
- CONCEDE(A) is moved if HTB(A) was moved previously, all attackers of A have been retracted, and this move was not yet played.
- RETRACT(A) is moved if **Con** made a CB(A) move in the past which has not yet been retracted, and A has an attacker B for which the move CONCEDE(B) was played.

An additional condition is that HTB and CB moves cannot be repeated (to prevent the dialogue going around in circles), and HTB and CB cannot be played for the same argument.

Pro wins the game if **Con** concedes the focal argument while **Con** wins if they make a CB move to which **Pro** cannot respond.

Caminada [3] demonstrates a strategy for this game which is sound and complete for the grounded semantics. That is, **Pro** will win the game if and only if the focal argument is in the grounded extension, and **Con** will win otherwise.

Example 4 Continuing our running example, a user might question whether the UAV ends up using reserve fuel (i.e., whether $A10$ is in the grounded extension). The dialogue could then proceed as illustrated in Table 9.3.

Comparing Table 9.3 with Fig. 9.2, the primary advantage of proof dialogues over the standard labelling-based approaches becomes apparent. Proof dialogues allow for the incremental presentation of arguments which are relevant to the user's interests, while ignoring arguments which the user accepts (by not having the user query such arguments), or are not central to the explanation. By operating in this way, proof dialogues mitigate against information overload and allow the user to "drill down" to where the explanation is necessary.

The dialogue above illustrates one weakness of many argumentation based explanation dialogues, namely that preferences are (normally) treated as meta features which induce defeats between arguments. The dialogue can therefore not explain these preferences directly. Techniques for overcoming this issue are discussed in Sect. 5.

Table 9.3 Sample dialogue for the UAV example

Pro	HTB(A10)	"The UAV uses reserve fuel as it lands on $rw1$"
	CB(A0)	"We know that by default, we can just land on rw2"
Pro	HTB(A3)	"But not being overweight means we must land on runway 1"
Con	CONCEDE(A3)	"I accept that"
	RETRACT(A0)	"And that for that reason, we can't land on runway 2"
	CB(A9)	"But could it not be the case that no reserve fuel is used as it doesn't land on $rw1$"?
	CB(A6)	"After all, by default, no reserve fuel is used"
Pro	HTB(A4)	"But we know that the UAV is not overweight, and therefore can't land on $rw2$". Not landing on $rw2$ means it lands on $rw1$, and therefore uses reserve fuel"
Con	CONCEDE(A4)	"I accept that line of argument"
	RETRACT(A9)	"And retract what I said"
	RETRACT(A6)	

4 Putting it all Together: The SAsSy Demonstrator

Figure 9.3 shows a screenshot of the prototype plan explanation tool developed as part of the Scrutable Autonomous Systems project. In this section, we provide a brief overview of this tool, its strengths, and its limitations.

Underpinning the tool were plans expressed as YAML workflows.[2] Such workflows contain choice points with regards to actions, and decisions as to which action to pursue were made—by the system—through argument-based reasoning. More specifically, a domain model made up of strict and defeasible rules was constructed. From this model, arguments could be generated, and extensions computed. The conclusions of arguments within the extension then identified which actions should be selected when choices existed (c.f., the UAV running example).

As illustrated in the screenshot, the tool's user interface consisted of three main portions. At the top, a visual display of the plan was shown. A textual summary of the plan (or portions of the plan) appears on the bottom left, while an area wherein a user can interact with the system via dialogue appears on the bottom right.

[2]https://www.commonwl.org/user_guide/.

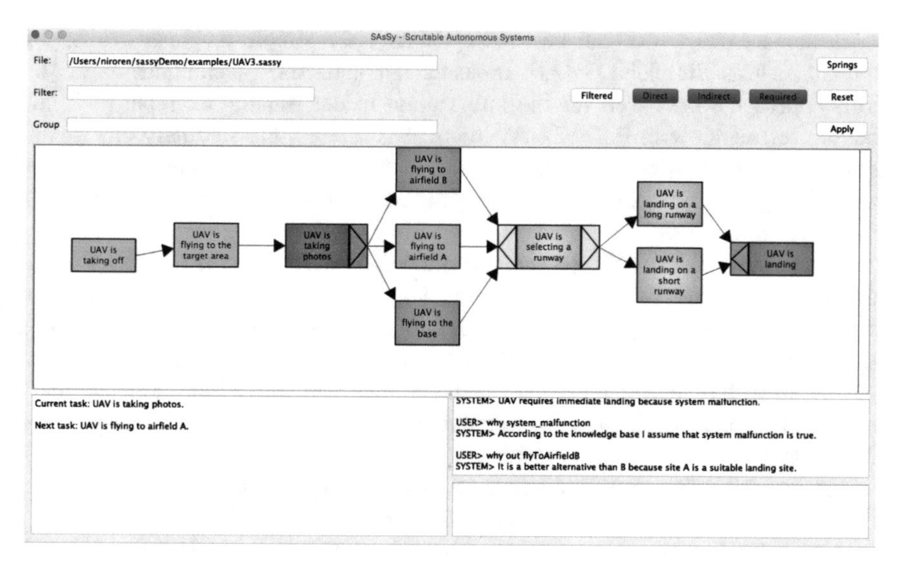

Fig. 9.3 A screenshot from the plan explanation tool

4.1 Plan Visualisation

The plan visualisation window provides a simple view of the plan, showing the ordering between tasks, actions which can be executed in parallel, and the like.

Actions within a plan are executed by different entities, and may affect various resources. Different filtering options were provided to the user, allowing them—for example—to highlight or hide only those actions which affect a specific resource. In [26], the authors show that such highlighting techniques reduce the number of errors and improve response times for users of the system when considering small and medium sized plans. Surprisingly, however, while highlighting relevant portions of the plan led to improved performance, the hiding of unimportant parts of the plan did not lead to improved performance by the user. In addition, questions remain as to whether these results carry through to larger plans than those investigated by [26].

4.2 Natural Language Generation

While plan visualisation using action labels may provide users with important insight, natural language descriptions of these actions can sometimes be easier for such users to understand. Such descriptions may be offered in isolation or—as is often preferable—in addition to graphs, and this is where Natural Language Generation (NLG) comes in.

In some cases, NLG can be accomplished via a simple language realisation toolkit such as SimpleNLG [12], or using template-based techniques as in [8]. This approach works well for the information in our running example (e.g., the bottom left window in Fig. 9.3). As the plan is filtered, the summary changes to reflect only the portions relevant to the user. Similarly, as the plan executes, the natural language summary describes only the current action to be executed, omitting irrelevant information.

Complicated tasks such as planning, however, pose difficult additional challenges, particularly when plans become large (i.e., containing many steps) or structurally complex (e.g., with choices or parallel paths), in which cases NLG needs to find suitable ways to summarise what would otherwise become an unwieldy list of lists. In such cases, the NLG-generated text may start with a high-level summary saying "This plan consist of a large number of actions, which need to be performed in parallel," before going into further detail.

Furthermore, the generator needs to avoid misunderstandings. For example, the text "But could it not be the case that no reserve fuel is used as it doesn't land on rw1"?" contains some syntactic ambiguities that might be misconstrued. Misunderstandings are also known to arise when English expressions such as "if .. then" are employed to express a logical construct such as the material implication (i.e., the standard "arrow" of FOPL), and finding better alternatives automatically is not always easy.

Finally, plans are often the result of automated theorem proving, where formulas in First-Order Predicate Logic (FOPL), or more complicated logics, are manipulated to find the solution for a planning problem. In these cases, the dialogue needs to inform the user that some action a was chosen (or that some action b, which the user may have suggested, is not feasible) because of some logic proposition p. The problem is that, frequently, the system expresses p in a form that may seem unwieldy to human users, for instance because of background knowledge that they possess. (For example, p may say that a crane is at location loc_1 and *not* at location loc_2, where the latter part is redundant because a crane can only be at one location at a time.) Thus, NLG faces the challenge of having to *"optimise"* p before any standard NLG techniques can be applied.

4.3 Dialogue Based Plan Explanation

The bottom right portion of the prototype allows users to interact with the system through dialogue, asking why the system believes certain facts do, or do not hold. If a query asks why some conclusion holds, the system initiates a proof dialogue taking on the role of **Pro**, while if the user asks why some conclusion does not hold, then the system takes on the role of **Con**. A simple domain specific language allows the user to participate in the dialogue, and arguments are presented to the user as natural language rather than logical formulae.

The dialogue language also allows the user to assert or delete facts within the knowledge base, enabling them to update the system with new knowledge. If the user changes the facts within the knowledge base, the system will determine if any of its actions need to be changed, and will allow the dialogue to restart. The system can change the user's beliefs (by presenting them with justified arguments), and allows the user to change the system's beliefs (via assertions and deletions) through the same natural dialogue based interface.

It should be noted that within the prototype, the grounded persuasion game of [5] was used as the proof dialogue. However, this latter game is not sound, and—unlike the proof dialogue described above—does not allow both participants to introduce arguments, a feature which is undesirable in some instances [3].

5 Discussion and Related Work

This chapter described an argument-based system for explaining plans. The planning domain was encoded using YAML, and argumentation was used to select actions where choices existed. Other researchers have described how classical planning techniques can be recreated using argumentation. For example, [10] provided an algorithm for performing partial order planning in defeasible logic programming, while Pardo [20] and others [19] described how dialogue can be used to perform multi-agent planning. Several researchers [14, 27] have examined how BDI agent programs (which bear strong similarities to HTN planning) can be explained via simple dialogues. The focus of this strand of work involves using argument and dialogue to drive the planning process. The techniques described in this paper can then be used to explain how the plan was generated, and why other plans were not selected, by advancing arguments for the plan and demonstrating attacks against other plans' arguments.

The focus of this work was on dialogue games for the grounded semantics. This semantics is sceptical, selecting arguments which—in some sense—must be justified, and can be contrasted with the preferred semantics, which identify sets of arguments that could be justified together. In the context of explaining planning, selecting a sceptical semantics appears correct, as it is only possible to have selected a single alternative for execution [21]. However, other sceptical semantics do exist, such as the sceptical preferred semantics, which select those arguments lying in the intersection of all preferred extensions. It has been argued that this latter semantic is more natural for humans [22], and dialogue games for identifying the sceptical preferred extension have been proposed [24]. One downside of this semantics is however the computational complexity involved in computing it [2], which may make it unfeasible for large domains.

Argumentation and dialogue appear to be natural techniques for explanation, and is increasingly being used in the context of explainable AI research [13]. Psychologists have claimed that humans innately reason using argument [15], while computer scientists have shown that formal argumentation agrees with human intuition [6, 22] in most cases. As future work, we intend to evaluate the effectiveness of our approach to explaining plans through human experimentation. In addition, we intend to overcome two of the current limitations of our approach, namely the meta-logical nature of preferences, and the lack of temporal concepts in our argumentation system.

One approach to addressing this first limitation could be to use an extended argumentation system [16] which encodes preferences as arguments. Addressing the second limitation may be possible through the use of timed argumentation frameworks [7], which explicitly include temporal concepts. However, both these systems have been described only in abstract terms, and it will therefore be necessary to create a structured instantiation of them. Doing so will allow us to provide more refined explanations about more aspects of the plans. Finally, we will also consider how explanations can be provided for richer planning languages such as PDDL.

6 Conclusions

In this chapter we described how plans can be explained through the use of a dialogue game, where participants take turns to make utterances which are used to establish whether some argument (and therefore its conclusions) is justified. To use our technique, domain rules describing the plan must be transformed to arguments, which we achieved through the use of the ASPIC− formalism. Finally, we described how these plan visualisation and natural language generation, together with dialogue based explanation, can be used to create a tool to explain plans to non-technical users.

While our approach appears promising, a complete evaluation with users is still required. In addition, it suffers from shortcomings with regards to explaining preferences and temporal concepts, suggesting a clear path for future work.

Acknowledgements This work was supported by the Engineering and Physical Sciences Research Council (EPSRC, UK), grant ref. EP/J012084/1 ("Scrutable Autonomous Systems").

References

1. P. Baroni, M. Caminada, and M. Giacomin. An introduction to argumentation semantics. *The Knowledge Engineering Review*, 26(4):365–410, 2011.
2. P. Baroni and M. Giacomin. *Semantics of Abstract Argument Systems*, pages 25–44. Springer US, Boston, MA, 2009.
3. M. Caminada. A discussion game for grounded semantics. In *International Workshop on Theory and Applications of Formal Argumentation*, pages 59–73. Springer, 2015.
4. M. Caminada, S. Modgil, and N. Oren. Preferences and unrestricted rebut. In *Proceedings of the 2014 conference on Computational Models of Argument*, pages 209–220, 2014.

5. M. Caminada and M. Podlaszewski. Grounded semantics as persuasion dialogue. In *Proceedings of the 4th International Conference on Computational Models of Argument (COMMA 2012)*, volume 245, pages 478–485. IOS Press, 2012.
6. F. Cerutti, N. Tintarev, and N. Oren. Formal arguments, preferences and natural language interfaces to humans: an empirical evaluation. In *Proc. ECAI*, pages 207–212, 2014.
7. M. L. Cobo, D. C. Martínez, and G. R. Simari. On admissibility in timed abstract argumentation frameworks. In *ECAI*, volume 215, pages 1007–1008, 2010.
8. K. V. Deemter, M. Theune, and E. Krahmer. Real versus template-based natural language generation: A false opposition? *Computational Linguistics*, 31(1):15–24, 2005.
9. P. M. Dung. On the acceptability of arguments and its fundamental role in nonmonotonic reasoning, logic programming and n-person games. *Artificial Intelligence*, 77(2):321–357, 1995.
10. D. R. García, A. J. García, and G. R. Simari. Defeasible reasoning and partial order planning. In *Proceedings of the 5th International Conference on Foundations of Information and Knowledge Systems*, FoIKS'08, pages 311–328, Berlin, Heidelberg, 2008. Springer-Verlag.
11. A. Gatt and E. Krahmer. Survey of the state of the art in natural language generation: Core tasks, applications and evaluation. *Journal of Artificial Intelligence Research*, 61:65–170, 2018.
12. A. Gatt and E. Reiter. SimpleNLG: A realisation engine for practical applications. In *Proceedings of the 12th European Workshop on Natural Language Generation (ENLG 2009)*, pages 90–93, 2009.
13. D. Gunning, Explainable artificial intelligence (XAI). *Defense Advanced Research Projects Agency, DARPA/I20*, (DARPA, 2017).
14. V. Koeman, L. A. Dennis, M. Webster, M. Fisher, and K. Hindriks. The "Why did you do that?" Button: Answering Why-questions for end users of Robotic Systems. In *Proceedings of the 7th International Workshop in Engineering Multi-Agent Systems*, Montreal, Canada, 2019.
15. H. Mercier and D. Sperber. *The enigma of reason.* Harvard University Press, 2017.
16. S. Modgil. Reasoning about preferences in argumentation frameworks. *Artificial Intelligence*, 173(9–10):901–934, 2009.
17. S. Modgil and M. Caminada. *Proof Theories and Algorithms for Abstract Argumentation Frameworks*, chapter 6. Springer, 2009.
18. S. Modgil and H. Prakken. The ASPIC+ framework for structured argumentation: a tutorial. *Argument and Computation*, 5(1):31–62, 2014.
19. S. Pajares and E. Onaindia. Temporal defeasible argumentation in multi-agent planning. In *Proceedings of the Twenty-Second International Joint Conference on Artificial Intelligence - Volume Three*, IJCAI'11, pages 2834–2835. AAAI Press, 2011.
20. P. Pardo, S. Pajares, E. Onaindia, L. Godo, and P. Dellunde. Multiagent argumentation for cooperative planning in DeLP-POP. In *The 10th International Conference on Autonomous Agents and Multiagent Systems-Volume 3*, pages 971–978. International Foundation for Autonomous Agents and Multiagent Systems, 2011.
21. H. Prakken. Combining sceptical epistemic reasoning with credulous practical reasoning. *COMMA*, 144:311–322, 2006.
22. I. Rahwan, I. Madakkatel, M., J. Bonnefon, R. N. Awan, and S. Abdallah. Behavioral experiments for assessing the abstract argumentation semantics of reinstatement. *Cognitive Science*, 34(8):1483–1502, 2010.
23. E. Reiter and R. Dale. Building applied natural language generation systems. *Natural Language Engineering*, 3(1):57–87, 1997.
24. Z. Shams and N. Oren. A two-phase dialogue game for skeptical preferred semantics. In *JELIA*, volume 10021 of *Lecture Notes in Computer Science*, pages 570–576, 2016.
25. N. Tintarev, R. Kutlak, J. Masthoff, K. Van Deemter, N. Oren, and W. W. Vasconcelos. Adaptive visualization of plans. In *UMAP Workshops*, 2014.
26. N. Tintarev and J. Masthoff. Effects of individual differences in working memory on plan presentational choices. *Frontiers in Psychology*, 7:1793, 2016.

27. M. Winikoff. Debugging agent programs with why? Questions. In *Proceedings of the 16th Conference on Autonomous Agents and MultiAgent Systems*, pages 251–259. International Foundation for Autonomous Agents and Multiagent Systems, 2017.
28. Y. Wu, M. Caminada, and M. Podlaszewski. A labelling-based justification status of arguments. *Studies in Logic*, 3(4):12–29, 2010.

Chapter 10
Interactive Planning-Based Hypothesis Generation with LTS++

Shirin Sohrabi, Octavian Udrea, Anton Riabov, and Oktie Hassanzadeh

Abstract We present LTS++, an interactive development environment for planning-based hypothesis generation motivated by applications that require multiple hypotheses to be generated in order to reason about the observations. Our system uses expert knowledge and AI planning to reason about possibly incomplete, noisy, or inconsistent observations derived from data by a set of analytics, and generates plausible and consistent hypotheses about the state of the world. Planning-based reasoning is enabled by knowledge models obtained from domain experts that describe entities in the world, their states, and relationship to observations. To address the knowledge engineering challenge, we have developed a language, also called LTS++ that allows the domain expert to specify the state transition model and encoding of the observations without any knowledge of AI planning or existing planning languages (i.e., PDDL). LTS++ integrated development environment facilitates model testing and debugging, generating, and visualizing multiple hypotheses for user-provided observations, and supports model deployment for online observation processing, publishing generated hypotheses for analysis by experts or other systems. To compute hypotheses we use an efficient planner that finds a set of high-quality plans. We experimentally evaluate our planning algorithm and conduct empirical evaluation to demonstrate the feasibility of our approach and the benefits of using planning-based reasoning. In this chapter we focus on describing the modeling and the knowledge engineering challenges of our system.

S. Sohrabi (✉) · O. Udrea · O. Hassanzadeh
IBM Research, Yorktown Heights, NY, USA
e-mail: ssohrab@us.ibm.com; udrea@us.ibm.com; hassanzadeh@us.ibm.com

A. Riabov
Logitech Inc., Newark, CA, USA
e-mail: ariabov@logitech.com

M. Vallati, D. Kitchin (eds.), *Knowledge Engineering Tools and Techniques for AI Planning*, https://doi.org/10.1007/978-3-030-38561-3_10

1 Introduction and Motivation

The set of planning-based tools, collectively called LTS++, address the hypothesis generation problem that arises in applications that require multiple hypotheses to be generated in order to reason about possibly incomplete or inconsistent sequences of observations received from external sources. For example, when analyzing observations derived from sensor data in intensive care, the goal can be to generate plausible hypotheses about the condition of the patient. The resulting hypotheses can then be further refined and analyzed to create a recovery plan for the patient. In another application, decisions aimed to prevent malware spread in computer networks can be based on hypotheses about change in behavior of individual hosts generated by reasoning about observations of network traffic over time.

The core idea of the approach to planning-based hypothesis generation we implement in LTS++ is the following. Modeling the hypothesis generation problem as one of inferring a sequence of state transitions from a sequence of observations and transforming the sequence of observations together with the state transition model into a planning task. In particular, we extend the work of Sohrabi et al., [19] to address unreliable observations and generate multiple near-optimal lowest-cost plans, mapping the generated plans to hypotheses [17, 27]. This mapping ensures that lower cost plans are mapped to more plausible hypotheses; hence, finding a number of lowest-cost plans results in the same number of most plausible hypotheses.

Our LTS++ implementation uses an efficient planner that finds top-k plans, i.e., k plans such that no valid plans with lower cost exist [11, 16, 24]. We have evaluated several algorithms for this purpose, and currently use the k-shortest path algorithm K* [1]. More details can be found in [16].

Knowledge engineering requirements come to the forefront in designing a system like LTS++, where domain knowledge is encoded and maintained directly by the domain experts, such as clinicians or network security engineers. To address these requirements, we developed the LTS++ language that allows the domain experts to easily describe the state transition models and observations specific to their domain, without requiring the experts to learn about the underlying planning technologies or Planning Domain Definition Language (PDDL) [13]. The LTS++ browser-based Integrated Development Environment (IDE) includes an editor with syntax highlighting and static error checking, as well as integrated tools for interactive model testing and debugging, generating, and visualizing multiple hypotheses for user-provided observations. Models created in the IDE can then be deployed to LTS++ servers to generate hypotheses automatically as observations are received, generating alerts based on hypotheses for further analysis by experts or other systems.

We build upon a significant body of prior research. While expert judgment is the primary method used for generating hypotheses and evaluating their plausibility, automated methods have been proposed, to assist the expert, and help improve accuracy and scalability. Notably, model-based diagnosis methods can determine whether observations can be explained by a model (e.g., [3]). Also, several researchers have proposed use of automated planning technology to address several related classes of problems including diagnosis (e.g., [7, 18]), plan recognition (e.g., [14, 15, 22]), and finding excuses [5]. These problems share a common goal of finding a sequence of actions that can explain the set of observations given the model-based description of the system. However, most of the existing literature makes an assumption that the observations are reliable and should all be explainable according to the model. But that is not true in general; as a further complication, we cannot assume the system model is complete. The hypothesis generation approach we propose handles the unreliable observations and incomplete models by offering multiple alternative hypotheses explaining each given observation sequence. Our LTS++ tool automates the generation and evaluation of hypotheses in addition to addressing the knowledge engineering challenges of encoding and maintaining models.

While we have performed experimental evaluation and conducted empirical evaluation to demonstrate the feasibility of our approach and the benefits of using planning-based reasoning, in this chapter, we focus on describing the modeling and the knowledge engineering challenges of our system. In particular, in Sect. 2, we describe our two applications, early detection of complications in ICU and early detection of malware in computer networks. In Sect. 3, we describe the hypothesis generation problem and its relationship to planning. In Sect. 4, we describe our proposed language LTS++ and its main elements as well as its relationship to a planning problem. We will then discuss the LTS++ IDE and provide a number of example hypotheses in Sect. 5. We will conclude with a discussion of related work and summary.

2 Application Description

In this section we describe two real-world applications that motivate our approach: the early detection of patient complications in Intensive Care Units (ICUs) and suspicious behavior of hosts (computers) in computer networks. A key characteristic of these applications is that the true state of monitored patients, network hosts, or other entities, while essential for timely detection and prevention of critical conditions, is not directly observable. Furthermore, there are several ways of analyzing the raw data to create observations about the entity, and there are multiple potentially ambiguous observations, each of which can have differing interpretations. We must then analyze the sequence of available observations to reconstruct or estimate the entity state, and use that to drive further analysis or take specific actions. To make this possible, our approach relies on a model of the entity consisting of states, transitions between states, and a many-to-many correspondence between states, observations, and actions. The model is a representation of the knowledge a domain expert uses to perform the corresponding monitoring and

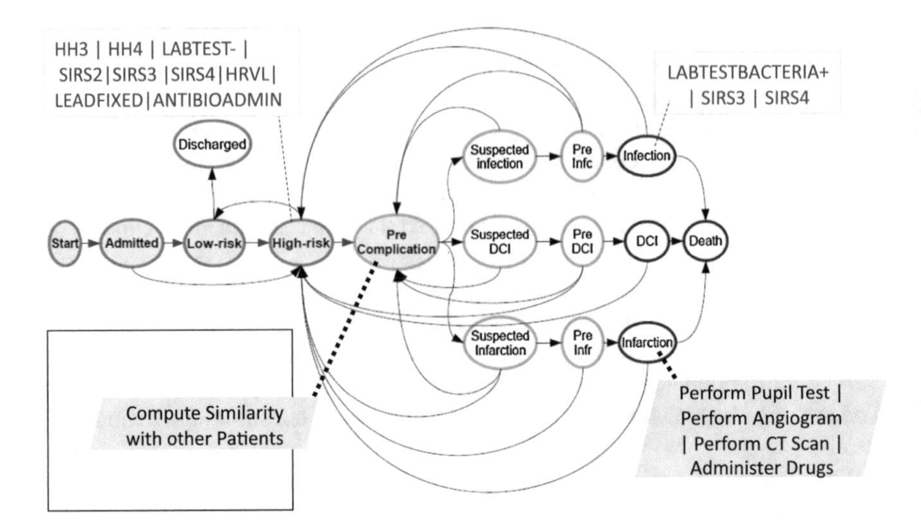

Fig. 10.1 Patient complication detection

diagnosis task. We next describe two real-world models for patient monitoring and cybersecurity analysis derived from experts, and the encoded LTS++ language.

To help describe the patient monitoring application, we describe a state transition model that was drawn by talking with neuro ICU physicians from the Columbia University Neurological unit. Figure 10.1 shows this state transition model. The states have types, *good* drawn with a green outline or *bad*, drawn with an orange or red outline. Each state has a name, and has associated observations that it explains, and actions that it triggers. Note, these actions are analytic actions not to be confused with planning actions. In the figure, the bad states correspond to critical states of a patient such as *Infection*, *DCI*, *Infarction*, or sometimes even a terminal state such as *Death*. The good states are the non-critical states. Upon admission the patient is classified as either in *Lowrisk* or in *Highrisk*. From a *Highrisk* state, they may get to the *Infection*, *Infarction*, or the *DCI* state through intermediate precomplication states. These intermediate states represent states where some clinical signals are present, but before the appearance of definitive symptoms. The patient's condition may improve; hence, the patient's state may move back to the *Lowrisk* state from for example the *Infection* state, based on interventions that the physicians can perform.

Observations in the model are computed from raw data captured by patient monitoring devices (e.g., the patient's blood pressure, heart rate, temperature) as well as other measurements and computations provided by doctors and nurses. In the figure, a subset of all possible observations is shown in light rectangular boxes—attached to the state that explains them. Examples of observations include measures computed from physiological parameters, such as the Systemic Inflammatory Response Syndrome (SIRS) score, may be provided by doctors, such the Hunt and Hess score (HH), and may be the result of performing lab-tests on the patient, such as LabTestBacteria+ (positive test for Bacteria). As is shown, the same observation

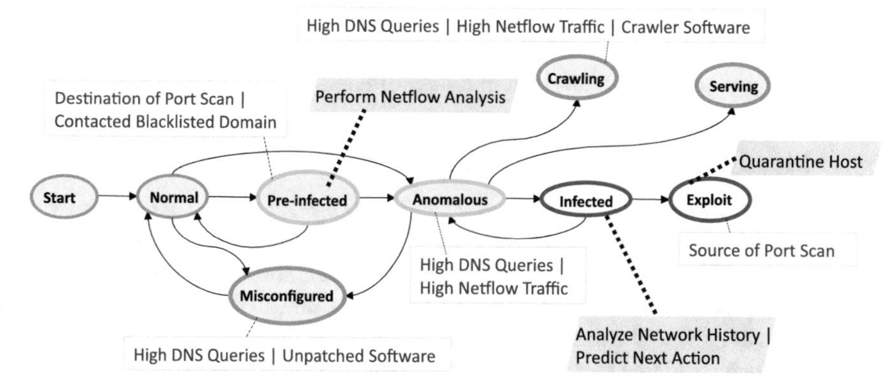

Fig. 10.2 Malware detection

can have multiple interpretations. For instance, the patient may have a SIRS score 3 or 4 both in *Highrisk* as well as in *Infection* state. Also shown in the figure—within the shaded parallelograms—are corresponding actions that may be taken when the patient is identified to be in that state. For instance, when the patient is estimated to be in the *Precomplication* state, it is recommended to look at similarity with other patients to further diagnose the patient's condition. Such similarity can be computed using different analytics. Alternately, when the patient is identified as being in a complication state, the actions can correspond to interventions that the physicians need to perform, e.g., *Perform Pupil Test*.

Figure 10.2 shows a state transition model for network host (entity) monitoring in a computer network. This model is derived through consultation with cybersecurity and network monitoring analysts. Bad states in this model correspond to the malware lifecycle, with the host becoming infected with malware. These states include the *Infection* and *Exploit* states shown in the figure. Good states include states associated within the normal modes of operation of a host such as *Crawling*, *Normal*, *Serving* (server behavior), etc. There are also intermediate states that are not completely indicative of infection, but may be pre-cursors to bad states. These include the *Anomalous* and *Pre-Infected* states. Observations for this model are computed from the raw network traffic, including measurements from Domain Name System (DNS) queries, from Netflow measurements, and from Firewall alerts, as well as looking at the network behavior of the hosts. As with the ICU model the same observation of *High Number of DNS Queries* may be associated with either *Anomalous* behavior for end-user machines or *Crawling* behavior. Actions in this case represent analytic tasks, such as *Analyze Network History* to identify how a machine got infected, or physical tasks, such as *Quarantine Host* when the host is identified as being in the *Exploit* state.

While the complexity of the analysis involved to derive observations from the raw data can vary, it is important to note that observations, in both cases described earlier, are by nature *unreliable*:

The Set of Observations Can Be Incomplete Operational constraints will prevent us running in-depth analysis on *all* of the data all the time. However, observations are typically time stamped, and can be temporally ordered.

Observations May Be Ambiguous This is depicted in multiple examples in Figs. 10.1 and 10.2, where the same observation may be explained by many states.

Not All Observations Are Explainable, Given Other Observations There are several reasons while some observations may remain unexplained: (1) observations are (sometimes weak) indicators of a behavior, rather than authoritative measurements; (2) the model description is by necessity incomplete, unless we are able to design a perfect model; (3) in the case of malware detection, malware could try to confuse detectors by either hiding in normal traffic patterns or originating extra traffic.

For instance, consider the cybersecurity model, and a sequence of observations: *Monitoring On*, *High DNS Queries*. This can be explained by the following state sequences:

- *Normal* → *Misconfigured*
- *Normal* → *Anomalous* → *Misconfigured*
- *Normal* → *Misconfigured*
- *Normal* → *Anomalous* → *Crawling*

and other observations are required to disambiguate these states. Some of these may be pre-cursors to infection (e.g., *Anomalous*) and so require careful analysis. Given a sequence of observations and the model, the hypothesis generation task infers a number of plausible hypotheses about the evolution of the entity. In practice this requires a high degree of human skill to perform. By encoding domain knowledge using a simple model of the form described, and coupling with an automated technique that allows for incomplete state transition models, and unreliable observations, we can provide action decision support to human experts. The result of our automated technique is presented as recommendations to physicians or network analysts, or may be used automatically to drive additional analyses.

3 Hypothesis Generation Problem

In this section, we formally define the hypothesis generation problem. To do so, we first define a dynamical system that can model the system behavior. We then define a notion of a hypothesis and hypothesis "plausibility."

A dynamical system is a tuple $\Sigma = (F, A, I)$, where F is a finite set of fluent symbols, A is a set of actions, and $I \subseteq F$ defines the initial state. Actions are defined by their precondition and effects, over the set of fluents F. The set of actions A includes both actions that account for the possible transitions in the model as well as the *discard* actions, one per each observation o with precondition ¬o and no effect. The "*discard*' actions to simulate the "explanation" of an unexplained observation.

That is, the instances of the *discard* action add transitions to the system that account for leaving an observation unexplained. The added transitions by the *discard* action help us define the satisfaction of observations as we will discuss next.

A *system state* s is a set of fluents which defines all that is true in a particular state of a dynamical system. For a state s, let $M_s : F \rightarrow \{true, false\}$ be a truth assignment that assigns $true$ to f if $f \in s$ and $false$ otherwise. An action a is *executable* in a state s if all of its preconditions are met by the state s or $M_s \models c$ for every c in the precondition of a. We define the successor state as $\delta(a, s) = (s\backslash$ delete effects of $a) \cup$ (add effects of action a) for the executable actions. The sequence of actions $[a_1, \ldots, a_n]$ is executable in s if the state $s' = \delta(a_n, \delta(a_{n-1}, \ldots, \delta(a_1, s)))$ is defined; henceforth, is executable in Σ if it is executable from the initial state.

Let $T \subseteq F$ be the set of fluents that are observable. An *observation* is a fluent in T. Observation formula φ or what we call a *trace* is a sequence of observations. While in general the observation formula φ can be expressed as an Linear Temporal Logic (LTL) formula [4], we consider the trace φ to have the form $\varphi = [o_1, \ldots, o_m]$, where $o_i \in T$, with the following standard LTL interpretation[1]:

$$o_1 \wedge \bigcirc\Diamond(o_2 \wedge \bigcirc\Diamond(o_3 \ldots (o_{n-1} \wedge \bigcirc\Diamond o_n) \ldots))$$

Note that the observations are totally ordered in the above formula. It is typical for the applications we consider to have observations that are timestamps and hence are considered to be totally ordered.

Intuitively, not all observations can be explained; hence, we define the notion of satisfaction of a trace which considers an observation *satisfied* if it is explained or discarded as long as the order of which observations are considered is met by the action sequence. More formally, we define the satisfaction of a trace φ by an action sequence π in Σ as follows.

Definition 1 A trace $\varphi = [o_1, \ldots, o_m]$ is satisfied by an action sequence $\pi = [a_1, \ldots, a_n]$ if π is executable from the initial state and there is a non-decreasing function f that maps the observation indices $j = 1, \ldots, m$ into action indices $i = 1, \ldots, n$, such that for all $0 \leq j \leq m$, either $o_j \in s$, where s is the state reached after execution of action $a_{f(j)}$, or $discard_{o_j} = a_{f(j)}$

Consider the following set of actions: A_{o_1} with effect o_1, A_{o_2} with effect o_2, A_{o_3} with effects o_2 and o_3, and action A_{o_4} with effects o_1, o_2, and o_4. Then the trace $[o_1, o_2]$ is satisfied by action sequence $[A_{o_1}, A_{o_2}]$ $(f(1) = 1, f(2) = 2)$, $[A_{o_4}]$ $(f(1) = 1, f(2) = 1)$,$[discard_{o_1}, A_{o_2}]$ $(f(1) = 1, f(2) = 2)$, but not by the action sequence $[A_{o_2}, discard_{o_1}]$ or $[A_{o_1}, discard_{o_1}]$. This is because the order of observation must be met by the function f. No such function would exist for $[A_{o_2}, discard_{o_1}]$. Additionally, an action may explain multiple observation. For example, action A_{o_4} explains both o_1 and o_2; hence, the function f maps both observations to the same action index.

[1] $\bigcirc$ is a symbol for next, $\Diamond$ is a symbol for eventually.

Definition 2 Given the dynamical system description $\Sigma = (F, A, I)$, and a trace $\varphi = [o_1, \ldots, o_m]$, an observation $o_i \in \varphi$ is said to be *ambiguous* if there are at least two actions in A that have the fluent o_i as part of their effects. Further, if φ is satisfied by an action sequence $\pi = [a_1, \ldots a_n]$, an observation o is said to be *missing* from the trace if (1) o is observable (i.e., $o \in T$); (2) $o \notin \varphi$; and (3) o is part of an effect of at least one action a_i in the action sequence π, and $o \in \varphi$ is said to be *noisy* if o is never added by any of the actions $a_i \in \pi$.

According to the above definition, observation o_1 is ambiguous because both action A_{o_1} and action A_{o_4} may explain it. Also given a trace $\varphi = [o_1, o_2]$, φ is satisfied by the action sequence $[A_{o_1}, A_{o_3}]$ and in that case, observation o_3 is said to be missing from the trace φ because o_3 is part of the effect of A_{o_3}, but not in the given trace. Furthermore, o_1 is said to be noisy given the action sequence $[discard_{o_1}, A_{o_2}]$ because o_1 is not added by any of the actions in the plan.

A hypothesis is the sequence of actions that explains the given trace. In the case of unreliable observations, a hypothesis may not explain all the observations by discarding some. Hence, we use our definition of a trace satisfied by an action sequence to formally define a hypothesis as follows.

Definition 3 Hypothesis generation problem is a tuple $HG = (\Sigma = (F, A, I), \varphi)$, where Σ is a dynamical system and φ is the given trace. A hypothesis for HG is a sequence of actions $\pi = [a_1, \ldots, a_n]$, $1 \leq i \leq n$, $a_i \in A$ such that the trace φ is satisfied by the sequence of actions π.

Given a trace, there are many possible hypotheses, but some could be stated as more plausible than others. Hence, we define a notion of plausibility of a hypothesis. A hypothesis π is said to be at least as plausible as hypothesis π', stated as $\pi \preceq \pi'$, where $\preceq$ is assumed to be a reflexive and transitive plausibility relation.

Definition 4 Given a hypothesis generation problem $HG = (\Sigma = (F, A, I), \varphi)$, π is the most plausible hypothesis for HG if and only if π is a hypothesis for HG and there does not exists another hypothesis π' for HG such that π' is more plausible or $\pi' \preceq \pi$ and $\pi \npreceq \pi'$.

Next, we define a few cases for the notion of plausibility between hypothesis. A hypothesis π is at least as plausible as hypothesis π', $\pi \preceq \pi'$, if one or more of the following statements hold: π can explain more observations than π', π is a shorter hypothesis, π has minimum number of designated "unlikely" or "bad" actions. The third criteria is similar to the notion of minimum number of "faulty" actions in a diagnostic setting, based on having an optimistic view on what can go wrong.

Back to our example, the hypothesis $[A_{o_4}]$ is more plausible than for example, $[A_{o_1}, A_{o_2}]$ because it is shorter, and the hypothesis $[A_{o_1}, A_{o_2}]$ is more plausible than the hypothesis $[discard_{o_1}, A_{o_3}]$ because it explains both observations. The third criteria is similar to the notion of minimum number of "faulty" actions in a diagnostic setting, based on having an optimistic view on what can go wrong. Note that a hypothesis may be shorter but have more discard actions or more unlikely actions. We address combining the above plausibility relations using numerical cost values of the underlining planning domain. Therefore, plans with smaller costs are more plausible.

4 Model Description in LTS++

In this section, we will describe our proposed language LTS++, derived from LTS (Labeled Transition System) [12] that can be used to define the domain knowledge by a domain expert. As described in the application section, encoding the domain knowledge is itself a challenge specially if the domain expert is not familiar with AI planning. Hence, we also discuss our knowledge engineering effort that can guide the domain expert in describing their knowledge about a particular application. This knowledge is implicitly the same knowledge captured theoretically by the dynamical system. Furthermore, the LTS++ model description together with a trace encodes the hypothesis generation problem we are trying to solve.

Note, LTS++ does not have a full expressive power of PDDL since it encodes state transitions in a simple "next-state" predicate model. A PDDL encoding allows encoding of richer actions with preconditions and effects. Hence, while we can express the LTS++ language into PDDL, we cannot go from a PDDL encoding of the domain to the LTS++ encoding.

We propose a process that further helps the domain experts in creating a model. Figure 10.3 shows our 7-step creation process for an LTS++ model. The arrows are intended to indicate the most typical transitions between steps. This process is meant to help provide guidance to the new users in developing an LTS++ model. While this process is geared towards our applications, we believe that it also provides insight and inspiration into creation of a practical planning problem. Next, we will describe the basic elements in the description of a model in LTS++ following the steps in the model creation process.

1. **Entity**: The domain expert needs to identify the entity which is what the system monitors. This depends on the objective of the hypotheses generator, the available

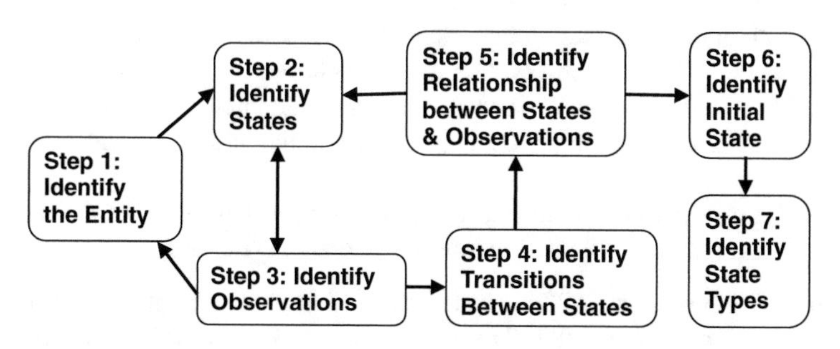

Fig. 10.3 Process for LTS++ model creation

data, and the available actions. The entity could be a patient or a host or other objects in the application.

2. **States**: The domain expert needs to identify the possible states of the entity (different from a planning state). States are not directly observable but can be hypothesized. The states of patient for example could be Delayed Cerebral Ischemia (*DCI*), *SuspectedDCI*, *Infection*, *Precomplication* or *Highrisk*. The states could form a hierarchy, in which case all non-child states are called *hyperstates*. For example, there could be multiple precomplication states, each a child of *Precomplication* hyperstate. Designating a state as a hyperstate is useful when it comes to modeling incomplete model and unreliable observations. For example, if a transition through one or several states of the hyperstate is required, but no specific observation is associated with the transition, the hyperstate itself is included as part of the hypothesis, indicating that the model may have a missing state within the hyperstate, and that state in turn may need a new observation type associated with it.
3. **Observations**: The domain experts need to identify a set of observation types that the system needs to reason about. Since observations are received from analytics as a result of analyzing raw data, the available data and analytics may limit the space of observations. Heart Rate Variability Low (*OHRVL*), is an example of an observation. It is important to note that observations are by nature unreliable: the set of observations will be incomplete, observations may be ambiguous, and not all observation will be explainable.
4. **State Transitions**: The domain expert has to describe possible transitions between states. An example transition is going from state *Infection* to *Highrisk*. This transition reflects an improvement in patient state, without describing the cause of this transition. Enumerating all possible transitions may be a tedious task, depending on the number of states. However, one can use hyperstates to help manage these transitions. Any transition into (or out of) a hyperstate is carried to every child of the hyperstate.
5. **Association between States and Observations**: The domain expert has to associate observations to states meaning that this observation *can* be explained by this state. Note, this association can be many-to-many as observations could be ambiguous or indicative of more than one state, and each state can be associated with multiple observations. The observation *OHRVL* is an example of an ambiguous observation because it can be associated with multiple states. Note we add "*O*" to the observation as a convention.
6. **Initial State**: The domain expert can also define the initial state if the initial state is known. For example in the case of healthcare, the initial state can be the state *Admitted*.
7. **State Types**: States could also have types such as, unlikely, or "bad" states. This maps to the notion of "faulty" or "unlikely" planning actions.

Figure 10.4 shows an LTS++ model description for our healthcare application. The states are shown in blue. The observations are specified within the curly brackets and are shown in green. Multiple observations can be separated by

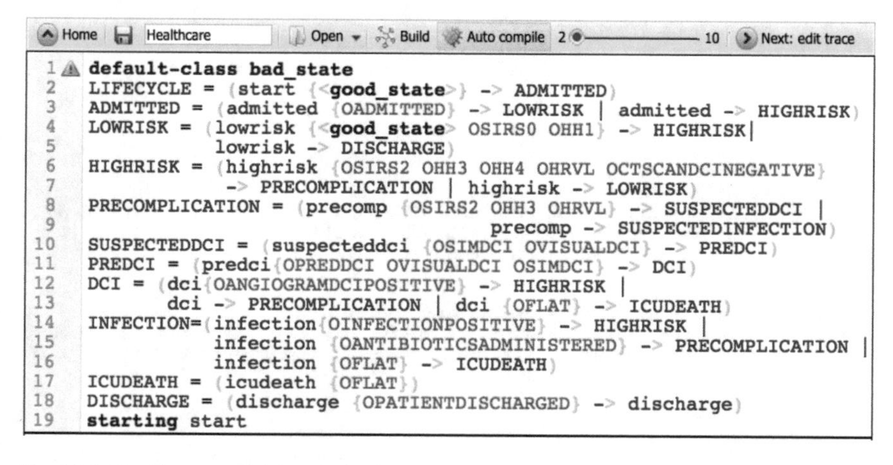

Fig. 10.4 Healthcare model description in LTS++

whitespace or a comma. The state types are specified within angle brackets with a default state type shown in the first line. The transitions between states are specified using arrows. Multiple transitions between states can be specified using a vertical bar. The starting state is specified in the last line.

The knowledge encoded in the LTS++ model is implicitly the same knowledge in the theory of the dynamical system. Informally, each state can be thought of as a label for a subset of planning states of interest and therefore be modeled using a special fluent such as "(at-state)." Each observation belongs to the set T of observable fluents. The state transitions together with the relationship between states and observations define the set of actions A such that the specified state transition is ensured and the observations are part of the effect of the actions. The initial state can also map directly to I. The state types can also map to fluents. Hence, the hypothesis generation problem can now be captures using the LTS++ model description together with a provided trace.

4.1 From LTS++ to a Planning Problem in PDDL

In this section, we describe the planning problem using one fixed encoding of the planning domain, (i.e., description of planning actions, predicates), but varied the planning problem/instance (i.e., initial state, goal state, and variables) based on the given LTS++ model and the given observations. The planning domain is shown in Fig. 10.5 and the planning problem is shown in Fig. 10.6.

The planning domain in Planning Domain Definition Language (PDDL) [13] includes a total of 6 actions. In short, we use two phases, state transitions and explaining or discarding observations and switch between these two using the "ready" predicate. Each transition is followed by either explaining, explain-

```
(:action explain-observation
  :parameters(?x - state ?obs1 - obs ?obs2 - obs ?cat - obs-type)
  :precondition (and (is-next-obs ?obs1 ?obs2)(matches ?obs1 ?cat)
                     (explains ?x ?cat)(at-state ?x)(at-obs ?obs1))
  :effect (and  (not (at-obs ?obs1)) (at-obs ?obs2) (ready)
                     (increase (total-cost) 0)))

 (:action discard-observation
  :parameters(?x - state ?obs1 - obs ?obs2 - obs)
  :precondition (and (is-next-obs ?obs1 ?obs2)(at-state ?x)(at-obs ?obs1))
  :effect (and (not (at-obs ?obs1))(at-obs ?obs2)(ready)
                  (increase (total-cost) 2000)))

 (:action state-change
  :parameters(?x - state ?y - state ?obs - obs)
  :precondition (and (is-next-state ?x ?y)(at-obs ?obs)(at-state ?x)(ready))
  :effect (and (not (at-state ?x))(not (ready))(entering-state ?y)
                    (increase (total-cost) 0)))

 (:action enter-state-good
  :parameters(?y - state ?obs - obs)
  :precondition (and (at-obs ?obs) (entering-state ?y) (good-state ?y))
  :effect (and (at-state ?y)(not (entering-state ?y))
               (increase (total-cost) 1)))

 (:action enter-state-bad
  :parameters(?y - state ?obs - obs)
  :precondition (and (at-obs ?obs)(entering-state ?y)(bad-state ?y))
  :effect (and (at-state ?y)(not (entering-state ?y))
               (increase (total-cost) 10)))

 (:action allow-unobserved
  :parameters(?x - state ?obs - obs)
  :precondition (and (at-obs ?obs)(at-state ?x))
  :effect (and (ready)(increase (total-cost) 1100)))
```

Fig. 10.5 Partial encoding of our sample PDDL domain

```
(:init
  (at-state admitted) (at-obs o_1)(ready)

  (matches o_1 OHH1)(matches o_2 OSIRS0)(matches o_3 OSIRS2)

  (explains lowrisk OSIRS0) (explains highrisk OSIRS2)
  (explains precomp OSIRS2) (explains lowrisk OHH1)
  (explains dci OANGIOGRAMDCIPOSITIVE)
  (explains highrisk OHRVL) (explains precomp OHRVL)

  (is-next-state admitted highrisk) (is-next-state admitted lowrisk)
  (is-next-state lowrisk highrisk) (is-next-state highrisk lowrisk)
  (is-next-state highrisk precomp) (is-next-state dci highrisk)
  (is-next-state dci icudeath) (is-next-state dci precomp)

  (bad-state dci) (bad-state highrisk) (good-state lowrisk)

  (is-next-obs o_1 o_2)(is-next-obs o_2 o_3) (is-next-obs o_3 o_end))

 (:goal (and (at-obs o_end) (ready)))
```

Fig. 10.6 Partial encoding of our sample PDDL problem for the intensive care application

observation, or discarding, discard-observation, an observation or moving to the next state transition without explaining or discarding any observation, allow-unobserved which is useful in order to allow missing observations. The explain action has a cost of 0, the discard-observation action has a high cost (e.g., 2000), and the unobserved transition has a cost of 1100 in this encoding. These numbers were set arbitrary here to show the relative comparison between the different costs set for each different action. In practice these numbers can be learned from data but we found that the number we used modeled the behavior as expected.

The predicates “is-next-obs,” and “at-obs” are used to keep track of observation order. Observation categories (i.e., ?cat) defines the possible observations in the domain. The predicate “matches” together with the “is-next-obs” defines the current trace or the sequence of observation. The predicate “explains” is used to connect states and observations; “(explains ?x ?cat)” means that state ?x can explain observation of category ?cat. The action explain-observation can explain an observation if the resulting state can explain the observation category of the observation in the trace.

We also had one action, state-change that represents the transitions defined by the actions A. This action had a cost of 0 and the predicate “is-next-state” is used to encode the transitions between states. Two additional actions, enter-state-good and enter-state-bad, are used to associate different costs for good and bad states. The predicates “(bad-state)” and “(good-states)” are used to define the good and bad states in the problem. We used a cost of 1 for good states and a cost of 10 for bad states.

This encoding of the domain allowed us to automatically generate multiple problem sets that include different number of observations as well as different transitions. Partial encoding of our sample PDDL problem is shown in Fig. 10.6. This encoding matches our LTS++ description shown in Fig. 10.4 with the following trace [OHH1, OSIRS0, OSIRS2]. The initial state is a special state ‘admitted’ with transitions to highrisk and lowrisk states. The goal state is encoded by two predicates “(ready)” and the “(at-obs o_end)” predicate to ensure the last observation is considered. The last observation is only considered if all other observations are considered in the order in which they are given.

Theorem 1 *Let P' be a planning problem constructed as above for a given LTS++ model and a trace φ and HG be the corresponding hypotheses generation problem; HG has only state transition actions in which observations are part of their effects and the discard actions. If π is a plan for P' then there exists a hypotheses π' for HG that can be constructed from π by considering only the state transition actions and the discard actions. On the other hand, if π' is a hypotheses for HG, then there exists a plan π for P' that can be constructed from π' by adding the extra actions explain, enter, and allow-unobserved and by modifying the state transition action.*

Proof If π is a plan for P', therefore it is executable from the initial state and the goal is satisfied; each observation is either explained or discarded and the ordering is preserved. Therefore, there is a non-decreasing function that maps the observation indices into the action indices: if the observation is satisfied it maps to the state-change action and if it is discarded, it maps to the "discard" action. Therefore, if only the state transition and discard actions are kept, then the trace is still satisfied. If the state-change action is modified to include the observation fluent as part of its effect, then this is a hypotheses for HG. On the other hand, if π' is a hypothesis for HG, then we can add the extra actions to π' and modify the state-change action to remove the explicit mention of the observation and the trace would still be satisfied. The result is a plan for π.

Note, the exact PDDL encoding of the planning problem P' determines if for each found plan for P' there would be exactly one corresponding hypotheses or multiple. If we used the encoding shown earlier, then for each plan there could be multiple possible hypotheses because of the positioning of the explain action. It is possible to have a more complex planning domain and force a one-to-one relationship between hypotheses and plans. Nevertheless, the above theorem shows that a hypothesis can be found by translating the hypothesis generation problem into a planning problem and using an AI planner to find a plan. The resulting plan can be turned into a hypotheses by a post-processing step that removes the extra actions from the plan. Furthermore, assuming that the costs of the actions in P' model the plausibility notion correctly, then the lowest-cost plan maps to the most plausible hypotheses. More formally,

Corollary 1 *Let P' be a planning problem constructed as above for a given LTS++ model and a trace φ and HG be the corresponding hypotheses generation problem. Further, let π_1 and π_2 be two plans for P', and π_1' and π_2' be the corresponding hypotheses for HG. Then π_1' is at least as plausible as π_2' if and only if $cost(\pi_1) < cost(\pi_2)$.*

Given the association between plans and hypotheses we use top-*k* planning to find a set of plans with low cost. These plans can be translated to hypotheses to find the most plausible hypotheses to the hypotheses generation problem. For details on top-*k* please see [11, 16, 24].

5 LTS++ Integrated Development Environment

LTS++ Integrated Development Environment (IDE) is a web-based tool that helps the domain experts to create model descriptions by describing LTS++ models and to generate hypotheses. LTS++ IDE consists of an LTS++ editor, graphical view of the transition system, specification of the trace, and generation of hypotheses. The tool automatically generates planning problems from the LTS++ specification and the entered trace. The generated hypotheses are the result of running our planner and presenting the result from top-most plausible hypothesis to the least plausible hypothesis.

Model Editor The top part of the model editor screen (Fig. 10.7) is the LTS++ language editor with syntax highlighting and the bottom part is the automatically

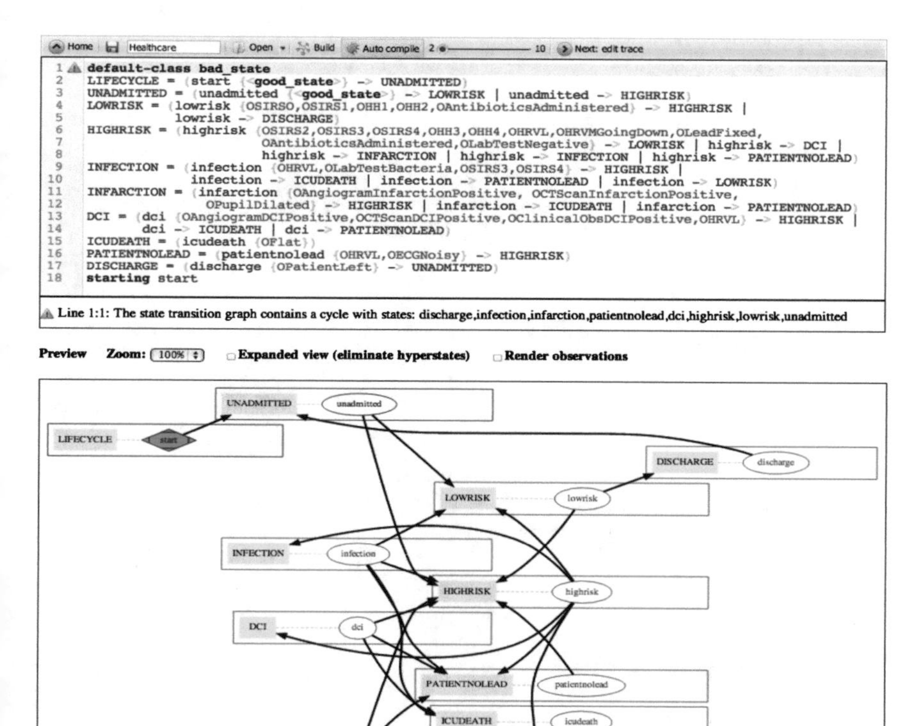
Home
Healthcare
Open
Build
Auto compile
2
10
Next: edit trace
1 default-class bad_state
2 LIFECYCLE = (start {<good_state>} -> UNADMITTED)
3 UNADMITTED = (unadmitted {<good_state>} -> LOWRISK | unadmitted -> HIGHRISK)
4 LOWRISK = (lowrisk {OSIRS0,OSIRS1,OHH1,OHH2,OAntibioticsAdministered} -> HIGHRISK |
5 lowrisk -> DISCHARGE)
6 HIGHRISK = (highrisk {OSIRS2,OSIRS3,OSIRS4,OHH3,OHH4,OHRVL,OHRVMGoingDown,OLeadFixed,
7 OAntibioticsAdministered,OLabTestNegative} -> LOWRISK | highrisk -> DCI |
8 highrisk -> INFARCTION | highrisk -> INFECTION | highrisk -> PATIENTNOLEAD)
9 INFECTION = (infection {OHRVL,OLabTestBacteria,OSIRS3,OSIRS4} -> HIGHRISK |
10 infection -> ICUDEATH | infection -> PATIENTNOLEAD | infection -> LOWRISK)
11 INFARCTION = (infarction {OAngiogramInfarctionPositive, OCTScanInfarctionPositive,
12 OPupilDilated} -> HIGHRISK | infarction -> ICUDEATH | infarction -> PATIENTNOLEAD)
13 DCI = (dci {OAngiogramDCIPositive,OCTScanDCIPositive,OClinicalObsDCIPositive,OHRVL} -> HIGHRISK |
14 dci -> ICUDEATH | dci -> PATIENTNOLEAD)
15 ICUDEATH = (icudeath {OFlat})
16 PATIENTNOLEAD = (patientnolead {OHRVL,OECGNoisy} -> HIGHRISK)
17 DISCHARGE = (discharge {OPatientLeft} -> UNADMITTED)
18 starting start
Line 1:1: The state transition graph contains a cycle with states: discharge,infection,infarction,patientnolead,dci,highrisk,lowrisk,unadmitted
Preview
Zoom: 100%
Expanded view (eliminate hyperstates)
Render observations
UNADMITTED
unadmitted
LIFECYCLE
start
DISCHARGE
discharge
LOWRISK
lowrisk
INFECTION
infection
HIGHRISK
highrisk
DCI
dci
PATIENTNOLEAD
patientnolead
ICUDEATH
icudeath
INFARCTION
infarction

Fig. 10.7 LTS++ IDE

generated transition graph. In the editor, the states appear in blue. The observations are specified within the curly brackets and appear in green. You can specify multiple observations by using space or comma between observations. The transitions between states are specified using arrows. Multiple transitions between states can be specified using a vertical bar. The LTS++ model editor automatically detects errors in LTS++ language and shows them below the text editor.

Model Testing To test the model, a sequence of observations can be entered by clicking on "Next: edit trace" from the LTS++ IDE main page. The tool automatically generates planning problems from the LTS++ specification and entered trace. The generated hypotheses are the result of running a planner and finding the most plausible hypotheses ranked by plausibility from highest to lowest. Figure 10.8 shows an example of hypotheses generated for the critical care model; the result is automatically generated by our tool. Each hypothesis is shown as a sequence of states matched to an observed event sequence. The observations that are explained by a state are shown in green ovals, and unexplained observations are shown in purple. The arrows between the observations show the sequence of

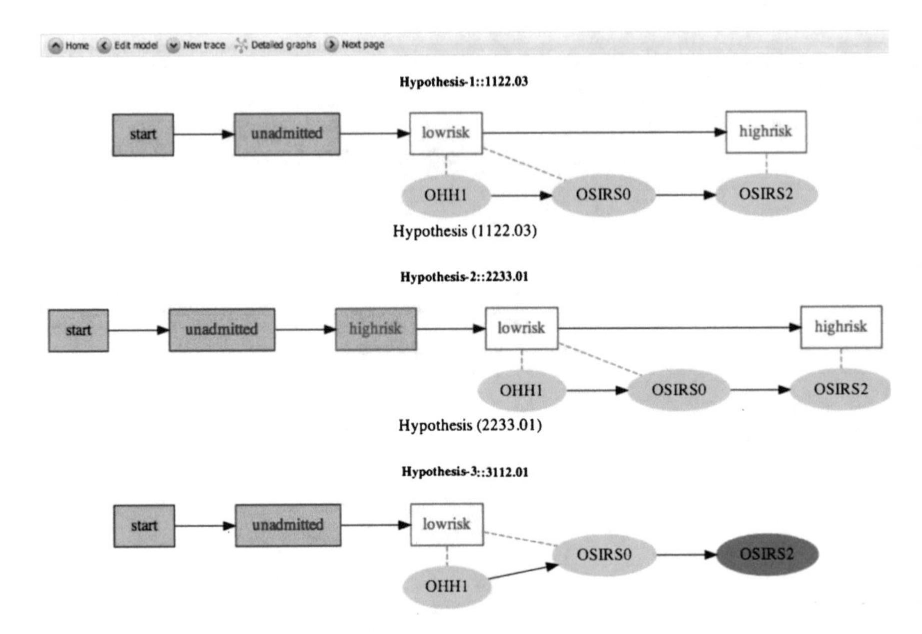

Fig. 10.8 Sample healthcare example

observations in the trace. Each hypothesis is associated with a cost. The lower the cost value, the more plausible is the hypothesis.

Model Discovery and Update Our tool uses a simple bootstrapping technique to discover an initial model given a set of historical observations. Several candidate models will be presented to the domain expert who can choose one as an initial LTS++ model and further improve it. We also implement automated model updates to produce better quality hypotheses as we do not assume the model will be accurate in perpetuity. To do this, we use an aggregate measure of the plausibility of top-N hypotheses as our optimization criteria. Using a genetic algorithm, we attempt small atomic changes to the model (e.g., addition and deletion of states, observations and transitions) and measure the increase in aggregate hypothesis plausibility as a result. In subsequent generations, we combine the promising atomic changes and repeat until we can no longer increase hypothesis plausibility.

Model Composition A single LTS++ model describes a state transition system for a single type of entity, such as a patient. Given multiple entities, each with their own associated model, our tool also allows automated composition of multiple models. It does so by considering a cross product of all possible joint states while paying special attention to the association between observations and the combined states.

Hypothesis Clustering Many of the generated hypotheses are only slightly different from each other. That is, they do seem to be duplicates of each other, except for one or more states or actions that are different. To consolidate similar plans produced by the planner, we apply a clustering algorithm to cluster similar plans and present clusters of plans, where each cluster can be replaced by its representative plan.

6 Related Work

There are several approaches in the diagnosis literature related to the hypothesis generation problem in which use of planners as well as SAT solvers are explored (e.g., [2, 6, 7, 18]). The hypothesis generation problem is also related to the plan recognition problem and the use of AI planning have been explored in that space as well [14, 15, 22]. In particular, Sohrabi et al., explored the same ideas as discussed here with respect to the notion of unreliable observations for the several related problem such as future state projection problem [23] and enterprise risk management [20, 21, 25, 26] that have a corresponding a plan recognition problem. It is important to note that these papers also discuss and address the knowledge engineering challenge through what is called a Mind Map. A Mind Map is a graphical representation of the concepts and relations. The domain knowledge can be encoded by one or more Mind Maps connected by the same concept used in multiple Mind Maps. The Mind Maps can be created in a tool such as FreeMind that produces an XML representation of the Mind Maps and be provided to a system. The system then translates the Mind Maps into an AI planning problem automatically. It is also possible to learn the causal relation between the concepts in order to build the Mind Maps automatically from scratch or augment or validate existing ones [8]. Then similarly, a set of top-k or top-quality plans are found through top-k planning [11, 16, 24]. Diverse planning [9] or top-quality [10] planning can also be explored to compute such a set of plans.

7 Summary

We presented LTS++, an interactive development environment for planning-based hypothesis generation. To enable our planning-based reasoning, we proposed a characterization of the hypothesis generation problem and showed its correspondence to an AI planning problem. To address the knowledge engineering challenge, we have developed a language, also called LTS++ that allows the domain expert to specify the state transition model and encoding of the observations without any knowledge of AI planning or existing planning languages. LTS++ IDE facilitates model testing and debugging, generating, and visualizing multiple hypotheses for user-provided observations. The tool automatically generates planning problems from the LTS++ specification and the entered trace. The generated hypotheses are the result of running our planner that computes a set of high-quality plans. The hypotheses can be visualized and shown to the analyst or can be further investigated automatically.

References

1. Aljazzar, H., Leue, S.: K*: A heuristic search algorithm for finding the k shortest paths. Artificial Intelligence **175**(18), 2129–2154 (2011)
2. Bauer, A., Botea, A., Grastien, A., Haslum, P., Rintanen, J.: Alarm processing with model-based diagnosis of discrete event systems. In: Proceedings of the 22nd International Workshop on Principles of Diagnosis (DX). pp. 52–59 (2011)
3. Cassandras, C., Lafortune, S.: Introduction to discrete event systems. Kluwer Academic Publishers (1999)
4. Emerson, E.A.: Temporal and modal logic. Handbook of theoretical computer science: formal models and semantics **B**, 995–1072 (1990)
5. Göbelbecker, M., Keller, T., Eyerich, P., Brenner, M., Nebel, B.: Coming up with good excuses: What to do when no plan can be found. In: Proceedings of the 20th International Conference on Automated Planning and Scheduling (ICAPS). pp. 81–88 (2010)
6. Grastien, A., Anbulagan, Rintanen, J., Kelareva, E.: Diagnosis of discrete-event systems using satisfiability algorithms. In: Proceedings of the 22nd National Conference on Artificial Intelligence (AAAI). pp. 305–310 (2007)
7. Haslum, P., Grastien, A.: Diagnosis as planning: Two case studies. In: International Scheduling and Planning Applications Workshop (SPARK). pp. 27–44 (2011)
8. Hassanzadeh, O., Bhattacharjya, D., Feblowitz, M., Srinivas, K., Perrone, M., Sohrabi, S., Katz, M.: Answering binary causal questions through large-scale text mining: An evaluation using cause-effect pairs from human experts. In: Proceedings of the 28th International Joint Conference on Artificial Intelligence (IJCAI) (2019)
9. Katz, M., Sohrabi, S.: Reshaping diverse planning. In: Proceedings of the 34th Conference on Artificial Intelligence (AAAI-20) (2020)
10. Katz, M., Sohrabi, S., Udrea, O.: Top-quality planning: finding practically useful sets of best plans. In: Proceedings of the 34th Conference on Artificial Intelligence (AAAI-20) (2020)
11. Katz, M., Sohrabi, S., Udrea, O., Winterer, D.: A novel iterative approach to Top-k planning. In: Proceedings of the 28th International Conference on Automated Planning and Scheduling (2018)
12. Magee, J., Kramer, J.: Concurrency - state models and Java programs (2. ed.). Wiley (2006)
13. McDermott, D.V.: PDDL—The Planning Domain Definition Language. Tech. Rep. TR-98-003/DCS TR-1165, Yale Center for Computational Vision and Control (1998)
14. Ramírez, M., Geffner, H.: Plan recognition as planning. In: Proceedings of the 21st International Joint Conference on Artificial Intelligence (IJCAI). pp. 1778–1783 (2009)
15. Ramírez, M., Geffner, H.: Probabilistic plan recognition using off-the-shelf classical planners. In: Proceedings of the 24th National Conference on Artificial Intelligence (AAAI). pp. 1121–1126 (2010)
16. Riabov, A., Sohrabi, S., Udrea, O.: New algorithms for the top-k planning problem. In: Proceedings of the Scheduling and Planning Applications Workshop (SPARK) at the 24th International Conference on Automated Planning and Scheduling (ICAPS). pp. 10–16 (2014)
17. Riabov, A.V., Sohrabi, S., Sow, D.M., Turaga, D.S., Udrea, O., Vu, L.H.: Planning-based reasoning for automated large-scale data analysis. In: Proceedings of the 25th International Conference on Automated Planning and Scheduling (ICAPS). pp. 282–290 (2015)
18. Sohrabi, S., Baier, J., McIlraith, S.: Diagnosis as planning revisited. In: Proceedings of the 12th International Conference on the Principles of Knowledge Representation and Reasoning (KR). pp. 26–36 (2010)
19. Sohrabi, S., Baier, J.A., McIlraith, S.A.: Preferred explanations: Theory and generation via planning. In: Proceedings of the 25th National Conference on Artificial Intelligence (AAAI). pp. 261–267 (2011)
20. Sohrabi, S., Katz, M., Hassanzadeh, O., Udrea, O., Feblowitz, M.D.: IBM scenario planning advisor: Plan recognition as AI planning in practice. In: Proceedings of Demonstration Track at the 27th International Joint Conference on Artificial Intelligence (IJCAI-18) (2018)

21. Sohrabi, S., Katz, M., Hassanzadeh, O., Udrea, O., Feblowitz, M.D., Riabov, A.: IBM scenario planning advisor: Plan recognition as AI planning in practice. AI Commun. **32**(1), 1–13 (2019)
22. Sohrabi, S., Riabov, A., Udrea, O.: Plan recognition as planning revisited. In: Proceedings of the 25th International Joint Conference on Artificial Intelligence (IJCAI). pp. 3258–3264 (2016)
23. Sohrabi, S., Riabov, A., Udrea, O.: State projection via AI planning. In: Proceedings of the 31st Conference on Artificial Intelligence (AAAI-17). pp. 4611–4617 (2017)
24. Sohrabi, S., Riabov, A., Udrea, O., Hassanzadeh, O.: Finding diverse high-quality plans for hypothesis generation. In: Proceedings of the 22nd European Conference on Artificial Intelligence (ECAI). pp. 1581–1582 (2016)
25. Sohrabi, S., Riabov, A., Udrea, O., Yuan, F.: Using lightweight semantic models to assist risk management in a large enterprise. In: Proceedings of the 16th International Semantic Web Conference - Industry Track (ISWC-17) (2017)
26. Sohrabi, S., Riabov, A.V., Katz, M., Udrea, O.: An AI planning solution to scenario generation for enterprise risk management. In: Proceedings of the 32nd National Conference on Artificial Intelligence (AAAI). pp. 160–167 (2018)
27. Sohrabi, S., Udrea, O., Riabov, A.: Hypothesis exploration for malware detection using planning. In: Proceedings of the 27th National Conference on Artificial Intelligence (AAAI). pp. 883–889 (2013)

Chapter 11
Web Planner: A Tool to Develop, Visualize, and Test Classical Planning Domains

Maurício C. Magnaguagno , Ramon Fraga Pereira , Martin D. Móre , and Felipe Meneguzzi

Abstract Automated planning tools are complex pieces of software that take declarative domain descriptions and generate plans from domains and problems. New users often find it challenging to understand the plan generation process, while experienced users often find it difficult to track semantic errors and efficiency issues. In response, we develop a cloud-based planning tool with code editing and state-space visualization capabilities that simplifies this process. The code editor focuses on visualizing the domain, problem, and resulting sample plan, helping the user see how such descriptions are connected without changing context. The visualization tool explores two alternative visualizations aimed at illustrating the operation of the planning process and how the domain dynamics evolve during plan execution.

Keywords Classical planning · STRIPS · PDDL · State-space visualization

1 Introduction

Classical planning algorithms typically require a declarative domain specification describing action schemata, which, in turn, define the dynamics of the underlying domain. Since the inception of the International Planning Competition (IPC) [24], the standard specification language for classical planning is the Planning Domain Definition Language (PDDL) [3, 15]. Given the declarative nature of PDDL, planning algorithm implementations are often opaque regarding the intermediate steps between reading the formalism and generating a plan. This creates a twofold problem for domain engineers that wish to use automated planning to solve

M. C. Magnaguagno (✉) · R. F. Pereira · M. D. Móre · F. Meneguzzi
Pontifical Catholic University of Rio Grande do Sul (PUCRS), School of Technology, Porto Alegre, RS, Brazil
e-mail: mauricio.magnaguagno@acad.pucrs.br; ramon.pereira@acad.pucrs.br; martin.more@acad.pucrs.br; felipe.meneguzzi@pucrs.br

M. Vallati, D. Kitchin (eds.), *Knowledge Engineering Tools and Techniques for AI Planning*, https://doi.org/10.1007/978-3-030-38561-3_11

problems: ensuring the correctness of each domain description, and optimizing the efficiency of a planning algorithm for each domain description.

First, regarding correctness, writing PDDL specifications may be a challenging task for new users even for simple domains, while detecting semantic mistakes in complex domains is always non-trivial. Even when the user successfully compiles and executes a planning instance with the chosen heuristic function, the planner may fail to find a correct plan for the intended domain. In these cases, virtually no planning algorithm offers extra information, and the user only knows that either the domain has some kind of description error or that specific problem supplied to the planner is unsolvable, such that the planner cannot find a correct plan.

Second, practical applications of classical planners require not only a formalization of the domain in PDDL that is correct, but also exploit the search mechanisms employed by the underlying planners to find solutions efficiently. Most modern classical planning solvers [8, 9, 11, 19] use heuristic functions to estimate which states are likely to be closer to the goal state and save time and memory during the planning process. Different planning domains may require different heuristic functions to focus the search on promising branches and be solved within a reasonable time with little memory footprint. Thus, key to understanding the efficiency of a domain formalization is its impact on the heuristic function used by the underlying planner.

In order to address these challenges, we developed WEB PLANNER, an online tool aimed at helping domain engineers to tune a formalization to a number of common planning heuristics and spotting semantic errors in planning domains. Our tool, which we describe in Sect. 3, includes a PDDL code editor with syntax highlight and auto-complete aimed at helping users to efficiently develop PDDL domains in a similar workflow to many popular integrated development environments (IDEs). Importantly, we integrate the editor to two visualization tools, described in Sect. 2, developed to help users cope with the declarative nature of PDDL and explore the effects of changes to the domain in solving concrete problems. First, we use a visual metaphor from the literature to see how a plan execution achieves (or does not) a goal state from an initial state [14]. Second, we develop a new state-space search visualization that uses tree drawing (in both Cartesian and radial layouts) in conjunction with heatmaps to represent how the distance (e.g., how colder or warmer) to the goal state changes during search. We conducted a structured case study (described in Sect. 4.1) to illustrate how our approach works and validate from user tests, which we describe in Sect. 4.2 showing the results we obtained from employing the tool in a planning course. WEB PLANNER has been deployed for 2.5 years as openly available tool for the planning community, which allowed us to collect anonymous usage statistics. In Sect. 5, we survey related work on planning tools and data visualization, and conclude the paper in Sect. 6 discussing our conclusions and future work.

2 Background

2.1 *Planning*

Planning is the problem of finding a sequence of actions (i.e., plan) that achieves a particular goal from an initial state [4]. A state is a finite set of facts that represent logical values according to some interpretation. Facts are divided into two types: positive and negated facts. Predicates are denoted by an n-ary predicate symbol applied to a sequence of zero or more terms. An operator is represented by: a name that represents the description or signature of an action; a set of preconditions, i.e., a set of facts or predicates that must be true in the current state to be executed; a set of effects, which has an add-list of positive facts or predicates, and a delete-list of negative facts or predicates. An action is an instantiated operator over free variables. A planning instance is represented by: a domain definition, which consists of a finite set of facts and a finite set of actions; and a problem definition, which consists of an initial state and a goal state. The solution of a planning problem is a plan, which is a sequence of actions that modifies the initial state into one in which the goal state holds by the successive execution of actions in a plan. To formalize planning instances, we use the STRIPS [2] fragment of PDDL [15], which contains domain and problem definition in different files.

Heuristic functions are used to estimate the cost of achieving a particular goal [4]. In classical planning, this estimate is often the number of actions to achieve the goal state from a particular state by exploring only promising states. Estimating the number of actions is a NP-hard problem [1]. In automated planning, heuristics can be domain-dependent or domain-independent, and a well-tuned heuristic can result in a substantial reduction in search time by pruning a vast part of the state-space.

2.2 *Data Visualization*

Visualization techniques aim to convey some kind of information using graphical representation [26]. The use of data visualization techniques is often associated to a set of data with the aim of communicating a particular information clearly and efficiently via graphical representation.

Data visualization techniques are concerned with what is the best way to display a dataset, for instance, how to display relation information. Relation information can be displayed efficiently by using hierarchies that convey relation information. Edges in a hierarchical tree represent a relation between nodes. A Cartesian tree visualization is a way to display hierarchical trees as a coordinate system. A radial tree visualization is a way to display a hierarchical tree structure in which such tree expands outwards and radially. In Sect. 3.2 we explore such tree visualizations. Besides hierarchical visualization, we highlight other visualization methods that are closely related to the ones we develop in this work, such as *Gantt charts* [27], which are used to show how tasks are correlated and how much time is expected to complete them, *Waveforms* [6, Chapter 1—page 2] are used to express the behavior of analog or digital data through time, and heatmap visualization [26], which uses a color scheme to illustrate values in a graphic in which each color in the scheme represents one limit value and the many values in the interval are represented by the mix of such colors.

3 Web Planner Architecture

We designed our tool envisioning a development process centered around two tasks by the domain developer. In the first task, the user aims to describe both domain and problem correctly. In the second task, the user tries to identify details of the description (in terms of predicate use) that impact performance and how these predicates appear during the planning process. The domain designer is free to move between these tasks and repeat until satisfied with the results. Once a planning instance is described it is possible to visualize its explored state-space, even when the planning process fails. When the planning process returns a plan the user is able to visualize how predicates were added or deleted by each action in the plan. Such interface could also help planning system developers to explore how planners in development behave.

To avoid the considerable setup time of some planner implementations and maintain a consistent interface across platforms, we use a web interface. The planner is executed in a server, while the editor, output and visualizations are displayed and executed in the browser. The communication between the two sides uses JSON.[1] Figure 11.1 shows the architecture of Web Planner.

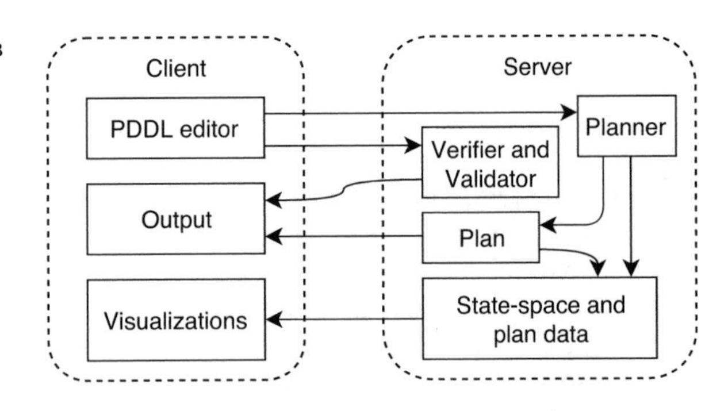

Fig. 11.1 Overview of Web Planner architecture

[1] JSON (JavaScript Object Notation) is an open-standard format for structuring data.

3.1 Domain Development Interface

To better describe planning domains and problems, we identified three key requirements to improve editing such descriptions. First, we required our tool to provide the two common IDE features of syntax highlighting, code auto-completion, and templates (PDDL snippets) to streamline the editing process. For example, to define a new action, our PDDL editor provides an action template (an auto-complete function of our editor, pressing *CTRL+Space* after typing the word *action*) that shows how an action is defined in PDDL, as illustrated in Fig. 11.2. Besides templates for PDDL actions, our editor also provides templates for domain and problem description, just pressing *CTRL+Space* after typing the word *domain* or *problem*, respectively. Second, we the interface must show both domain and problem simultaneously, to avoid forcing the user to go back and forth between descriptions or browser tabs. This interface arrangement improves the designer's awareness of the interactions between a domain and instances of its problems and minimizes user effort in terms of required interface actions (i.e., key presses and mouse clicks). Finally, our interface must include a visualization and an action button in the same context as the editors, allowing the designer to execute the current planning instance without a changing context.

To meet such requirements, we split the editor interface horizontally in three parts: domain, problem, and output. The ability to see input alongside output is very important for both advanced users that are modifying or extending legacy PDDL, and new users, such as students, that are not used with the domain and problem distinction. Instead of starting with a blank planning instance we opted for a simple but complete Towers of Hanoi example to be loaded by default.

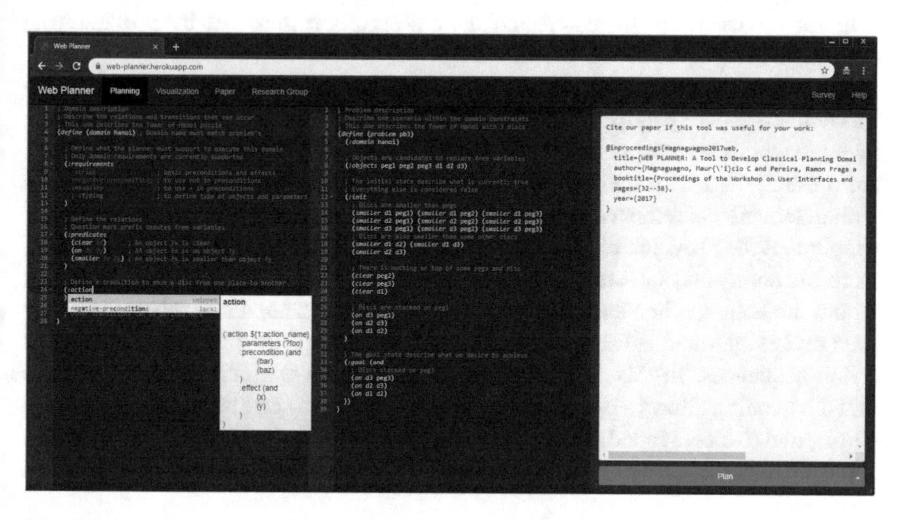

Fig. 11.2 WEB PLANNER editor interface with domain editor (left), problem editor (center), and plan output (right). Action template is provided by auto-complete shortcuts. Verification and validation tools available through caret button

The **Plan** button sends the planning instance to the server to obtain an output based on the domain and problem descriptions contained in the editor. Our editor uses *brace*,[2] a variant of the *ace editor* without server-side processing to highlight programming language elements. In our case most PDDL elements are highlighted, some of which are currently not supported by the back-end planner. The output provided by the planner contains the plan and execution time when successful, error messages when the parser fails, or a failure message when no plan is found. Due to screen space limitations, the visualizations were left to a secondary interface, as users can only visualize after an initial description step. Our goal is to make clear that domain and problem are described together, while planning insights and optimization steps can be obtained later, if required, without overloading the user with information.

Verifier and Validator Plan output alone is not enough to identify errors in a planning description. The declarative nature of PDDL obscures the intermediate structures of the planner for novice users (or users without working knowledge of planner implementation), requiring further modification of the chosen planner to log such information. To address this problem we provide two extra tools to find description errors and mistakes in their domain and problem. The first is a *verifier*, a tool that finds common mistakes in both domain and problem descriptions. The second is a *validator*, a tool that tries to execute a plan provided by the user in the domain and problem previously described, and reports any errors found while doing so. Tests cover only atomic or conjunctive preconditions and effects, limited to *:strips*, *:negative-preconditions*, *:equality*, and *:typing* requirements. Our verifier includes different test cases for domain, Table 11.1, and problem, Table 11.2. Some verifier tests refer to uncommon but valid PDDL, and can be seen as warnings for new users, such as actions with empty preconditions. Our verifier offers substantial help for novice users to understand their description mistakes by providing an automated analysis of the PDDL encoding.

Our validator applies each plan action, testing if such action exists (i.e., the action was defined and all parameters are defined objects/constants), is applicable (all positive preconditions are present in the current state, while no negative preconditions are present), and with their effects generate each intermediate state (current state with delete effects removed and add effects added). Note that the validator ensures simply that the provided plan is a solution to the problem, regardless of optimality, therefore empty and sub-step-optimal plans can also be used, as some problems may require no action, when an initial state satisfies a goal state, while other plans may even revisit intermediate states or simply take more steps than required using other action sequences. In this way, validators help domain engineers verify that their PDDL encoding allows a planner to generate valid plans, and that these plans indeed correspond to the intended semantics of the planning domain. Nevertheless, verifiers and validators tools are often separated from the actual planner software [12], which

[2]https://github.com/thlorenz/brace.

Table 11.1 Rules used by verifier in the domain description

Domain rule	Description
Predicate defined	Every predicate must be defined in *:predicates*
Predicate with valid name	Predicates must contain only valid characters, starting with *a-z*
Predicate arity	Predicates must have the same amount of parameters
Action redefined	Each action must have a unique name
Action parameter unused	One or more parameters of an action are unused
Action parameter repeated	One or more parameters of an action are repeated
Parameter with valid name	Parameters must contain valid characters, starting with *?*
Predicate repetition	Each predicate must appear only once in preconditions and effects
Empty precondition	Preconditions contain no predicates
Null effect	Effect is either empty or does modify state based on preconditions
Unnecessary equality	Preconditions contain *(= ?x ?y)*
Equality contradiction	Preconditions contain *(not (= ?x ?x))*
Precondition contradiction	Preconditions contain *(pre ?a)* and *(not (pre ?a))*
Effect contradiction	Effects contain *(pre ?a)* and *(not (pre ?a))*
Effect contains equality	Equality is only supported in preconditions
Missing/extra requirements	Requirements must match what is used in the description

Table 11.2 Rules used by verifier in the problem description

Problem rule	Description
Predicate repetition	Each predicate must appear only once in initial or goal states
Object with valid name	Objects must contain only valid characters, starting with *a-z*
Object unused	Objects must appear as constant terms in actions, initial, or goal states
Forced equality	Initial or goal states contain *(= a b)*
Goal contradiction	Goal state contain *(pre ?a)* and *(not (pre ?a))*
Rigid goal	Rigid goal predicate is unachievable unless present in initial state
Empty goal state	No planning is required for an empty goal state

makes reviewing and revising domain formalizations less straightforward. Thus, our coupling of the validator and verifier with the editor streamlines the domain formalization process by providing immediate feedback to the domain engineer.

3.2 Visualization Interface

We currently support two visualizations, one focusing on the explored state-space and the other on the execution of the first plan found.The impact of heuristics in the state-space is often introduced in AI lectures using images, such as the ones from

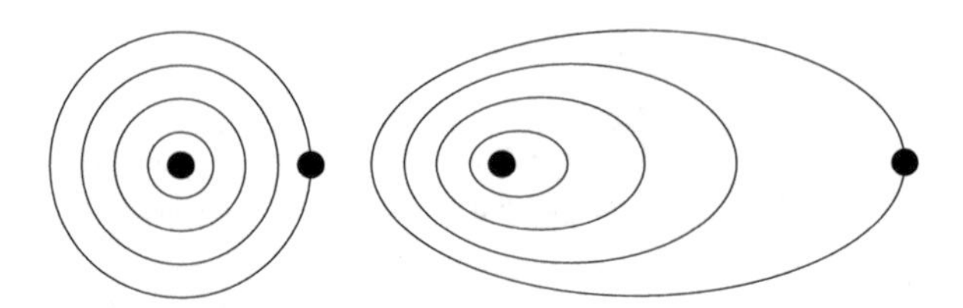

Fig. 11.3 Search contours are defined by search mechanism and heuristic function, either equally exploring in all directions (left) or giving priority towards the goal state (right)

Fig. 11.3, to show how the contour of the explored states grows in all directions on blind search and towards the goal state in informed search (using heuristics) [20, Chapter 3—page 97]. Such images target an audience new to the concept of using a computed auxiliary function to speed-up search. Different from textbooks, implementations that target the same audience use dynamic grids to show both how the state-space is explored and how the heuristic is computed in an Euclidean space. Such examples show the step-by-step process of search. Since not all domains can be mapped to a grid, the visualization process is often limited to path-finding domains. To generate such contours we opted for a tree-based visualization, as they better represent state relations while ignoring repeated states by not expanding a previously found state. If we also added connections to previously found states, a cyclic graph would be obtained and the contours would not be visible.

Heuristic Visualization The heuristic visualization we developed takes advantage of interactive elements to avoid information overload while providing alternative layouts, Cartesian and radial tree visualizations. The radial layout matches the abstraction used by heuristic examples, while the Cartesian layout generates a more compact visualization. In practice, we use the Reingold–Tilford algorithm [18][3] to display both tree layouts. Using tree visualizations we aim to show how planning heuristics explore the state-space to achieve a particular goal.

To compare and explore the state-space of a planning instance, we implemented two planning methods. The first method is based on breadth-first search, and thus uses no heuristic, exploring the state-space in the order of distance from the initial state. The second method implements greedy best-first search using Hamming distance [7] as a heuristic. While we selected these two methods as examples to show the impact of no heuristic vs a generic distance metric for states, our visualization tool supports other search mechanisms and heuristic functions as long as such mechanisms search through the state-space.

To represent the data obtained from the planning process, we use a tree containing the explored state-space and heuristic information about each state. In this tree, each node represents a state (i.e., a set of instantiated predicates), an edge represents a state-transition (i.e., the execution of an action), and the root node represents the

[3]Reingold–Tilford is an algorithm for an efficient tidy arrangement of layered nodes. We use an implementation based on a D3 example available at: http://bl.ocks.org/mbostock/4063550.

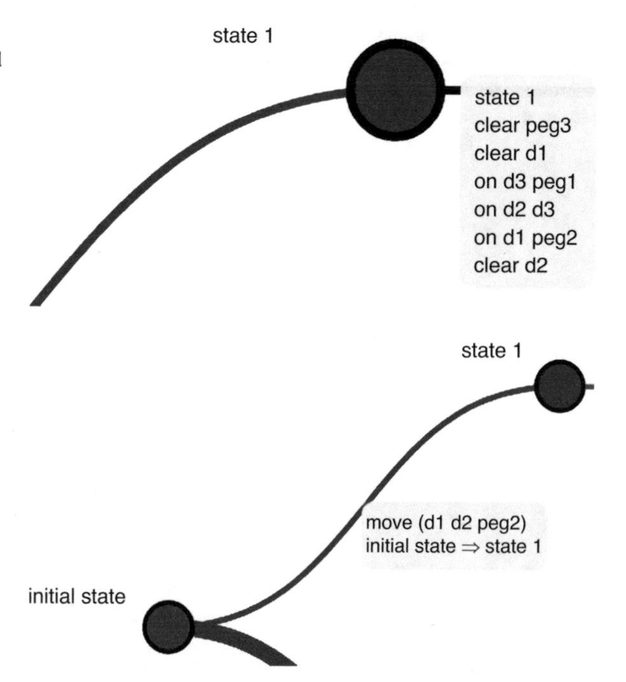

Fig. 11.4 Tooltip that displays the set of instantiated predicates in a state. This figure illustrates state 1 and its predicates for a planning instance of the Hanoi domain

Fig. 11.5 Tooltip that displays the instantiated action applied between two states. This figure illustrates state 1 and its predicates for a planning instance of the Hanoi domain

initial state. The information about the set of predicates in the states (nodes) and the applied actions in such states (edges) are hidden in our heuristic visualization. Such information about states and actions can be seen when the user hovers the cursor over nodes and edges, which then shows, the state's and action's detail as a tooltip. Figures 11.4 and 11.5, respectively, illustrate how our visualization tool show the information about states and actions.

Our visualization tool displays the state-space of a planning heuristic by coloring the estimated distance between states using a heatmap, as in Fig. 11.6. Red nodes represent the states closer to the goal state, i.e., warmer, while distant nodes are represented by blue, i.e., colder. Nodes and edges are colored according to the estimated distance to the goal state. We illustrate the heuristic gradient as a heatmap in Fig. 11.7. Other heuristic functions could generate not only other distance estimations for each state (visible through colors in the graph), but also a different graph, as states would be explored in a different order, as in Figs. 11.10 and 11.11. Here, the radial layout of Fig. 11.11 provides a visualization of the search contours of the heuristic, provided a large enough sample of the total number of states has been explored. Edges between initial and goal state are emphasized (in bold, and as a thick line) to show which path contains the actions that constitute the plan. Such emphasized path is only available when planning is successful for the give planning task. Failed planning cases still obtain data to draw the explored state-space as a tree, which can be used as an interactive debug tool.

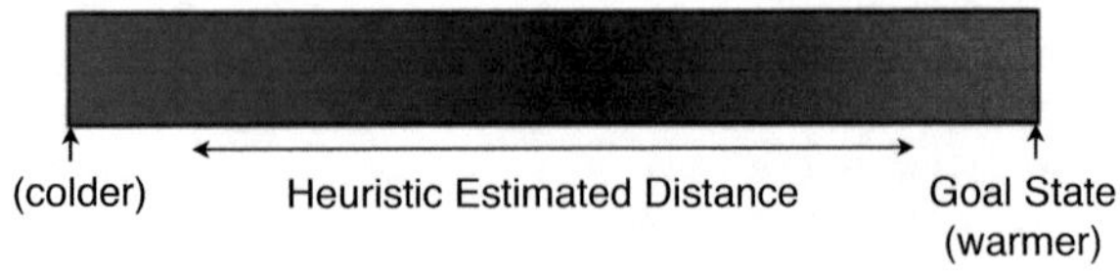

Fig. 11.6 Color scheme that our visualization tool uses to represent the estimated distance

(colder)
Heuristic Estimated Distance
Goal State
(warmer)

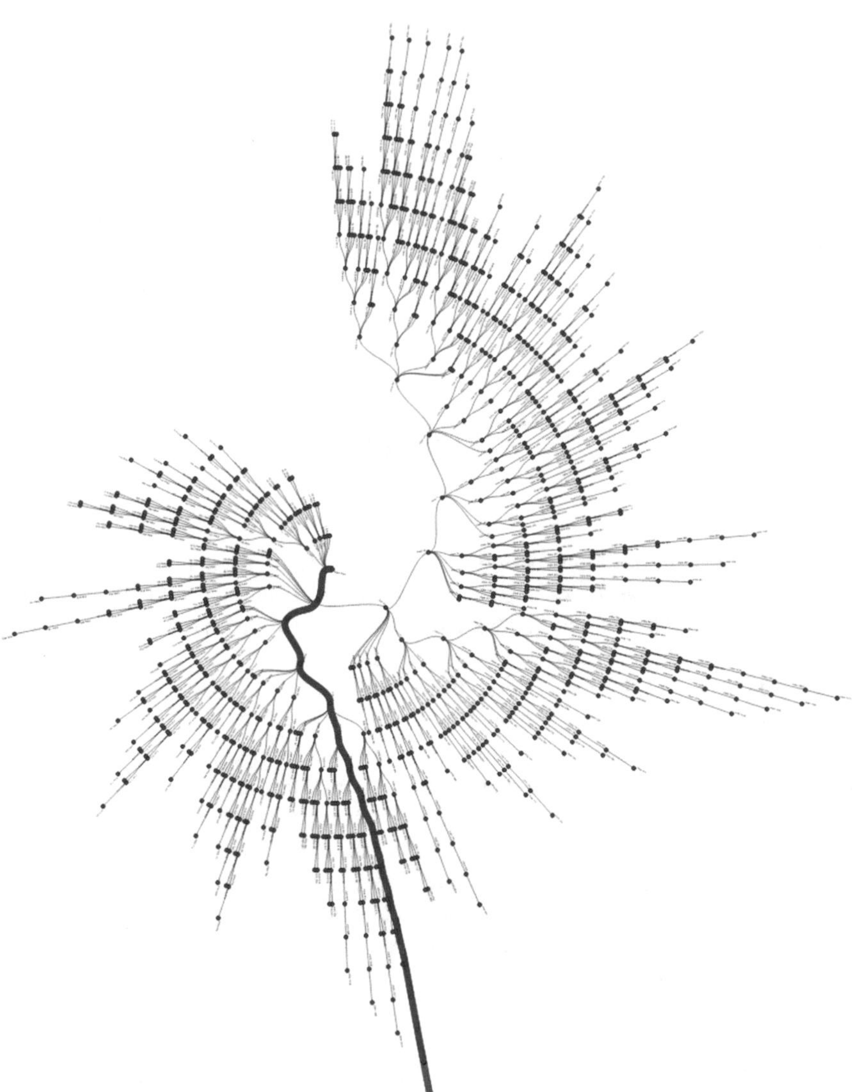

Fig. 11.7 Search contours become visible as more states are explored. This planning instance obtain all goal predicates at the same time, which makes the heatmap mostly blue (colder), while the goal state is located at the bottom in red (warmer)

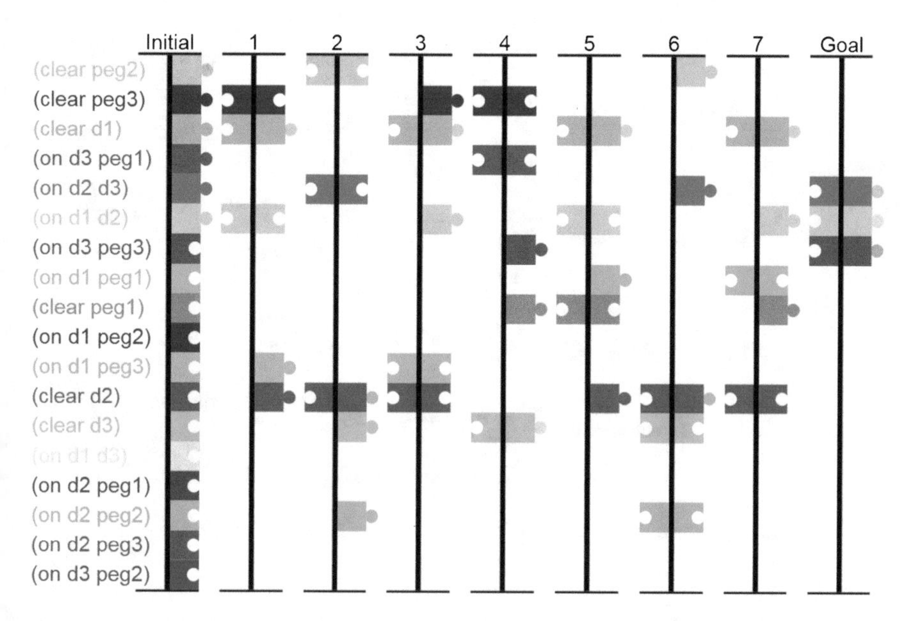

Fig. 11.8 Dovetail plan visualization of Hanoi domain with 3 discs and a plan of size 7

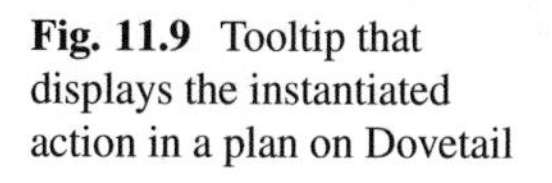
Fig. 11.9 Tooltip that displays the instantiated action in a plan on Dovetail

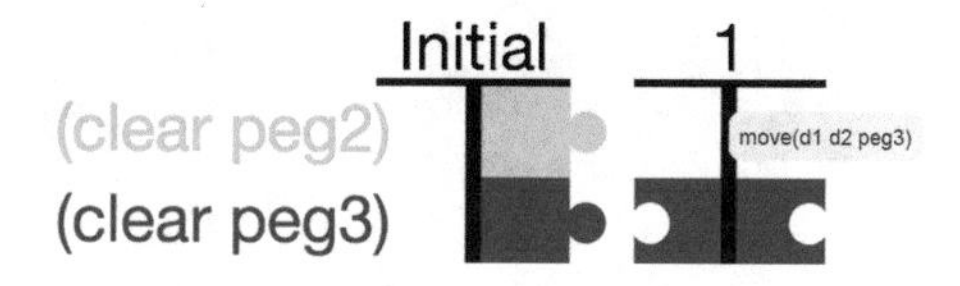

Dovetail Metaphor Visualization The second visualization we implemented is a visual metaphor called Dovetail [14], which is useful to see how predicates change along the plan execution. Each ground predicate that appears in an action effect is represented as one line while both initial state, goal state, and actions are represented as columns. Our interface allows a user to move and zoom to parts of this visualization (illustrated in Fig. 11.8), with tooltips providing extra information as shown in Fig. 11.9 for the domain of the case study of Sect. 4.1. The use of this visual abstraction (Dovetail) aims to improve the learning curve for defining and debugging planning domains and problems.

4 Deployment and Evaluation

4.1 Case Study

In order to validate our visualization tool, we now present a case study we carried out to show a planning instance using different planning heuristics displaying the state-space. To do so, we selected the Tower of Hanoi domain to illustrate our

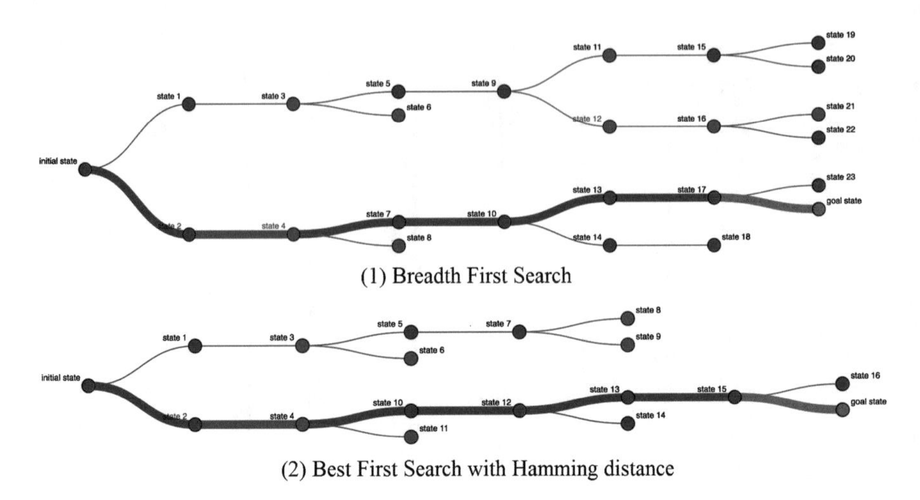

Fig. 11.10 Cartesian tree visualizations of the state-space of Hanoi with 3 discs

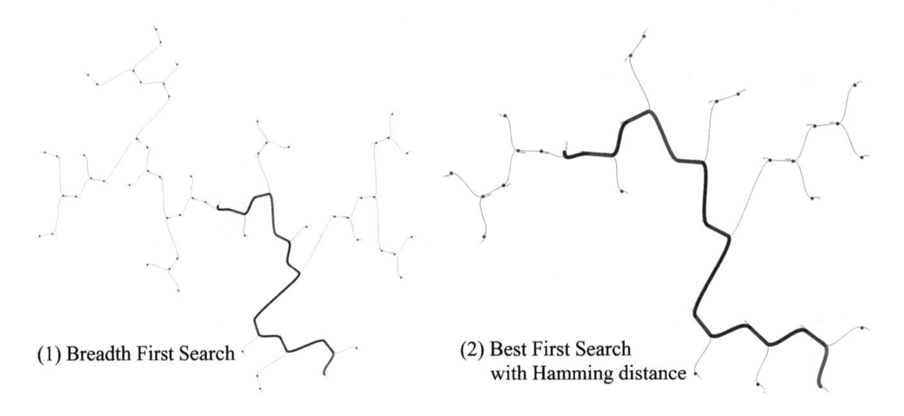

Fig. 11.11 Radial tree visualizations of the state-space of Hanoi with 3 discs

heuristic visualizations. In this domain, one must move a stack of discs from one peg to another without stacking a larger disc onto a smaller one, three pegs are available in total. Problem instances for this domain show that the goal state cannot be achieved in an incremental way, requiring a plan to build and destroy partial towers several times, and then obtain the complete tower in the final peg. Domains with such particular behavior are not pruned as much as others by the Hamming distance as a heuristic function and have a visible color fluctuation between the gradient limits instead of a clear movement towards red, as seen in the Cartesian tree of Fig. 11.10. The Cartesian tree generates a more compact representation, while the radial tree highlights the side to which the heuristic gave priority during search, as seen in Fig. 11.11, where the top-left branch was not explored. Other domains may suddenly achieve a goal state from a mostly blue colored graph, in which all states are far away from the goal, as seen in Fig. 11.7, or incrementally achieving the goal clearly going from one extreme of the gradient to the other, as in the Logistics domain.

To better understand how the predicates are affected by the plan we use the Dovetail [14] metaphor. This particular Hanoi planning instance is solved by a 7-step plan, represented by the pieces labeled with numbers at the top, Fig. 11.8. Each piece has preconditions represented on the left side and effects represented on the right side. In this case we can see the first action, *move(d1 d2 peg3)*, moving a clear disc *d1* that starts on disc *d2* to a clear peg *peg3*, leaving *d2* clear and *peg3* not clear. We can see the predicate *clear d1* being tested by each odd-index action, revealing the pattern of movements related with the disc *d1*.

4.2 Case Study Survey Results

To evaluate WEB PLANNER, a group of four users from our automated planning course[4] were asked to fill a survey after using the tool to describe the *RPG* domain from the International Competition on Knowledge Engineering for Planning and Scheduling.[5] The survey contained the following questions and answers:

- How familiar are you with automated planning languages and algorithms?
  - Only 2 users have used PDDL before.
  - Did WEB PLANNER visualizations help you to find any bugs/errors/interesting points during the course of your task?
  - One user found missing preconditions.
  - Mark other planners/tools you used in your experiments:
  - Fast-Downard (1), JavaFF (1), JavaGP (3), Planning.domains (3), STRIPS-Fiddle (1)
  - Which features you missed the most?
  - Support more requirements (2), Auto-complete (1), Option to clear console (1), Find (common) errors in PDDL (1).

Results of system reaction show evidence of the utility of our tool, albeit with many suggested improvements, in Fig. 11.12 with minimum, maximum, and average represented. The current planning output must be improved in order to provide more meaningful messages about errors while taking advantage of the integrated editor to draw attention to specific lines where parsing errors were detected. Other improvements are more related to the editor itself, making it more

[4]https://github.com/pucrs-automated-planning/syllabus.

[5]https://ickeps2016.wordpress.com.

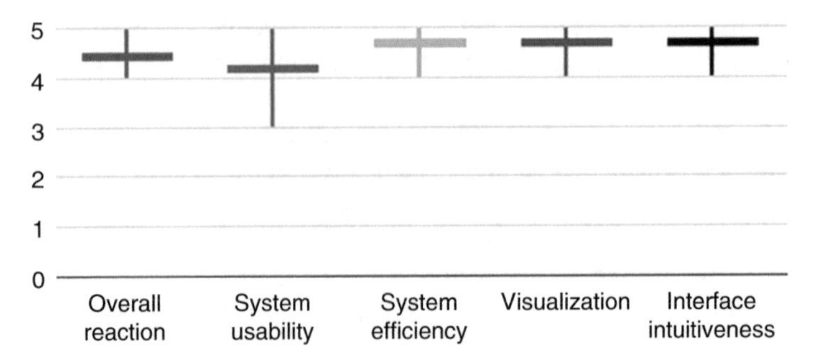

Fig. 11.12 Survey results, users were asked to evaluate the system between frustrating (0) and satisfying (5)

flexible to attend different user needs, such as theme, font size, and the ability to re-size each part of the editor. Users also asked for more planners/requirements to be supported.

4.3 General Public Usage Statistics

We collected anonymous data in WEB PLANNER from January 1 to May 30, 2019 to verify user habits. We identified users from multiple countries with varying session durations, with most users being in Brazil, where it was proposed as a classroom tool. World usage can be seen in Fig. 11.13

5 Related Work

We now discuss related work and tools to formalize and validate planning domains, visualize changes on a large amount of hierarchical data, and visualize state-space search algorithms.

`Planning.Domains`[6] is a collection of web tools for automated planning. These web tools provide a PDDL editor, an API that contains a wide collection of PDDL benchmark domain and problem files (most of them used on the International Planning Competition), and a planner in the cloud that allows using not only a planning solver, but also debugging tools, such as TorchLight [10], and even WEB PLANNER visualizations as plugins. Similar to our approach, `Planning.Domains` provides a PDDL editor, however, our approach provides

[6]http://planning.domains.

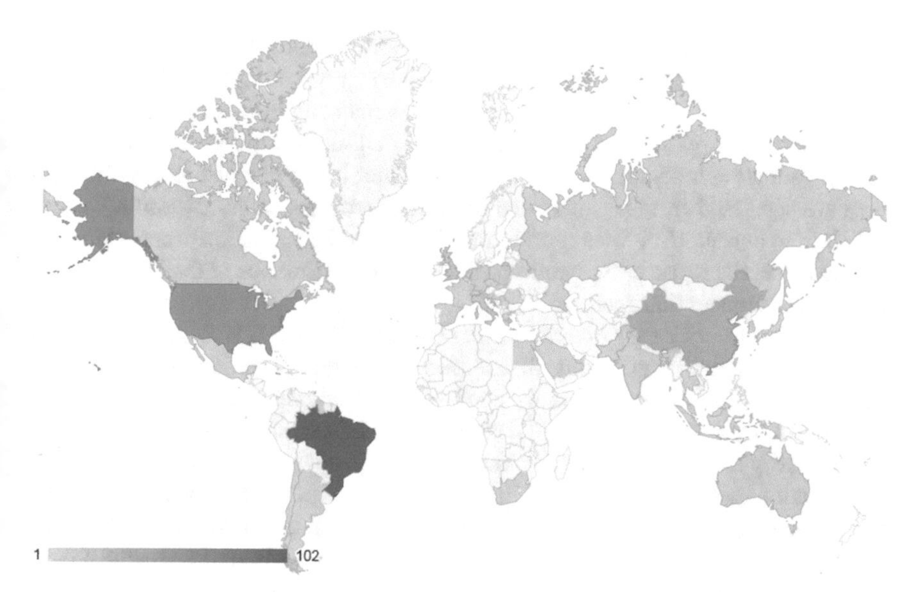

Fig. 11.13 User sessions per country during the first 4 months of 2019

not only a web editor with syntax highlighting, but also a set of tools to develop, analyze, and visualize planning domains using metaphors and alternative data visualization methods.

We consider two offline tools for PDDL file editing as related enough to our approach for comparison: `myPDDL`; and PDDL Studio. `myPDDL`[7] [22] is an editor extension for Sublime Text, which provides PDDL syntax highlighting, snippets, and domain visualization (e.g., diagram types). PDDL Studio [16] is an IDE to edit PDDL domains and problems. This IDE provides syntax highlighting, code completion, and context hints specifically designed for PDDL. Both tools have editor capabilities similar to ours, with `myPDDL` being able to generate type diagram and calculate distances automatically, two unique features that benefit only users that are either debugging typing errors or avoiding calculating distances in problems. PDDL Studio is able to list description errors and integrates with external planners using a command-line interface, leaving the user responsible for installation and call to each planner. While `myPDDL` and PDDL Studio are more flexible than our approach, being open or able to use any local planner, respectively, they need an initial setup phase that consumes valuable classroom time. One of our goals was to minimize the time spent to go from planning description explanation to planner call.

To validate a domain description one can follow the steps of a known valid plan to solve one problem and either achieve a goal state or discover errors in the domain

[7] https://github.com/Pold87/myPDDL.

description. An entire branch of plan validation tools was created from VAL[8] [12] to do this job automatically. More recent implementations, InVAL[9] and ReviVAL,[10] try to complement VAL, being independent implementations that can increase trust in domain descriptions and warn ambiguous PDDL descriptions to users. More PDDL validation tools means more interest in their usage in real-life activities, yet they are separated from planners and domain description tools. By adding a plan validator to our interface we expect to make not only the validation process simpler, but also essential to a user that wants an automated confirmation of their work, while bringing awareness that such tools exist.

Graphical Interface for Planning with Objects (GIPO) [21] is a tool for planning domain knowledge engineering that allows the textual specification of domains in PDDL and *Hierarchical Task Network* (HTN), like other code editors. Besides domain knowledge engineering, GIPO provides an animator tool to graphically inspect the plans produced by the internal planner, given a domain and problem specification. Like our approach, GIPO can use a set of plans to validate domain and problem specification, indicating whether the specification do support the given plans. Similar to Dovetail metaphor we implemented in WEB PLANNER, GIPO also provides an animator tool to visualize how a sequence of actions (i.e., a plan) connects to form a plan that achieves a goal state from an initial state. *VisPlan* [5] is an interactive tool to visualize and verify plans' correctness. This tool is closely related to Dovetail metaphor in the sense of helping planning users to better understand how a sequence of actions achieve a goal from an initial state. *VisPlan* identifies possible flaws (i.e., incorrect actions) in a plan, allowing users to manually modify this plan by repairing these identified flawed actions.

PDVer [17] is a methodology and tool that verifies if a PDDL domain satisfies a set of requirements (i.e., planning goals). This tool allows an automatic generation of these requirements from a *Linear Temporal Logic* (LTL) specification into a PDDL description. This tool is concerned with how the corresponding PDDL action constraints are translated from an LTL specification. *PDVer* provides a summary of test cases (positive and negative) indicating why a PDDL domain specification does not satisfy a set of requirements to achieve a goal. Our verification tests are only based on common PDDL mistakes and lack domain-dependent constraints.

itSIMPLE[11] [25] is concerned with domain modeling, using steps to guide the user from informal requirements (UML) to an objective representation (Petri Nets). *itSIMPLE* features a visualization and simulation tool to help understanding planning domains through diagrams. *itSIMPLE* uses UML diagrams to model planning instances and Petri Nets for validating planning instances. WEB PLANNER

[8]https://github.com/KCL-Planning/VAL.

[9]https://github.com/patrikhaslum/INVAL.

[10]https://github.com/guicho271828/ArriVAL.

[11]https://github.com/tvaquero/itsimple.

does not provide an incremental formalization approach to domain engineering, requiring users to start with PDDL descriptions and, once done, able to generate visualizations from it.

Magnaguagno et al. [14] developed a visual metaphor to help users visualize and learn how the planning process works. Dovetail results suggest that this visual metaphor can be useful to define and debug the planning process. We have applied this visual metaphor in WEB PLANNER by using colors each instantiated predicate in the state along a plan execution.

We found two approaches to data visualization suitable for heuristics. In [13], Kuwata and Cohen develop visualization methods to understand and analyze the search-space and behavior of heuristic functions, by exploring the usefulness of these methods on shaping state-space search. The heuristic functions they explore are A* and IDA*. Tu and Shen [23] propose a set of strategies to visualize and compare changes in hierarchical data using treemaps. We currently only support state-space non-cyclic graphs obtained from the planning process and no graph comparison, as abrupt layout changes would impact a side-by-side comparison as perceived by Tu and Shen. We opted for the current tree structure to obtain a visible contour visualization that better matches abstract explanations.

6 Conclusions

In this paper, we describe WEB PLANNER, a cloud-based planning tool we developed that consists of a PDDL editor to formalize planning domains and problems, and visualizations to help understand the effect of planning heuristics in the domains. This work aims to simplify the setup process required to execute planners while providing visualizations to better understand how domain differences and heuristics can impact the performance of the planner. Our small-scale survey indicated promising results with user-feedback pointing towards improvements and new features already in development.

As future work, we intend to support user-defined heuristics in our planner along with alternative options to the user, such as selectable color schemes for the visualization and a side-by-side state-space view for comparison. WEB PLANNER has being used in the lectures from the Artificial Intelligence and Automated Planning courses since August 2016 to explain planning concepts using both PDDL and visualizations to around 50 students every year while being available to anyone online, reaching over 300 accesses in the first quarter of 2019.

We believe that such web tool can help new heuristics to be developed and tested, providing to users a better grasp of the impact of heuristics to the state-space exploration, which is usually an invisible entity. WEB PLANNER tool is available online at: https://web-planner.herokuapp.com.

Acknowledgements We acknowledge the support given by CAPES/Pro-Alertas (88887.115590/2015-01) and CNPQ within process number 305969/2016-1 under the PQ fellowship.

This research was achieved in cooperation with HP Brasil Indústria e Comércio de Equipamentos Eletrônicos LTDA using incentives of Brazilian Informatics Law (Law n° 8.2.48 of 1991).

Part of this research was also financed by the Coordenação de Aperfeiçoamento de Pessoal de Nivel Superior—Brasil (CAPES)—Finance Code 001.

References

1. Bylander, T.: The Computational Complexity of Propositional STRIPS Planning. Journal of Artificial Intelligence Research (JAIR) **69**, 165–204 (1994)
2. Fikes, R.E., Nilsson, N.J.: STRIPS: A new approach to the application of theorem proving to problem solving. Journal of Artificial Intelligence Research (JAIR) **2**(3), 189–208 (1971)
3. Gerevini, A., Long, D.: Plan Constraints and Preferences in PDDL3. The Language of the Fifth International Planning Competition. Technical Report, Department of Electronics for Automation, University of Brescia, Italy (2005)
4. Ghallab, M., Nau, D.S., Traverso, P.: Automated Planning—Theory and Practice. Elsevier (2004)
5. Glinskỳ, R., Barták, R.: VisPlan–Interactive Visualisation and Verification of Plans. Proceedings of the Workshop on Knowledge Engineering for Planning and Scheduling (KEPS) pp. 134–138 (2011)
6. Ha, T.T.: Theory and design of digital communication systems. Cambridge University Press (2010)
7. Hamming, R.W.: Error detecting and error correcting codes. Bell System Technical Journal **29**(2), 147–160 (1950)
8. Helmert, M.: The Fast Downward Planning System. Journal of Artificial Intelligence Research **26**, 191–246 (2006)
9. Hoffmann, J.: The Metric-FF Planning System: Translating "Ignoring Delete Lists" to Numeric State Variables. Computing Research Repository (CoRR) **abs/1106.5271** (2011), http://arxiv.org/abs/1106.5271
10. Hoffmann, J.: The TorchLight Tool: Analyzing Search Topology Without Running Any Search. In: Proceedings of the System Demonstrations, in the 21th International Conference on Automated Planning and Scheduling. pp. 37–41 (2011)
11. Hoffmann, J., Nebel, B.: The FF Planning System: Fast Plan Generation Through Heuristic Search. Journal of Artificial Intelligence Research (JAIR) **14**(1), 253–302 (May 2001)
12. Howey, R., Long, D., Fox, M.: VAL: automatic plan validation, continuous effects and mixed initiative planning using PDDL. In: 16th IEEE International Conference on Tools with Artificial Intelligence (ICTAI 2004), 15–17 November 2004, Boca Raton, FL, USA. pp. 294–301 (2004)
13. Kuwata, Y., Cohen, P.R.: Visualization Tools for Real-Time Search Algorithms. Computer Science Technical Report (1993)
14. Magnaguagno, M.C., Pereira, R.F., Meneguzzi, F.: DOVETAIL - An Abstraction for Classical Planning Using a Visual Metaphor. In: Proceedings of FLAIRS, 2016. (2016), http://www.aaai.org/ocs/index.php/FLAIRS/FLAIRS16/paper/view/12966
15. McDermott, D., Ghallab, M., Howe, A., Knoblock, C., Ram, A., Veloso, M., Weld, D., Wilkins, D.: PDDL – The Planning Domain Definition Language. Technical Report – Yale Center for Computational Vision and Control (1998)
16. Plch, T., Chomut, M., Brom, C., Barták, R.: Inspect, edit and debug PDDL documents: Simply and efficiently with PDDL Studio. In: Proceedings of ICAPS'09. pp. 15–18 (2012)

17. Raimondi, F., Pecheur, C., Brat, G.: PDVer, a Tool to Verify PDDL Planning Domains. In: Proceedings of ICAPS'09 Workshop on Verification and Validation of Planning and Scheduling Systems, Thessaloniki, Greece (2009)
18. Reingold, E.M., Tilford, J.S.: Tidier drawings of trees. IEEE Transactions on Software Engineering (2), 223–228 (1981)
19. Richter, S., Westphal, M.: The LAMA planner: Guiding cost-based anytime planning with landmarks. Journal of Artificial Intelligence Research (JAIR) **39**(1), 127–177 (2010)
20. Russell, S., Norvig, P.: Artificial Intelligence: A Modern Approach. Prentice Hall Press, Upper Saddle River, NJ, USA, 3rd edn. (2009)
21. Simpson, R.M., Kitchin, D.E., McCluskey, T.L.: Planning domain definition using GIPO. Knowledge Eng. Review **22**(2), 117–134 (2007). https://doi.org/10.1017/S0269888907001063
22. Strobel, V., Kirsch, A.: Planning in the Wild: Modeling Tools for PDDL. In: Joint German/Austrian Conference on Artificial Intelligence. pp. 273–284. Springer (2014)
23. Tu, Y., Shen, H.W.: Visualizing Changes of Hierarchical Data using Treemaps. IEEE Transactions on Visualization and Computer Graphics **13**(6), 1286–1293 (Nov 2007). https://doi.org/10.1109/TVCG.2007.70529
24. Vallati, M., Chrpa, L., McCluskey, T.L.: What you always wanted to know about the deterministic part of the International Planning Competition (IPC) 2014 (but were too afraid to ask). Knowledge Engineering Review **33**, 383 (2018)
25. Vaquero, T., Tonaco, R., Costa, G., Tonidandel, F., Silva, J.R., Beck, J.C.: itSIMPLE 4.0: Enhancing the modeling experience of planning problems. In: Proceedings of ICAPS'12. pp. 11–14 (2012)
26. Ward, M.O., Grinstein, G., Keim, D.: Interactive Data Visualization: Foundations, Techniques, and Applications, Second Edition - 360 Degree Business. A. K. Peters, Ltd., Natick, MA, USA, 2nd edn. (2015)
27. Wilson, J.M.: Gantt charts: A centenary appreciation. European Journal of Operational Research **149**(2), 430–437 (2003)

Part III
Case Studies and Applications

Chapter 12
Design of Timeline-Based Planning Systems for Safe Human-Robot Collaboration

Andrea Orlandini, Marta Cialdea Mayer, Alessandro Umbrico, and Amedeo Cesta

Abstract During the last decade, industrial collaborative robots have entered assembly cells supporting human workers in repetitive and physical demanding operations. Such human-robot collaboration (HRC) scenarios entail many open issues. The deployment of highly flexible and adaptive plan-based controllers is capable of preserving productivity while enforcing human safety is then a crucial requirement. The deployment of plan-based solutions entails knowledge engineers and roboticists interactions in order to design well-suited models of robotic cells considering both operational and safety requirements. So, the ability of supporting knowledge engineering for integrating high level and low level control (also from non-specialist users) can facilitate deployment of effective and safe solutions in different industrial settings. In this chapter, we will provide an overview of some recent results concerning the development of a task planning and execution technology and its integration with a state of the art Knowledge Engineering environment to deploy safe and effective solutions in realistic manufacturing HRC scenarios. We will briefly present and discuss a HRC use case to demonstrate the effectiveness of such integration discussing its advantages.

1 Introduction

During the last decade, *industrial robotic* systems have entered assembly cells supporting human workers in repetitive and physical demanding operations. The co-presence of robots and humans in a shared environment entails many issues to be

A. Orlandini (✉) · A. Umbrico · A. Cesta
Institute of Cognitive Science and Technology, CNR – National Research Council of Italy, Rome, Italy
e-mail: andrea.orlandini@istc.cnr.it

M. Cialdea Mayer
Department of Engineering, University "Roma TRE", Rome (RM), Italy
e-mail: cialdea@ing.uniroma3.it

M. Vallati, D. Kitchin (eds.), *Knowledge Engineering Tools and Techniques for AI Planning*, https://doi.org/10.1007/978-3-030-38561-3_12

properly addressed requiring *robust controllers* capable of preserving *productivity* and enforcing *human safety* [1]. Namely, Human-Robot Collaboration (HRC) scenarios entail challenges on *physical interactions*, to always guarantee *safety* of human operators, and activities *coordination*, to improve cell productivity. Thus, the deployment of highly flexible and adaptive controllers, capable of preserving productivity while enforcing human safety, is a crucial requirement.

Flexible plan-based solutions are a key enabling feature of HRC controllers where robot motions must be continuously adapted to the presence of humans, which act as uncontrollable "agents" in the environment. Their presence entails the ability of evaluating robot execution time variability and, in this sense, standard control methods are not fully effective. Moreover, the integration of Planning and Scheduling (P&S) technology with Knowledge Engineering solutions and, more specifically, with Verification and Validation (V&V) techniques is a key element to synthesize safety critical systems in robotics [2]. Indeed, the deployment of plan-based solutions requires domain experts (i.e., production engineers), knowledge engineers, and roboticists to deeply interact in order to design well-suited models of robotic cells considering both operational and safety requirements [3]. Therefore, the ability of supporting knowledge engineering for integrating high level and low level control (also from non-specialist users) can facilitate deployment of effective and safe solutions in different industrial settings.

Since a decade, a research initiative has been started to investigate the possible integration of a timeline-based planning framework (APSI-TRF [4]) and V&V techniques based on Timed Game Automata (TGA) [5] to automatically synthesize a robot controller that guarantees robustness and safety properties [6, 7]. Indeed, some plan-based controllers rely on temporal planning mechanisms capable of dealing with coordinated task actions and temporal flexibility (e.g., [8, 9]) that leverage temporal planners (e.g., [10, 11]). Unfortunately, these systems do not allow an explicit representation of *uncontrollability* features. Consequently, the resulting controllers are not endowed with the *robustness* needed to deal with the *temporal uncertainty* of HRC scenarios and controllability issues [12]. These system usually rely on *replanning mechanisms* that may however strongly penalize the production performance. The long-term research goal is to realize a robust task planning system enabling flexible, safe, and efficient HRC. In [13], the general pursued approach is presented aiming at realizing controllers capable to dynamically coordinate tasks according to the behaviors of human workers.

This chapter provides an overview of the results collected in the last decade concerning the development of a task planning and execution framework (PLATINUM) [14, 15] and its integration with a Knowledge Engineering ENvironment (KEEN) describing its deployment in safe and effective solutions for manufacturing HRC scenarios. KEEN [16] is a knowledge engineering environment with Verification and Validation features based on Timed Game Automata model checking. PLATINUM is a timeline-based planning systems capable of supporting flexible temporal planning and execution with uncertainty. Also, more recent results on the development of a task planning and execution technology deployed in realistic manufacturing scenarios will be presented. Specifically, the paper presents an Engineering & Control Environment which integrates a task planning system with an *engineering environment* taylored to support robust human-robot collaboration.

2 Fostering Autonomy via Timeline-Based Planning and Execution

Planning for real world problems with explicit temporal constraints is a challenging problem. Among different approaches, the use of flexible timelines in Planning and Scheduling (P&S) has been shown to be successful in a number of concrete applications, such as, for instance, autonomous space systems [17–19]. Timeline-based planning has been introduced by Muscettola [19], under a modeling assumption inspired by classical control theory. A planning problem is modeled by identifying a set of relevant components whose temporal evolution must be controlled to obtain a desired behavior. Components represent logical or physical subsystems whose state may vary over time. The behavior of the domain features under control is modeled as temporal functions whose values have to be decided over a temporal horizon. Such functions are synthesized during problem solving by posting planning decisions. The evolution of a single temporal feature over a temporal horizon is called the *timeline* of that feature.

In general, plans synthesized by temporal P&S systems may be (1) temporally flexible and (2) not fully controllable. Time flexibility reflects on modeling plans as made up of *flexible* timelines, describing transition events that are associated with temporal intervals (with given lower and upper bounds), instead of exact temporal occurrences. In other words, a flexible plan describes an envelope of possible solutions with the aim of facing uncertainty during actual execution. As a matter of fact, many P&S architectures return flexible plans, which are commonly accepted to be less brittle than fully specified plans, when coping with execution. The second above-mentioned property is due to the fact that not every value transition in a plan is under the system control, as events exist that depend on the *environment*. The execution of a flexible plan is usually under the responsibility of an executive system that forces value transitions over the timelines dispatching commands to the concrete system, while continuously accepting feedback and, thus, monitoring plan execution. In such cases, the execution time of controllable tasks should be chosen so that they can face uncontrollable events. This is known as the *controllability problem* [20].

2.1 A Theoretical Framework

After several attempts, a formal framework has been presented to provide a unique theoretical background on timelines [21]. This section provides a brief informal overview of the basic notions regarding flexible timelines and plans.

A timeline-based planning domain contains the characterization of a set of *state variables*, representing the components of a system. A *state variable* x is characterized by the set of values it may assume, denoted by $\text{values}(x)$, possible upper and lower bounds on the duration of each value, and rules governing the correct sequencing of such values. A *timeline* for a state variable is made up of a finite sequence of valued intervals, called *tokens*, each of which represents a time slot where the variable assumes a given value. In general, timelines may be *flexible*, i.e., the start and end times of each of its tokens are not necessarily fixed time points, but may range in given intervals. For the sake of generality, temporal instants and durations are taken from an infinite set of non-negative numbers $\mathbb{T}$, including 0. The notation $\mathbb{T}^\infty$ will be used to denote $\mathbb{T} \cup \{\infty\}$, where $t < \infty$ for every $t \in \mathbb{T}$.

Tokens in a timeline for the state variable x are denoted by expressions of the form x^i, where the superscript indicates the position of the token in the timeline. Each token x^i is characterized by a value $v_i \in \text{values}(x)$, an end time interval $[e_i, e'_i]$ referred to as $\text{endtime}(x^i)$, and a duration interval $[d_i, d'_i]$ (as usual, the notation $[x, y]$ denotes the closed interval $\{t \mid x \leq t \leq y\}$). The start time interval $\text{starttime}(x^i)$ of the token x^i is $[0, 0]$ if x^i is the first token of the timeline (i.e., $i = 1$), otherwise, if $i > 1$, $\text{starttime}(x^i) = \text{endtime}(x^{i-1})$. So, a token has the form $x^i = (v_i, [e_i, e'_i], [d_i, d'_i])$ and a timeline is a finite sequence of tokens $x^1, \ldots, x^k$. The metasymbol FTL (FTL_x) will henceforth be used to denote a timeline (for the state variable x), and **FTL** to denote a set of timelines. Being tokens flexible, their exact start end times will be decided at execution time. Tokens can be either *controllable* (the controller can decide both their start and end time), or *uncontrollable* (both start and end time depend on the environment's choices), or *partially controllable* (the controller can decide when to start them , but their exact duration is outside the system's control). Each token is consequently equipped also with a *controllability tag*, identifying the class it belongs to.

A *scheduled timeline* is a particular case where each token has a singleton $[t, t]$ as its end time, i.e., the end times are all fixed. A *schedule* of a timeline FTL_x is essentially obtained from FTL_x by narrowing down token end times to singletons (time points) in such a way that the duration requirements are fulfilled. In a given timeline-based domain, the behavior of state variables may be restricted by requiring that time intervals with given state variable values satisfy some temporal constraints. Such constraints are stated as a set of *synchronization rules* which relate tokens on possibly different timelines through temporal relations between intervals or between an interval and a time point. These temporal relations refer to token start or end points that will henceforth be called *events*. If **FTL** is a set of timelines and $\text{tokens}(\mathbf{FTL})$ the set of the tokens in **FTL**, then the set $\Upsilon(\mathbf{FTL})$ of the *events* in **FTL** is the set containing all the expressions of the form $\text{starttime}(x^i)$ and $\text{endtime}(x^i)$ for $x^i \in \text{tokens}(\mathbf{FTL})$. A *temporal relation* on tokens has then one of the following forms:

$$p \leq_{[lb,ub]} p' \quad p \leq_{[lb,ub]} t \quad t \leq_{[lb,ub]} p$$

where $p, p' \in \Upsilon(\mathbf{FTL})$, $t, lb \in \mathbb{T}$ and $ub \in \mathbb{T}^\infty$.

Intuitively, $p \leq_{[lb,ub]} p'$ states that the token start/end point denoted by p occurs from lb to ub time units before that denoted by p'; $p \leq_{[lb,ub]} t$ states that the token start/end point denoted by p occurs from lb to ub time units before the time point t and the third relation that it occurs from lb to ub time units after t. Other relations between tokens [22] can be defined in terms of the primitive ones, e.g., x^i before$_{[lb,ub]}$ y^j is the same as endtime$(x^i) \leq_{[lb,ub]}$ starttime(y^j); x^i during$_{[lb_1,ub_1][lb_2,ub_2]}$ y^j can be defined as starttime$(y^j) \leq_{[lb_1,ub_1]}$ starttime(x^i) and endtime$(x^i) \leq_{[lb_2,ub_2]}$ endtime(y^j); a *contains* relation is its converse: x^i contains$_{[lb_1,ub_1][lb_2,ub_2]}$ y^j if and only if y^j during$_{[lb_1,ub_1][lb_2,ub_2]}$ x^i.

Temporal relations are also used to state the synchronization rules of the planning domain. Here, it is sufficient to say that such rules allow to state requirements of the following form: for every token x_0^i where the state variable x_0 assumes the value v_0, there exist tokens $x_1^{i_1}, \ldots, x_n^{i_n}$ where the state variables $x_1, \ldots, x_n$ hold some given specified values, and all these tokens are related one to another by some given temporal relations. Unconditioned synchronization rules are also allowed and are useful for stating both domain invariants and planning goals.

A *flexible plan* Π is a pair (**FTL**, $\mathcal{R}$), where **FTL** is a set of timelines and $\mathcal{R}$ is a set of temporal relations, involving tokens in some timelines in **FTL**. An *instance* of the flexible plan $\Pi = (\mathbf{FTL}, \mathcal{R})$ is any schedule of **FTL** that satisfies every relation in $\mathcal{R}$. In order for a flexible plan $\Pi = (\mathbf{FTL}, \mathcal{R})$ to satisfy a synchronization rule it must be the case that $\mathcal{R}$ contains temporal relations guaranteeing what the rule requires. For the formal definitions the reader is again referred to [21], where it is also proved that whenever a flexible plan satisfies (in this sense) all the synchronization rules of a domain, then also any of its instances does.

2.2 *PLATINUm: A Timeline-Based Planning and Acting Framework*

Born as a follow-up of APSI-TRF, PLATINUM[1] is a general-purpose timeline-based planning and execution framework capable of dealing with *temporal uncertainty* and *controllability issues* [14, 23] that complies with the formalization given in Sect. 2.1. PLATINUM is able to deal with *uncontrollable dynamics* at both planning and execution time. Its solving process pursues a *plan refinement approach* which consists in iteratively refining a partial plan by reasoning in terms of *flaws* that must be solved. Flaw selection is supported by dedicated heuristics that guide the planning procedure. A PLATINUM-based planner relies on a set of data structures and algorithms called, respectively, *components* and *resolvers*. Components model the types of features that may compose a planning domain. They specify the set of states and constraints that characterize the temporal behaviors of a particular type

[1] https://github.com/pstlab/PLATINUm.

of domain feature. Resolvers are dedicated algorithms that encapsulate the logic for building valid temporal behaviors of a particular component. The reader may refer to [14, 23] for a more detailed description of the framework and the solving approach. However, it is important to point out that resolvers are not responsible for making decisions during the search process. They are responsible for detecting flaws on a component and computing all possible solutions of such flaws in order to guarantee completeness of the search. Each solution of a flaw represents a branch in the search and it is up to the planner deciding which flaw to solve and which solution to apply for search expansion (i.e., plan refinement). The types of flaws a PLATINUM-based planner is capable to deal with depend on the set of components and resolvers available in the framework. PLATINUM provides *state variables components* and the related resolvers that allow a planner to build valid *timelines* according to the semantics proposed in [21]. Thus, PLATINUM has been extended by adding new components and new resolvers in order to properly deal with *discrete* and *reservoir* resources [15].

3 KEEN: Knowledge Engineering ENvironment

In order to foster the deployment of reliable timeline-based P&S applications, the development of a Knowledge Engineering ENvironment (KEEN) for timeline-based planning has been pursued [16]. Here, the context in which the needs for a new tool emerged is described, and the main requirements for the new environment are presented. Then, the core design choices that lead the development of KEEN are discussed, divided accordingly to the features that they are intended to support.

KEEN is an open source software released under the Eclipse Public License, version 1.0, and as such its source code can be downloaded from its GitHub repository.[2] There were no tools for timeline-based planning supporting graphical modeling of the solution, and neither domain validation and plan verification; this alone might justify the creation of a new tool to specifically fill this niche. However, the motivations that led to the development of KEEN were due to some specific needs that arose in the Planning and Scheduling Technology (PST) Laboratory at the National Research Council of Italy (CNR-ISTC). Specifically, some research done in the field of Verification and Validation (V&V) for timeline-based planning [2, 6, 24]: specifically, a formalism has been developed to translate planning domains in the form of Timed Game Automata (TGA) [5], so as to make it possible to employ existing model checkers to verify the translated domain. Regarding plan verification, other work was done [24, 25] to also encode flexible plans generated by a timeline-based planner in the form of Timed Game Automata, thus making it possible to verify generated plans in the same way as mentioned earlier. Additionally, at the

[2]https://github.com/ugilio/keen. The reader interested in knowing how to use KEEN can find a detailed description at https://ugilio.github.io/keen/userguide.

PST there was the need to more efficiently exploit the technologies underlying the planners and frameworks that have been developed there during the years; a common language to express domains and problems, DDL, did exist, but there were no integrated tools to ease the development of timeline-based systems; this situation was far less than ideal because the work practice required to write domains using standard text editors, then manually invoking a planner on the files, and possibly execute tools to translate domains and plans to TGA encoding and running other tools to verify them; this was cumbersome and time-consuming in the first place; maintaining systems or refining them after some time had passed was a daunting task; if this development style was already very complicated for planning experts, it was a real stopper for beginners who approached the field for the first time.

This suggested that a tool to make developing comfortable was an absolute minimum; moreover, given the availability of Knowledge Engineering tools with support for graphical modeling in other fields of Automated Planning and Scheduling, it would have been desirable to have one for timeline-based planning too. And finally, a truly integrated environment had to also incorporate the work done on validation and verification to make it accessible to the developer in a simple and productive form.

3.1 Knowledge Engineering and Verification and Validation Features in KEEN

Requirements for the new environment were elicited by interviewing the potential new users of the system: developers of planning domains that already have a deep knowledge of timeline-based planning, and their colleagues which were marginally interested in writing code in detail, but were more concerned about the ability of sketching domains to be further refined by planning experts, or which needed to visualize the high-level information about the work done by them. A list of broad, high-level requirements has been identified as the following: provide all the traditional features of a modern Integrated Development Environment, i.e., (a) syntax highlighting, (b) code assist, (c) tree-view of syntactic elements, (d) error detection; leverage existing, well-known tools already in use at PST; support graphical editing of domains. Support Round-Trip Engineering, i.e., (a) be able to edit both the textual and graphical representations in a synchronized way, (b) without having to import and export code to/from the graphical representation, (c) synchronization between the views should be automatic; make it possible to use existing planners from the environment; support code sharing using popular version control systems; enable users to easily validate domains and verify plans. These requirements all refer to the quite common and well-known field of integrated development environments, given that these tools are commonly used in their daily work; for these reason, it has not been necessary to conduct in-depth analyses to better explore the domain of application of some requirements: instead, it has been enough clarify some aspects via informal conversations and releasing often early versions of the product to have a continuous feedback on the work being done (Fig. 12.1).

KEEN, built around APSI-TRF and now applied to PLATINUM, was designed as a set of active services to support knowledge engineering. We distinguish between two different service layers, i.e., a set of *Knowledge Engineering Services* (upper part of Fig. 12.2) and a set of V&V *services* (lower part of Fig. 12.2). The *Knowledge Engineering Services* provide "classical" KE functionalities specifically developed for timeline-based planning. At present, this part is composed of (1) a *Domain/Problem Editing and Visualization* module that supports synthesis and modification of planning models and (2) a *Plan Editing and Visualization* module

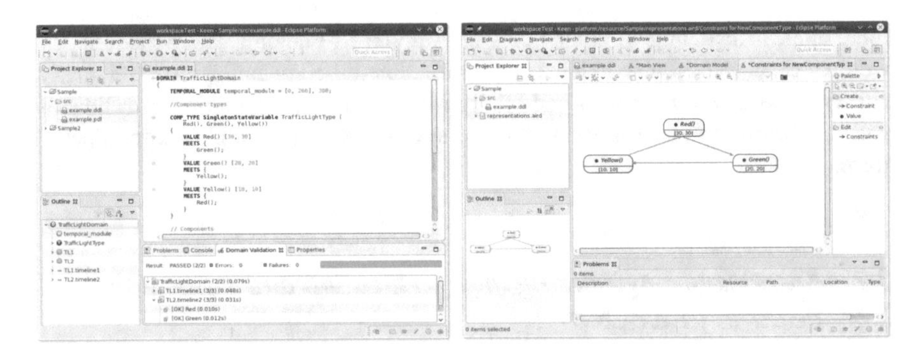

Fig. 12.1 KEEN graphical interface: eclipse-generated text highlights (left-side); a graphical view of the planning domain (right-side)

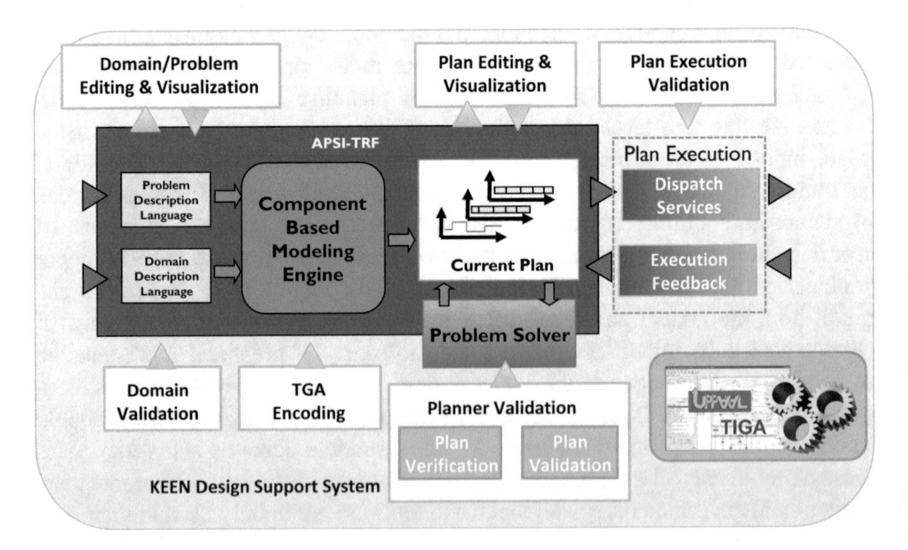

Fig. 12.2 The general V&V architecture in KEEN

that helps inspection, analysis, and direct manipulation of solution plans. The V&V *Services* contribute to the KEEN tool with a set of fully automated V&V features obtained by making operational some research results presented in [7, 25, 26]. The V&V functionalities are all based on model checking for Timed Game Automata (TGA) and rely on UPPAAL-TIGA [27] as the verification engine. UPPAAL-TIGA extends UPPAAL [28] by providing a toolbox for the specification, simulation, and verification of real-time games. Somehow such model checker constitutes an additional core engine for KEEN. The general KEEN concept is depicted in Fig. 12.2.

A *TGA Encoding* module provides the basic TGA automatic translation for P&S specification [25] which constitutes the basis for implementing the KEEN V&V services: (1) *Domain Validation* supports the model building activity allowing to check the P&S model with respect to system requirements; (2) *Planner Validation* assesses the P&S solver with respect to a given set of requirements. In this regard, two sub-modules are further deployed, i.e., *Plan Verification* to verify the correctness of the solution plans and *Plan Validation* to evaluate their adequacy; finally, (3) a TGA-based approach to *Plan Controller Synthesis* [7] is able to enforce robust execution of solution plans through the generation of robust plan controllers.

The pursued idea is the integration of KEEN with either an accurate simulator of a real environment or a real physical system (e.g., a robot). In this context, it is possible to take advantage of all KEEN functionalities during the different design phases, i.e., from initial design to actual solution execution and continuously exploiting the combination of "classical" KE and V&V functionalities. Users can also ask for solution plan generation by means of the KEEN functionalities. Indeed, exploiting the PLATINUM capabilities, this may be performed by means of specific planners. As for the domain, plan representation is completely handled by PLATINUM and a specific language generation component is deputed to the generation of a source file encoded with a *Problem Description Language* syntax. The user can then modify the generated solution plan and ask KEEN to perform V&V tasks.

4 Deploying Task Planning Solutions for Safe Human-Robot Collaboration

In order to deploy reliable task planning solutions for safe human-robot collaboration, we may consider a collaborative robotic workcell as a bounded connected space with two agents located in it, a human and a robot, and their associated equipment [29]. The robot system in a workcell consists of a robotic arm with its tools, its base, and possibly additional support equipment. In such a workcell four different degrees of interaction between a human operator and the robot can be defined [30] in which they may occupy the same spatial location and interact according to different modalities: *Independent*, the human and the robot operate on separate workpieces without collaboration, i.e., independently from each other; *Synchronous*, the human and the robot operate on sequential components of the same workpiece, i.e., one can start a task only after the other has completed a preceding task; *Simultaneous*, the human and the robot operate on separate tasks on the same workpieces at the same time; *Supportive*, the human and the robot work cooperatively to complete the processing of a single workpiece, i.e., they work simultaneously on the same task. Different interaction modalities entail the robot to be endowed with different safety settings while executing tasks.

4.1 A Specific Human-Robot Collaboration Case Study

In manufacturing, different production processes can be performed with HRC solutions, i.e., assembly/disassembly of parts, welding operations, large parts management, machine tending, etc. Among these, here we describe a specific case study considered in a research project named FourByThree[3] [31]. This case study corresponds to a real production industry with different relevant features for our perspective (e.g., space sharing, collaboration, interaction needs, etc.). The overall production process consists of a metal die which is used to produce a wax pattern in an injection machine. Once injected, the pattern is taken out of the die. Several patterns are assembled to create a cluster. The wax assembly is covered with a refractory element, creating a shell (this process is called investing). The wax pattern material is removed by the thermal or chemical means. The mould is heated to a high temperature to eliminate any residual wax and to induce chemical and physical changes in the refractory cover. The metal is poured into the refractory mould. Once the mould has cooled down sufficiently, the refractory material is removed by impact, vibration, and high pressure water-blasting or chemical dissolution. The casting is then cut and separated from the runner system. Other post-casting operations (e.g., heat treatment, surface treatment or coating, hipping) can be carried out, according to customer demands.

Here, we focus on the first step (preparation of the die for wax injection and extraction of the pattern from the die) which is a labor demanding operation that has a big impact on the final cost of the product. Specifically, the operation consists of the following steps: (1) mount the die; (2) inject the wax; (3) open the die and remove the wax; (4) repeat the cycle for a new pattern starting back from step (1). The most critical sub-operation is the opening of the die because it has a big impact on the quality of the pattern. In this context, the involvement of a collaborative robot has been envisaged to help the operator in the *assembly/disassembly* operation.

Due to the small size of the dies and the type of operations done by the worker to remove the metallic parts of the die, it is very complex for the robot and the worker to operate on the die simultaneously. However, both of them can cooperate

[3]http://www.fourbythree.eu.

in the assembly/disassembly operation. Once the injection process is finished, the die is taken to the workbench by the worker. The robot and the worker unscrew the bolts of the top cover. There are nine bolts, the robot starts removing those closer to it, and the worker the rest. The robot unscrews the bolts on the cover by means of a pneumatic screwdriver. The worker removes the top cover and leaves it on the assembly area (a virtual zone that will be used for the re-assembly of the die). The worker turns the die to remove the bottom die cover. The robot unscrews the bolts on the bottom cover by means of a pneumatic screwdriver. Meanwhile the operator unscrews and removes the threaded pins from the two lateral sides to release the inserts. The worker starts removing the metallic inserts from the die and leaves them on the table. Meanwhile, the robot tightens the parts to be assembled/reassembled together screwing bolts. The worker re-builds the die. The worker and the robot screw the closing covers. Thus the human and the robot must collaborate to perform assembly/disassembly on the same die by suitably handling different parts of the die and screwing/unscrewing bolts. Specifically, the human worker has the role of leader of the process while the robot has the role of subordinate with some autonomy. Moreover, the robot must be able to manage a screwdriver device and monitor the human location and its activities.

4.2 An Engineering and Control Architecture for HRC

Given the HRC scenarios described above, many features and constraints must be considered by the envisaged control architecture in order to realize an effective, robust, and safe collaboration. Such architecture must implement a well suited tradeoff among the requirements of the different stakeholders involved into the production process, i.e., a *Production Engineer*, a *Knowledge Engineer*, and a *Human Worker* in addition to the specific *Robot* requirements. The *Production Engineer* is the expert of the production needs and specifies operational requirements of the different processes that can be performed. The *Knowledge Engineer* knows the features of the robot and of the specific working environment and, therefore, is responsible to model the production processes according to specified operational requirements. The *Human Worker* and the *Robot* are the main actors that actually carry out the production tasks to achieve the production process. In general, several production processes can be performed within a factory. Each process consists of a set of *tasks* that must be executed according to some operational requirements. The perspective pursued here is the following: a *Worker* and a *Robot* represent two *autonomous agents* capable of executing different types of task. Some tasks can be executed only by the human, some tasks can be executed only by the robot, and some tasks can be executed both by the human and the robot. Thus, given a particular process, the control system is responsible for synthesizing the set of needed tasks to complete the working process, assigning tasks to the human and to the robot, and guaranteeing to *robustly and safely* executing them.

Figure 12.3 shows the envisaged FOURBYTHREE Engineering & Control Architecture [32] developed for flexible human-robot collaboration and implemented by means of PLATINUM and KEEN. The architecture shows the elements and the actors involved within the control loop as well as their relationships. Specifically, the labeled arrows describe all the phases of the *control process* starting from domain modeling up to physical task execution. The *FbT Engineering Environment* relies on KEEN (*Knowledge Engineering ENvironment*) [16] to support domain experts in the design of the control model exploited by the *FbT Controller* to coordinate the human and the robot tasks. Specifically, the *FbT Engineering Environment* allows the *Production Engineer* and/or the *FbT Knowledge Engineer* to model the working environment and the production processes without knowing in details the specific planning and execution technology utilized. Once the model is defined, the *FbT Task Planner* synthesizes a temporal flexible plan assigning tasks to the human and to the robot and the *FbT Plan Executive* executes such plans in order to achieve the production goals. Both the *FbT Task Planner* and the *FbT Plan Executive* rely on PLATINUm. Specifically, the developed task planner is capable of generating temporally robust plan by dealing with *temporal uncertainty* at solving time. This is crucial in the considered scenarios where a human must tightly cooperate with a robot. Indeed, a human is *uncontrollable* and his/her behavior may affect also the behavior of the robot from the control perspective. Thus, the *Human* is modeled as an autonomous and completely *uncontrollable* agent whose behavior may affect the behavior of the *Robot* which is modeled as a *partially controllable* agent.

A plan is executed by *dispatching* commands to the robot and to the human and by receiving *feedback* through dedicated communication channels implemented on

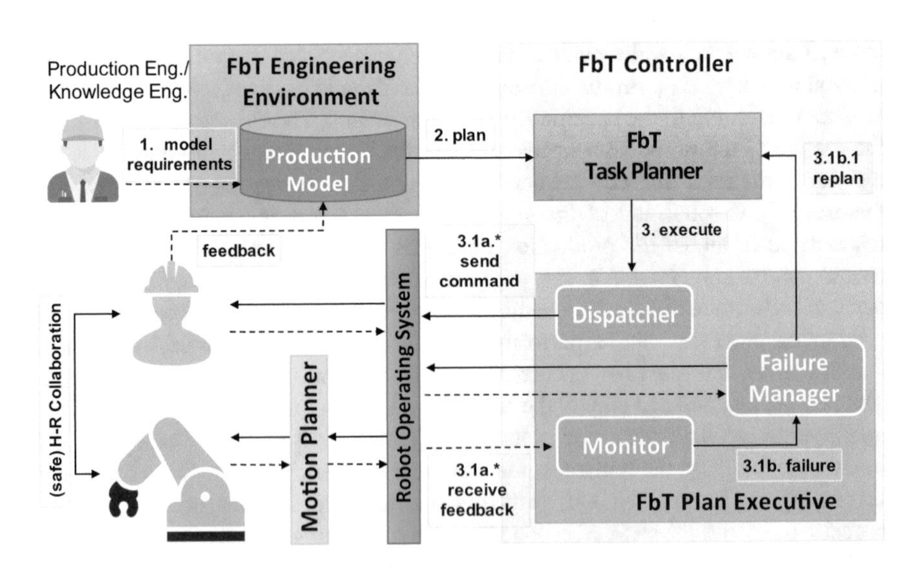

Fig. 12.3 The FOURBYTHREE engineering and control architecture

ROS.[4] The *FbT Plan Executive* realizes a closed-loop control process which puts the human-in-the-loop. Broadly speaking, the executive is capable of dynamically adapting a task plan (i.e., robot task execution) according to the detected behavior of the human. Thus, the executive can temporally adapt a task plan by *absorbing* execution delays and generate a new plan through *replanning* only if strictly needed. Replanning allows the executive to manage exogenous events the plan cannot capture, e.g., a failure of a robot actuator or a human task whose duration is longer than expected and synthesizes a new (adapted) plan which tries to complete the execution of the process. It is worth pointing out that the integration of *temporal uncertainty* at both planning and execution time makes the control process more *robust* than classical approaches in the literature, e.g., T-REX [8] or IXTET-EXEC [9], limiting the need for replanning.

4.3 *The* FOURBYTHREE *Controller*

The *FbT Controller* is the element responsible to actually carry out production processes and to coordinate the robot and the human. The synthesized tasks and the coordination of the human and the robot must follow the operational requirements specified by the *Production Engineer* and encoded into the *domain model* through KEEN. As Fig. 12.3 shows, the controller is composed by the *FbT Task Planner* and the *FbT Plan Executive* both relying on the timeline-based formalisms. The *FbT Task Planner* is responsible for generating the set of tasks needed to perform the production processes according to the desired requirements. In HRC scenarios, it is necessary to guarantee the safety of the human without penalizing the productivity of the factory. The task planner is in charge of finding a tradeoff between performance and safety and therefore there are several features to take into account when synthesizing plans.

The planning model can be characterized according to three different levels of abstraction: (1) the *supervision level*; (2) the *coordination level*; (3) the *implementation level*. At the *supervision level*, the task planner has to decide the set of tasks needed to execute the production process by modeling the operational requirements specified by the *Production Engineer*. At the *coordination level*, the task planner has to decide who, between the human and the robot, must perform each task harmonizing the activities of both. In this context, the human and the robot are modeled as *two autonomous agents* capable of executing some types of task. Given a production process, some tasks can be performed only by the human, some tasks can be performed only by the robot, and some tasks are *free* to be performed either by the human or by the robot. This choice-point represents the main *branching factor* of the task planning process. It can affect the *quality* of the collaboration and the efficiency of processes. Finally, at the *implementation level*, the task planner has to

[4]http://www.ros.org/.

decide the operations the robot must perform in order to execute the assigned tasks. According to the particular type of collaboration decided at coordination level, the task planner decides the most appropriate *execution modality* of the tasks of the robot in order to preserve the safety of the human.

Figure 12.4 (automatically generated by KEEN) shows an example of a timeline-based planning model for the collaborative assembly scenario. The model is hierarchically organized according to the three levels of abstraction identified (i.e., *supervision*, *coordination*, *implementation*). The *ALFA* (i.e., the name of the pilot plant) and *AssemblyProcess* state variables compose the *supervision level* of the model. These variables characterize the considered production context in terms of tasks that can be executed. The *AssemblyProcess* specifies the set of *high-level tasks* needed to complete the process and the related operational requirements. For example, the *RemoveTopCover* and *RemoveBottomCover* values in *AssemblyProcess* represent high-level tasks modeling part of the assembly/disassembly procedure. Notice that no task assignment is performed at this level of abstraction. The *Human*, *RobotController* and *CollaborationType* state variables compose the *coordination level* of the model. Specifically, the *Human* and *RobotController* state variables model the *low-level* tasks the human and the robot agents can perform over time. For example, the *Screw* or *Unscrew* values of *Human* and *RobotController* state variables model the capability of both *agents* of performing screwing operations. Instead, *RemovePart* or *Rotate* values of the *Human* state variables model *critical operations* that only the human is allowed to perform. The *CollaborationType* state variable models the possible types of human-robot collaboration within the

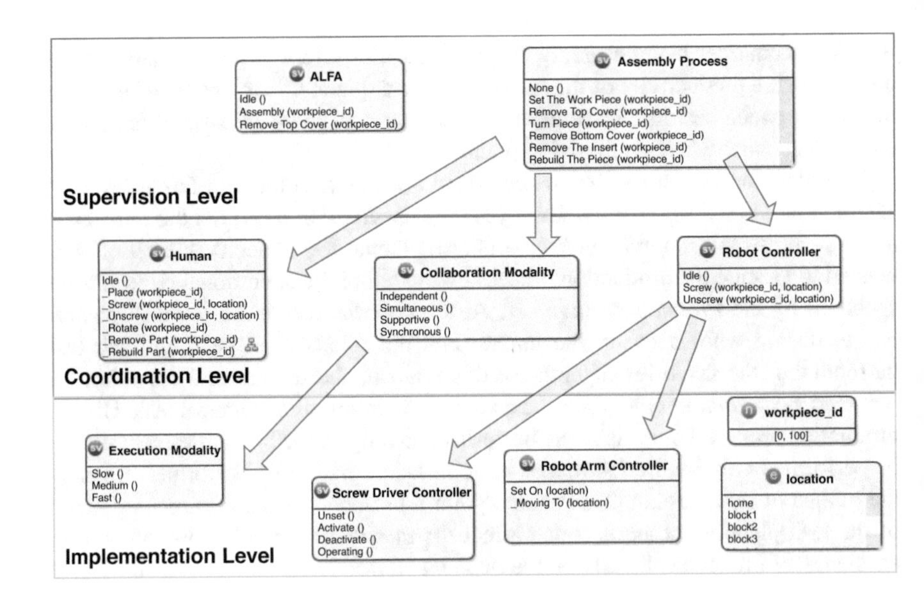

Fig. 12.4 A graphical overview of the hierarchical P&S model for collaborative assembly generated by KEEN

execution of the tasks of the desired process. The supervision and coordination layers are connected by a set of synchronization rules that specify decomposition constraints of *high-level tasks* in terms of *low-level tasks*. These rules specify how the tasks composing the process can be performed in collaboration by the human and the robot. Namely, these rules describe the possible task assignments between the human and the robot and specify the collaboration modalities suited for human–robot interactions.

The *RobotArmController*, *ScrewDriverController*, and *ExecutionModality* state variables constitute the *implementation level* of the model. These variables represent the physical and/or logical elements composing the production environment the system must directly interact with. The *RobotArmController* together with the *ExecutionModality* model the robotic arm. They represent the functional control interface of the robot provided by the integrated *motion planner* (see Fig. 12.3). Specifically, the *RobotArmController* models the motion tasks the robot can perform, while the *ExecutionModality* models the type of trajectory that must be used to perform the motion. The coordination and implementation layers are connected by another set of synchronization rules that specify how the robot must execute the assigned tasks. A particular execution modality of robot motions is selected according to the expected collaboration modality in the coordination layer.

4.4 Implementation with a Real Robot

The engineering & control architecture described above was deployed in a manufacturing case study integrating the task planning technology described above with a motion planning system for industrial robots [33]. The reference selected application is a human-robot collaborative environment for the preparation of the load/unload station (LUS) of a flexible manufacturing system (FMS). At the LUS, machined parts and raw parts have to be unmounted and mounted on ad-hoc fixturing system, called pallet, by a worker and a robot in order to be machined by the FMS. With the aim to increase productivity and grant human ergonomics and safety, robot trajectories and task allocation have to be, respectively, adequately designed and planned.

Thus, PLATINUM and KEEN features were leveraged to implement an integrated task and motion planning system capable of selecting different *execution modalities* for robot tasks according to the expected *collaboration* of the robot with a human operator. This is the result of a tight integration of PLATINUM with a motion planning system. Indeed, the pursued approach realizes an *offline analysis* of the production scenarios in order to synthesize a number of collision-free *robot motion trajectories* for each collaborative task with different *safety* levels. Each trajectory is then associated with an expected temporal execution bound and represents a tradeoff between "speed" of the motion and "safety" of the human. The integrated system has been deployed and tested in laboratory on an assembly case study similar to collaborative assembly/disassembly scenario described above. In [34], an empirical evaluation is provided in order to assess the overall productivity of the HRC cell while increasing the involvement of the robots. The idea is to gradually make free a set of tasks originally preallocated to the human, so to increase the number of degrees of freedom of the task planner during the minimization of the assembly time. The results show the effectiveness of the control architecture in finding well suited distribution of tasks between the human and the robot in different scenarios with an increasing workload for the control system. Indeed, the total assembly time was reduced of 65% (from 259 s to 169 s) and the percentage of tasks assigned by the controller to the robot moved from 25 to 65%. Thus, PLATINUM and KEEN have shown to be capable of increasing the productivity of the production process without affecting the safety of the operator. It is worth underscoring that the outcome of this integration constitutes the technological basis on which a new research project development is undergoing, i.e., the ShareWork project[5] funded by the European Commission within the Factories of the Future area.

5 Conclusions

A research initiative has been started to investigate the possible integration of a timeline-based planning framework (APSI-TRF [4]) and V&V techniques based on Timed Game Automata (TGA) [5] to automatically synthesize a robot controller that guarantees robustness and safety properties [6, 7]. This chapter provided an overview of the results collected in the last decade concerning the development of a task planning and execution framework (PLATINUM) [14, 15] and its integration with a Knowledge Engineering ENvironment (KEEN) describing its deployment for developing safe and effective solutions for manufacturing HRC scenarios. KEEN [16] is a knowledge engineering environment with Verification & Validation features based on Timed Game Automata model checking. PLATINUM is a timeline-based planning systems capable of supporting flexible temporal planning and execution with uncertainty. A brief overview on recent results about the development of a task planning and execution technology deployed in realistic manufacturing scenarios was presented. The research agenda is far from being completed. The long-term goal is to realize a robust task planning system enabling flexible, safe, and efficient HRC with a more tight integration of P&S technology with Knowledge Engineering solutions and V&V techniques is a key element to synthesize more effective, robust, and reliable safety critical systems in robotics [2].

Acknowledgements Amedeo Cesta, Andrea Orlandini, and Alessandro Umbrico wish to acknowledge the support by the European Commission and the ShareWork project (H2020—Factories of the Future—G.A. nr. 820807).

[5]https://sharework-project.eu/.

References

1. Freitag, M., Hildebrandt, T.: Automatic design of scheduling rules for complex manufacturing systems by multi-objective simulation-based optimization. {CIRP} Annals - Manufacturing Technology **65**(1) (2016) 433–436
2. Bensalem, S., Havelund, K., Orlandini, A.: Verification and validation meet planning and scheduling. International Journal on Software Tools for Technology Transfer **16**(1) (2014) 1–12
3. La Viola, C., Orlandini, A., Umbrico, A., Cesta, A.: ROS-TiPlEx: How to make experts in A.I. planning and robotics talk together and be happy. In: 28th IEEE International Conference on Robot and Human Interactive Communication (RO-MAN), New Delhi, India, 2019, pp. 1–6. http://dx.doi.org/10.1109/RO-MAN46459.2019.8956417
4. Cesta, A., Cortellessa, G., Fratini, S., Oddi, A.: Developing an End-to-End Planning Application from a Timeline Representation Framework. In: IAAI-09. Proc. of the 21st Innovative Application of Artificial Intelligence Conference, Pasadena, CA, USA. (2009)
5. Maler, O., Pnueli, A., Sifakis, J.: On the Synthesis of Discrete Controllers for Timed Systems. In: STACS. LNCS, Springer (1995) 229–242
6. Cesta, A., Finzi, A., Fratini, S., Orlandini, A., Tronci, E.: Validation and Verification Issues in a Timeline-Based Planning System. Knowledge Engineering Review **25**(3) (2010) 299–318
7. Orlandini, A., Suriano, M., Cesta, A., Finzi, A.: Controller synthesis for safety critical planning. In: IEEE 25th International Conference on Tools with Artificial Intelligence (ICTAI 2013), IEEE (2013) 306–313
8. Py, F., Rajan, K., McGann, C.: A systematic agent framework for situated autonomous systems. In: AAMAS. (2010) 583–590
9. Lemai, S., Ingrand, F.: Interleaving Temporal Planning and Execution in Robotics Domains. In: AAAI-04. (2004) 617–622
10. Barreiro, J., Boyce, M., Do, M., Frank, J., Iatauro, M., Kichkaylo, T., Morris, P., Ong, J., Remolina, E., Smith, T., Smith, D.: EUROPA: A Platform for AI Planning, Scheduling, Constraint Programming, and Optimization. In: ICKEPS 2012: the 4th Int. Competition on Knowledge Engineering for Planning and Scheduling. (2012)
11. Ghallab, M., Laruelle, H.: Representation and control in IxTeT, a temporal planner. In: 2nd Int. Conf. on Artificial Intelligence Planning and Scheduling (AIPS). (1994) 61–67
12. Morris, P.H., Muscettola, N.: Temporal Dynamic Controllability Revisited. In: Proc. of AAAI 2005. (2005) 1193–1198
13. Cesta, A., Orlandini, A., Bernardi, G., Umbrico, A.: Towards a planning-based framework for symbiotic human-robot collaboration. In: 21th IEEE International Conference on Emerging Technologies and Factory Automation (ETFA), IEEE (2016)
14. Umbrico, A., Cesta, A., Cialdea Mayer, M., Orlandini, A.: PLATINUM: A new Framework for Planning and Acting. In: AI*IA 2016 Advances in Artificial Intelligence: XVth International Conference of the Italian Association for Artificial Intelligence, Genova, Italy, November 29 – December 1, 2016, Proceedings, Springer International Publishing (2017) 508–522
15. Umbrico, A., Cesta, A., Cialdea Mayer, M., Orlandini, A.: Integrating resource management and timeline-based planning. In: The 28th International Conference on Automated Planning and Scheduling (ICAPS). (2018)
16. Orlandini, A., Bernardi, G., Cesta, A., Finzi, A.: Planning meets verification and validation in a knowledge engineering environment. Intelligenza Artificiale **8**(1) (2014) 87–100
17. Cesta, A., Cortellessa, G., Fratini, S., Oddi, A., Policella, N.: An Innovative Product for Space Mission Planning: An A Posteriori Evaluation. In: ICAPS. (2007) 57–64
18. Jonsson, A., Morris, P., Muscettola, N., Rajan, K., Smith, B.: Planning in Interplanetary Space: Theory and Practice. In: AIPS-00. Proceedings of the Fifth Int. Conf. on AI Planning and Scheduling. (2000)
19. Muscettola, N.: HSTS: Integrating Planning and Scheduling. In Zweben, M. and Fox, M.S., ed.: Intelligent Scheduling. Morgan Kauffmann (1994)

20. Vidal, T., Fargier, H.: Handling Contingency in Temporal Constraint Networks: From Consistency To Controllabilities. JETAI **11**(1) (1999) 23–45
21. Cialdea Mayer, M., Orlandini, A., Umbrico, A.: Planning and execution with flexible timelines: a formal account. Acta Informatica **53**(6–8) (2016) 649–680
22. Allen, J.F.: Maintaining knowledge about temporal intervals. Commun. ACM **26**(11) (1983) 832–843
23. Umbrico, A., Cesta, A., Cortellessa, G., Orlandini, A.: A goal triggering mechanism for continuous human-robot interaction. Lecture Notes in Computer Science (including subseries Lecture Notes in Artificial Intelligence and Lecture Notes in Bioinformatics) **11298 LNAI** (2018) 460–473
24. Cesta, A., Finzi, A., Fratini, S., Orlandini, A., Tronci, E.: Flexible Timeline-Based Plan Verification. In: KI 2009: Advances in Artificial Intelligence. Volume 5803 of LNAI. (2009)
25. Cesta, A., Finzi, A., Fratini, S., Orlandini, A., Tronci, E.: Analyzing Flexible Timeline Plan. In: ECAI 2010. Proceedings of the 19th European Conference on Artificial Intelligence. Volume 215., IOS Press (2010)
26. Orlandini, A., Finzi, A., Cesta, A., Fratini, S.: TGA-based controllers for flexible plan execution. In: KI 2011: Advances in Artificial Intelligence, 34th Annual German Conference on AI. Volume 7006 of Lecture Notes in Computer Science., Springer (2011) 233–245
27. Behrmann, G., Cougnard, A., David, A., Fleury, E., Larsen, K., Lime, D.: UPPAAL-TIGA: Time for playing games! In: Proc. of CAV-07. Number 4590 in LNCS, Springer (2007) 121–125
28. Larsen, K.G., Pettersson, P., Yi, W.: UPPAAL in a Nutshell. International Journal on Software Tools for Technology Transfer **1**(1–2) (1997) 134–152
29. Marvel, J.A., Falco, J., Marstio, I.: Characterizing task-based human-robot collaboration safety in manufacturing. IEEE Trans. Systems, Man, and Cybernetics: Systems **45**(2) (2015) 260–275
30. Helms, E., Schraft, R.D., Hagele, M.: rob@work: Robot assistant in industrial environments. In: Proceedings. 11th IEEE International Workshop on Robot and Human Interactive Communication. (2002) 399–404
31. Maurtua, I., Pedrocchi, N., Orlandini, A., Fernández, J.d.G., Vogel, C., Geenen, A., Althoefer, K., Shafti, A.: Fourbythree: Imagine humans and robots working hand in hand. In: 2016 IEEE 21st International Conference on Emerging Technologies and Factory Automation (ETFA). (Sept 2016) 1–8
32. Cesta, A., Orlandini, A., Umbrico, A.: Fostering robust human-robot collaboration through AI task planning. Procedia CIRP **72** (2018) 1045–1050 51st CIRP Conference on Manufacturing Systems.
33. Pellegrinelli, S., Moro, F.L., Pedrocchi, N., Tosatti, L.M., Tolio, T.: A probabilistic approach to workspace sharing for human–robot cooperation in assembly tasks. {CIRP} Annals - Manufacturing Technology **65**(1) (2016) 57–60
34. Pellegrinelli, S., Orlandini, A., Pedrocchi, N., Umbrico, A., Tolio, T.: Motion planning and scheduling for human and industrial-robot collaboration. CIRP Annals - Manufacturing Technology **66** (2017) 1–4

Chapter 13
Planning in a Real-World Application: An AUV Case Study

Lukáš Chrpa

1 Introduction

Automated planning deals with the problem of finding a (partially ordered) action sequence, a plan, transforming the environment from a given initial state to some required goal state [7]. In a nutshell, automated planning accounts for deliberative reasoning that intelligent entities leverage for finding strategies (plans) for their longer-term goals. There are many successful real-world applications ranging from space and planet observations [1], Urban Traffic Control [9] to narrative generation [8].

Domain-independent planning decouples a planning task descriptions (for example, in PDDL) and planning engines that can be understood as general solvers of planning tasks. The advantage of such approach is its modularity, i.e., the task description and the planning engine are independent components, that makes it easier to plug into larger systems. Planning task description can be further decoupled into a *domain model* that describes the environment and action in general level, and a *planning problem* that describes concrete objects, an initial state, and a goal. To communicate with the planning component, the system has to be able to generate planning problems (for example, in PDDL) and be able to interpret (and execute) generated plans.

This chapter summarises an experience of an automated planning expert with the application of task planning for autonomous underwater vehicles (AUVs) who went through the process of knowledge elicitation, domain modelling, and plan execution in real environment. The aim of the chapter is to give insights into how easy or hard

L. Chrpa (✉)
Faculty of Electrical Engineering, Czech Technical University in Prague, Prague, Czech Republic

Faculty of Mathematics and Physics, Charles University in Prague, Prague, Czech Republic
e-mail: chrpaluk@fel.cvut.cz

M. Vallati, D. Kitchin (eds.), *Knowledge Engineering Tools and Techniques for AI Planning*, https://doi.org/10.1007/978-3-030-38561-3_13

it might be for a planning expert to develop a reasonable domain model for a real-world application in which s/he is not an expert. On top of that the experience from the field experiments gives good insights into robustness of the planning approach and points out possible issues that arise from the discrepancy of the real environment and the (simplified) domain model. The chapter is based on the conference papers describing the "one-shot" model [3] and the "dynamic" model [2].

2 Background

While operating AUVs, human operators interact with a network of vehicles via NEPTUS, a graphical decision-support system with graphical user interface and analysis capabilities [4]. NEPTUS therefore allows users to view vehicle data and to define behaviours and tasks of the vehicles. NEPTUS is connected via inter-module communication protocol to DUNE that runs on board of each vehicle and is responsible for command execution and gathering of sensory data [10].

In a nutshell, an operator can specify and execute high-level commands in NEPTUS, for example, "move AUV1 to location X", or "sample AUV1 an object Y at location X". However, operating multiple (heterogeneous) AUVs to perform several tasks might be time-consuming and error prone for human operators even though the mission might not be very complex.

The idea how to automatise AUV operations is to leverage automated planning for generating plans for AUVs such that they eventually complete all tasks specified by operators [3]. An automated planning component can be embedded to the NEPTUS toolchain as depicted in Fig. 13.1. Intuitively, the high-level commands an

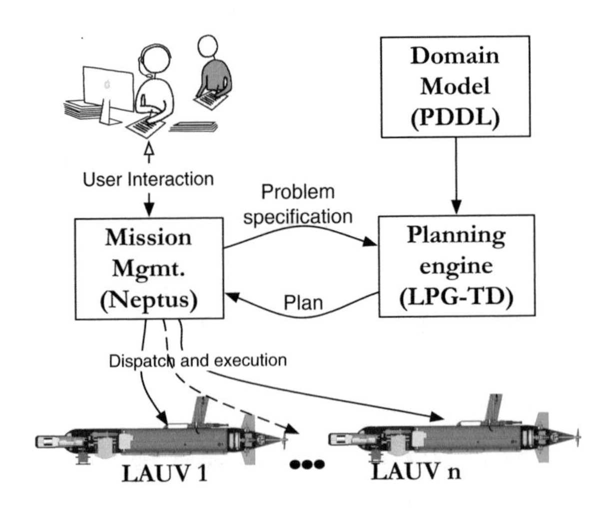

Fig. 13.1 A modular architecture of the system [3]

operator can specify in NEPTUS will correspond to actions specified in a planning domain model.

We use PDDL 2.1 [5] for representing our planning tasks. The environment is described by predicates and numeric fluents. Actions are specified via preconditions, effects, and the duration of their execution. Preconditions are sets of logical expressions that must be true in order to have the actions executable. In PDDL 2.1, these expressions can take place prior to starting action execution, prior to finishing action execution, or over the whole time period when the action is executed. Effects are sets of literals and function assignments that take place following action execution. In PDDL 2.1, effects can occur just after starting action execution, or just after finishing action execution.

As a planning engine, LPG-td [6] is used. LPG-td performs local search in a planning graph [7] which allows executing actions that do not interfere with each other in one step and thus LPG-td performs well in our domain.

3 One-Shot Planning

The initial idea of leveraging automated planning for task planning for AUVs can be summarised in three consecutive steps.

1. An operator specifies tasks
2. The planning component generates a plan
3. The plan is executed by AUVs

Henceforth, we will refer to such an approach as "one-shot planning" [3].

3.1 Requirements

Given a fleet of heterogeneous AUVs where each AUV has different payloads attached to it, a human operator specifies tasks in NEPTUS such that each task consists of an object or area of interest and a payload which has to be used to collect data about the object/area of interest. Noteworthy each task has to be fulfilled by a single AUV.

During operations none of the AUVs has to run out of power. Also, two or more AUVs cannot operate at the same location or in the same area. On the other hand, when moving between locations AUVs can be in different depths and hence they should not collide.

Each AUV can move between two locations if it has enough power. An AUV can sample an object of interest or survey an area of interest if it is in the required location, has enough energy, and has a required payload. If an AUV is close to the operation centre, it can transfer acquired data there.

The planning engine has to then find a plan that allocate all user-specified tasks to particular AUVs such that the above constraints are met (e.g., AUVs will not run out of power) and the AUVs return to their "safe spots" next to the operation centre and transmit acquired data.

3.2 Domain Model Specification

We have conceptualised the requirements in the form of a domain model specification. We consider three categories: object types, predicates and functions, and actions similar to [11]. Noteworthy such conceptualisation might be seen as biased towards PDDL as we use it for describing the domain model.

Object types refer to classes of objects that are relevant for the planning process, such as *vehicle* (V), *payload* (P), *phenomenon* (X), *task* (T), *location* (L). By "phenomenon" we mean either an object or an area of interest.

Predicates and numeric fluents describe states of the environment. In particular, predicates represent relationships between objects, and numeric fluents refer to quantity of resources related to the objects. We have defined the following predicates: $at \subseteq V \times L$—a location of the vehicle, $base \subseteq V \times L$—a location of the vehicle's *depot*, i.e., a safe location next to the control centre, $has \subseteq V \times P$—whether a payload is attached to the vehicle, $\textit{at-phen} \subseteq X \times L$—a location of the phenomenon, $task \subseteq T \times X \times P$—a task description of getting data about a phenomenon from a specific payload, $sampled \subseteq T \times V$—whether data of a given task has been acquired by the vehicle, $data \subseteq T$—whether the task data has been transmitted to the control centre. Then, we have defined the following numeric fluents: $dist : L \times L \rightarrow \mathbb{R}^+$—a distance between two locations, $speed : V \rightarrow \mathbb{R}^+$—speed over ground of the vehicle, $\textit{battery-level} : V \rightarrow \mathbb{R}_0^+$—the amount of energy in a vehicle's battery, $\textit{battery-use} : V \cup P \rightarrow \mathbb{R}^+$—battery consumption per distance unit (moving a vehicle) or per time unit (using a payload). Noteworthy, we assume a linear energy use both for moving or using a payload.

Actions modify the environment according to their effects. We consider "durative actions" that reason with time, i.e., action execution takes time. We have specified four actions (we denote t_s as time when an action is executed, and t_e as time when an action execution ends):

- $move(v, l_1, l_2)$—the vehicle v moves from its location of origin l_1 to a location of its destination l_2. As a precondition is must hold that in t_s: $(v, l_1) \in at$, $\textit{battery-level}(v) \geq dist(l_1, l_2) * \textit{battery-use}(v)$; and in t_e: $\neg\exists v_x \neq v : (v_x, l_2) \in at$. The effect is that in t_s: $(v, l_1) \notin at$, and $\textit{battery-level}(v) = \textit{battery-level}(v) - dist(l_1, l_2) * \textit{battery-use}(v)$,and in t_e: $(v, l_2) \in at$.
- $sample(v, t, x, p, l)$—the vehicle v samples a phenomenon x by payload p. As a precondition it must hold that in t_s: $\textit{battery-level}(v) \geq (t_e - t_s) * \textit{battery-use}(p)$, and in $[t_s, t_e]$: $(v, l) \in at$, $(x, l) \in \textit{at-phen}$, $(v, p) \in has$ and $(t, x, v) \in$

task. The effect is that in t_s: $battery\text{-}level(v) = battery\text{-}level(v) - (t_e - t_s) * battery\text{-}use(p)$, and in t_e: $(t, v) \in sampled$.

- $survey(v, t, x, p, l_1, l_2)$—the vehicle v surveys the area (between l_1 and l_2) of phenomenon x occurrence by the payload p. As a precondition is must hold that in t_s: $(v, l_1) \in at$, $battery\text{-}level(v) \geq dist(l_1, l_2) * battery\text{-}use(v) + (t_e - t_s) * battery\text{-}use(p)$, and in $[t_s, t_e]$: $(x, l_1) \in at\text{-}phen$, $(x, l_2) \in at\text{-}phen$, $(v, p) \in has$ and $(t, x, v) \in task$. Also, no other vehicle can perform the survey action over the phenomenon x in $[t_s, t_e]$. The effect is that in t_s: $(v, l_1) \notin at$, and $battery\text{-}level(v) = battery\text{-}level(v) - (dist(l_1, l_2) * battery\text{-}use(v) + (t_e - t_s) * battery\text{-}use(p))$, and in t_e: $(v, l_2) \in at$, $(t, v) \in sampled$.
- $collect\text{-}data(v, t, l)$—the data associated with a task t is collected by vehicle v. As a precondition it must hold that in $[t_s, t_e]$: $(v, l) \in at$, $(v, l) \in base$ and $(t, v) \in sampled$. The effect is that in t_e: $t \in data$.

Durations of the *move* and *survey* actions are determined from the distance between locations and speed of the vehicle. Durations of the *collect-data* and *sample* actions are constant. As an example, the `Sample` action is encoded as depicted in Fig. 13.2.

3.3 Problem Specification

The planning problem is specified by concrete objects (e.g., AUVs, phenomena, locations), an initial state and a goal. The goal is to have all the required data transmitted to the control centre and having AUVs back in their depots. The initial state consists of initial locations of AUVs and their depots, locations or areas of phenomena, vehicles' payloads, task descriptions (specified by a human operator in NEPTUS), i.e., which types of payloads is to be used to collect data about the phenomena, distance between the locations, vehicle speed, battery levels, and battery consumption values per vehicle/payload.

The planner decides which AUV does which task and in which order the tasks are performed. Plans follow the constraints specified in the action descriptions, i.e.,

```
(:durative-action sample
 :parameters (?v - vehicle ?l - location
              ?t -task ?o - phenomenon ?p - payload)
 :duration (= ?duration 60)
 :condition (and (over all (at-phen ?o ?l))
                 (over all (task ?t ?o ?p))
                 (over all (at ?v ?l))
                 (over all (having ?p ?v))
                 (at start (>= (battery-level ?v)
                               (* (battery-use ?p) 60))))
 :effect (and (at end (sampled ?t ?v))
              (at start (decrease (battery-level ?v)
                                  (* (battery-use ?p) 60)))) )
```

Fig. 13.2 The `Sample` action of the One-shot model in PDDL

collision avoidance and energy constraints with plans optimised for total mission time. If the planner is unable to find a plan, the user is notified of plan failure, requiring her/him to iteratively relax constraints (e.g., remove some tasks).

3.4 Field Experiment

The concept has been evaluated on a "mine-hunting" scenario in Porto Harbour [3]. We used three AUVs that in the first stage were set to perform several survey tasks, while in the second stage they were set to perform several sample tasks. Generated plans were successfully executed, hence AUVs successfully completed assigned tasks in both stages.

On the other hand, the results have shown considerable discrepancies between anticipated and actual action durations, especially for the move and survey actions. The reasons cover ocean currents, rough ocean floor, to mention a few, affecting actual duration of the survey and move actions. For our settings in which AUVs operate individually such discrepancies are not a major issue. For collaborative scenarios or longer-term missions, the discrepancies might be problematic. For deeper insights about the experiments, see [3].

4 Dynamic Planning, Replanning, and Plan Execution

The "one-shot" model [3] does not consider user changes during the mission execution. In other words, the two-stage mine-hunting scenario has to be planned and executed one by one, i.e., all the first-stage tasks have to finish before the operator can specify tasks of the second stage. In longer-term operations, such an approach is impractical as the mission operators cannot react to acquired data as soon as they get them. That said, mission operators should be able to add, modify, or remove tasks during the mission.

Another issue concerns possible vehicle failure while performing a given action. For example, an AUV might fail to complete a sample action if its payload is faulty. The "one-shot" model, however, does not reason with such cases and it is up to the operator to reinsert failed tasks into NEPTUS.

Also, while operating underwater AUVs might not be able to maintain reliable communication with the control centre. This is exacerbated in lager areas or higher depths where AUVs are often radio silent for several minutes. Lack of communication complicates monitoring plan execution as well as sending new or amending old plans to AUVs. Hence, AUVs have to establish a reliable communication with control centre from time to time to report completion/failure of tasks and receiving new ones. Details can be found in [2].

4.1 Requirements

To incorporate dynamic task allocation as well as recovery from task failures, the system has to be able to replan, i.e., to generate a new plan which replaces the old one. In particular, any user change might trigger replanning; however, for practical reason replanning is triggered (if a change occurs) periodically. Task failures can be directly reported to the system that reinserts the failed task back into the system and the task is considered for replanning.

Addressing the communication issue can be done by requesting AUVs to return to their "depots" and establish communication with the control centre in a given period of time. Specifically, each AUV can be away for at most a given period of time before returning to its "depot" to establish communication and then it can go away again (for at most the given period of time). That said, AUV will complete tasks in "rounds".

4.2 Domain Model Specification

In contrast to the "one-shot" model, the "dynamic" model [2] distinguishes between *Objects of Interest* (O) and *Areas of Interest* (A) instead of phenomenons.

In the dynamic model, we introduce a *can-move* $\subseteq V$ predicate determining whether a vehicle can perform a move action. The *at-phen* predicate is replaced by *at-obj* $\subseteq O \times L$ determining locations of objects of interest, and *at-entry* $\subseteq A \times L$ and *at-exit* $\subseteq A \times L$ determining entry and exit locations of areas of interest. The battery constraints are relaxed since vehicles have to regularly visit their depots and if their battery level is not high enough for the next "round", they are no longer considered for planning. On the other hand, we define two fluents determining how long the vehicles are away, *away* $: V \to \mathbb{R}_0^+$, and the maximum time they can be away from their depots, *max-away* $: V \to \mathbb{R}_0^+$.

We consider "durative actions" that reason with time, i.e., action execution takes time. We have specified six actions (we denote t_s as time when an action is executed, and t_e as time when an action execution ends). Noteworthy we split the move action into three variants depending where the vehicle goes, i.e., *move-to-sample*, *move-to-survey*, and *move-to-depot*:

- *move-to-sample*(v, l_1, l_2, o)—the vehicle v moves from its location of origin l_1 to a location of its destination l_2 in which it has to take a sample of o. As a precondition is must hold that in t_s: $(v, l_1) \in at$, $v \in$ *can-move*, $(o, l_2) \in$ *at-obj*, *away*$(v) \leq$ *max-away*$(v) - (t_e - t_s)$; and in t_e: $\neg\exists v_x \neq v : (v_x, l_2) \in at$. The effect is that in t_s: $(v, l_1) \notin at$, $v \notin$ *can-move*, and *away*$(v) =$ *away*$(v) + (t_e - t_s)$,and in t_e: $(v, l_2) \in at$
- *move-to-survey*(v, l_1, l_2, a)—the vehicle v moves from its location of origin l_1 to a location of its destination l_2 in which it has to survey a. As a precondition is must hold that in t_s: $(v, l_1) \in at$, $v \in$ *can-move*, $(a, l_2) \in$ *at-entry*, *away*$(v) \leq$

max-away$(v) - (t_e - t_s)$; and in t_e: $\neg\exists v_x \neq v : (v_x, l_2) \in at$. The effect is that in t_s: $(v, l_1) \notin at$, $v \notin$ *can-move*, and *away*$(v) =$ *away*$(v) + (t_e - t_s)$,and in t_e: $(v, l_2) \in at$

- *move-to-depot*(v, l_1, l_2)—the vehicle v moves from its location of origin l_1 to a location of its depot at l_2. As a precondition is must hold that in t_s: $(v, l_1) \in at$, $v \in$ *can-move*, $(v, l_2) \in$ *base*, *away*$(v) \leq$ *max-away*$(v) - (t_e - t_s)$; and in t_e: $\neg\exists v_x \neq v : (v_x, l_2) \in at$. The effect is that in t_s: $(v, l_1) \notin at$, and in t_e: $(v, l_2) \in at$ and *away*$(v) = 0$.
- *sample*(v, t, o, p, l)—the vehicle v samples an object of interest o by payload p. As a precondition it must hold that in t_s: *away*$(v) \leq$ *max-away*$(v) - (t_e - t_s)$;, and in $[t_s, t_e]$: $(v, l) \in at$, $(o, l) \in$ *at-obj*, $(v, p) \in$ *has* and $(t, o, v) \in task$. The effect is that in t_s: *away*$(v) =$ *away*$(v) + (t_e - t_s)$, and in t_e: $(t, v) \in sampled$, $v \in$ *can-move*.
- *survey*(v, t, a, p, l_1, l_2)—the vehicle v surveys the area a (between l_1 and l_2) by the payload p. As a precondition is must hold that in t_s: *away*$(v) \leq$ *max-away*$(v) - (t_e - t_s)$, $(v, l_1) \in at$ and in $[t_s, t_e]$:, $(a, l_1) \in$ *at-entry*, $(a, l_2) \in$ *at-exit*, $(v, p) \in$ *has* and $(t, a, v) \in task$. Also, no other vehicle can perform the survey action over the area a in $[t_s, t_e]$. The effect is that in t_s: $(v, l_1) \notin at$, and *away*$(v) =$ *away*$(v) + (t_e - t_s)$, and in t_e: $(v, l_2) \in at$, $(t, v) \in sampled$, $v \in$ *can-move*.
- *collect-data*(v, t, l)—the data associated with a task t is collected by vehicle v. As a precondition it must hold that in $[t_s, t_e]$: $(v, l) \in at$, $(v, l) \in base$ and $(t, v) \in sampled$. The effect is that in t_e: $t \in data$.

Durations of the *move* and *survey* actions are determined from the distance between locations and speed of the vehicle. Durations of the *collect-data* and *sample* actions are constant. As an example, the `Sample` action is encoded as depicted in Fig. 13.3.

```
(:durative-action sample
:parameters (?v - vehicle ?l - location ?t -task ?o -oi ?p - payload)
:duration (= ?duration 60)
:condition (and (over all (at-obj ?o ?l))
                (over all (task ?t ?o ?p))
                (over all (at ?v ?l))
                (over all (having ?p ?v))
                (over all (<= (from-base ?v)(max-to-base ?v)))
                )
:effect (and (at end (sampled ?t ?v))
             (at end (can-move ?v))
             (at start (increase (from-base ?v) 60))
        )
)
```

Fig. 13.3 The `Sample` action of the dynamic model in PDDL

4.3 Problem Specification

The *All-tasks* model [2] specifies the problem analogously to the "one-shot" model (defining the "away" constraints instead of the "battery" constraints). Plans allocate all specified tasks and are optimised for minimising makespan, i.e., the duration of plans' execution, and the number of *move-to-depot* actions. The latter is used to minimise the number of "rounds" vehicles have to take as well as the number of vehicles necessary to fulfil the tasks.

The *One round* model [2] plans only for the next round. Specifically, the *move-to-depot* action removes the *can-move* predicate and the *collect-data* action increments a fluent (additionally defined) *tasks-completed*. Goals, in contrast to the *All-Tasks* model, are to have all involved vehicles in their depots and a specified minimum number of tasks completed. Plans are optimised for maximising the number of completed tasks.

4.4 Planning and Execution

On top of the "one-shot" model, here large survey tasks are split into smaller ones (to satisfy the "away" constraints). The *move* actions are given more lenient durations as they are translated into timed waypoint behaviours which specify target location and absolute time of arrival. As a result, when vehicles move between locations they adapt their speed according to remaining time to arrive at the location of the task.

A new planning request is generated by operator changes as well as by reported failures of vehicles. Planning requests are considered periodically. If a new planning request arrives while a vehicle is executing the previous plan, the vehicle continues executing the former plan until its earliest arrival to its depot. Then, the new plan is assigned to the vehicle (notice that the planning request considers the estimated time when the vehicle will be available).

4.5 Field Experiment

Again, the dynamic models were evaluated on a "mine-hunting" scenario that incorporates a surveying and sampling stages. In contrast to "one-shot", the operator could add sampling tasks that were processed and allocated to AUVs before the stage-one survey tasks finished. Although in one case an AUV reported a temporary depth sensor fault and failed a given task, the task was reinserted to NEPTUS and later allocated to a different AUV that completed the task. In the remaining cases, plans were executed successfully.

Also, with the timed waypoint move actions, plans became more robust to discrepancies between planned and actual plan duration hence the vehicles always arrived to their depots in a planned time (or very close to it).

The "All-tasks" model generates better quality plans than the "one-round" model. On the other hand, we believe that for larger missions (dozens of tasks), the "All-tasks" model might struggle to generate plans in required time (plans will be longer) in contrast to the "one-round" model.

5 Conclusion

Domain-independent planning can be fruitfully exploited in application areas such as task planning for AUVs. The advantage is its modularity that makes it easy to plug into larger systems. The key part is developing of effective domain model that provides a symbolic description of the environment and (high-level) actions. This chapter described the knowledge engineering process of domain model development for the problem of task planning for AUVs. The field experiments indicate that the use of domain-independent planning can automatise task allocation for AUVs and that generated task allocation can be successfully executed in the real-world environment.

The future challenges concern robustness and adaptability of domain models. Collaborative as well as long-term missions require accurate and robust models as even small discrepancies might easily propagate and might cause mission failure. In particular, domain models should be able to adapt themselves according to current observations and generated plans must be safe or with a minimal risk of failure if environment unexpectedly changes. That would require to combine several research areas such as automated planning and machine Learning.

Acknowledgement This research was funded by the Czech Science Foundation (project no. 17-17125Y).

References

1. Ai-Chang, M., Bresina, J.L., Charest, L., Chase, A., Hsu, J.C., Jónsson, A.K., Kanefsky, B., Morris, P.H., Rajan, K., Yglesias, J., Chafin, B.G., Dias, W.C., Maldague, P.F.: MAPGEN: mixed-initiative planning and scheduling for the mars exploration rover mission. IEEE Intelligent Systems **19**(1), 8–12 (2004). https://doi.org/10.1109/MIS.2004.1265878
2. Chrpa, L., Pinto, J., Marques, T.S., Ribeiro, M.A., de Sousa, J.B.: Mixed-initiative planning, replanning and execution: From concept to field testing using AUV fleets. In: 2017 IEEE/RSJ International Conference on Intelligent Robots and Systems, IROS 2017, Vancouver, BC, Canada, September 24–28, 2017. pp. 6825–6830 (2017). https://doi.org/10.1109/IROS.2017.8206602

3. Chrpa, L., Pinto, J., Ribeiro, M.A., Py, F., de Sousa, J.B., Rajan, K.: On mixed-initiative planning and control for autonomous underwater vehicles. In: 2015 IEEE/RSJ International Conference on Intelligent Robots and Systems, IROS 2015, Hamburg, Germany, September 28 - October 2, 2015. pp. 1685–1690 (2015). https://doi.org/10.1109/IROS.2015.7353594
4. Dias, P.S., Gomes, R.M.F., Pinto, J., Gonçalves, G.M., de Sousa, J.B., Pereira, F.M.L.: Mission planning and specification in the Neptus framework. In: Proceedings of the 2006 IEEE International Conference on Robotics and Automation, ICRA 2006, May 15–19, 2006, Orlando, Florida, USA. pp. 3220–3225 (2006). https://doi.org/10.1109/ROBOT.2006.1642192
5. Fox, M., Long, D.: PDDL2.1: an extension to PDDL for expressing temporal planning domains. J. Artif. Intell. Res. (JAIR) **20**, 61–124 (2003)
6. Gerevini, A., Saetti, A., Serina, I.: Planning with numerical expressions in LPG. In: Proceedings of the 16th European Conference on Artificial Intelligence, ECAI'2004, including Prestigious Applicants of Intelligent Systems, PAIS 2004, Valencia, Spain, August 22–27, 2004. pp. 667–671 (2004)
7. Ghallab, M., Nau, D., Traverso, P.: Automated planning, theory and practice. Morgan Kaufmann (2004)
8. Haslum, P.: Narrative planning: Compilations to classical planning. J. Artif. Intell. Res. **44**, 383–395 (2012). https://doi.org/10.1613/jair.3602
9. McCluskey, T.L., Vallati, M.: Embedding automated planning within urban traffic management operations. In: Proceedings of the Twenty-Seventh International Conference on Automated Planning and Scheduling, ICAPS 2017, Pittsburgh, Pennsylvania, USA, June 18–23, 2017. pp. 391–399 (2017), https://aaai.org/ocs/index.php/ICAPS/ICAPS17/paper/view/15645
10. Pinto, J., Dias, P.S., Martins, R., Fortuna, J., Marques, E., Sousa, J.: The LSTS toolchain for networked vehicle systems. In: OCEANS-Bergen, 2013 MTS/IEEE. pp. 1–9. IEEE (2013)
11. Shah, M.M.S., Chrpa, L., Kitchin, D.E., McCluskey, T.L., Vallati, M.: Exploring knowledge engineering strategies in designing and modelling a road traffic accident management domain. In: IJCAI (2013)

Chapter 14
Knowledge Engineering and Planning for Social Human–Robot Interaction: A Case Study

Ronald P. A. Petrick and Mary Ellen Foster

Abstract The core task of automated planning is goal-directed action selection; this task is not unique to the planning community, but is also relevant to numerous other research areas within AI. One such area is interactive systems, where a fundamental component called the interaction manager selects actions in the context of conversing with humans using natural language. Although this has obvious parallels to automated planning, using a planner to address the interaction management task relies on appropriate engineering of the underlying planning domain and planning problem to capture the necessary dynamics of the world, the agents involved, their actions, and their knowledge. In this chapter, we describe work on using domain-independent automated planning for action section in social human–robot interaction, focusing on work from the JAMES (Joint Action for Multimodal Embodied Social Systems) robot bartender project.

1 Introduction

At a high level, automated planning can be viewed as a problem of context-dependent action selection: given a set of initial state conditions, action descriptions, and goals, the planner must generate a sequence of actions whose application to the initial state will bring about the goal conditions. However, this view of action selection is not unique to planning. One important area where this problem is also of primary concern is in **interactive systems**, a subfield of natural language dialogue that is focused on implementing tools and applications for interacting with human users.

R. P. A. Petrick (✉)
Department of Computer Science, Heriot-Watt University, Edinburgh, UK
e-mail: R.Petrick@hw.ac.uk

M. E. Foster
School of Computing Science, University of Glasgow, Glasgow, UK
e-mail: MaryEllen.Foster@glasgow.ac.uk

M. Vallati, D. Kitchin (eds.), *Knowledge Engineering Tools and Techniques for AI Planning*, https://doi.org/10.1007/978-3-030-38561-3_14

A fundamental component in the construction of an interactive system, such as a robot that is able to converse with a human using natural language, is the **interaction manager** [6], whose primary task is to carry out a form of action selection: based on the current state of the interaction and of the world, the interaction manager makes a high-level decision as to which spoken, non-verbal, and task-based actions should be taken next by the system as a whole. Compared with more formal, descriptive accounts of dialogue which aim to model the full generality of language use [3], work on interaction management has concentrated primarily on developing end-to-end systems that operate in specific task settings, and on evaluating them through interaction with human users [20, 27].

In contrast, the planning community has addressed the problem of high-level action selection through the development of domain-independent planners: systems that employ general-purpose problem-solving techniques that can be applied to a wide range of planning domains and problems, modelled in common representation language such as PDDL [26]. Action selection strategies are regularly compared within this common context, especially through events like the International Planning Competitions [16], while the representation languages themselves are often studied to better understand their expressiveness and applicability [36]. Applying planning tools to a complex scenario, therefore, involves appropriate engineering of the underlying planning domain and planning problem, to capture the necessary dynamics of the world, the agents involved, their actions, and their knowledge, for a suitable choice of planning system and representation language—and then often integrated as part of a larger system.

While the link between automated planning and natural language processing has a long tradition, the planning approach to natural language interaction has for the most part been largely overlooked more recently. In this chapter, we describe work on using domain-independent automated planning for action section in human–robot interaction, using an application from the JAMES (Joint Action for Multimodal Embodied Social Systems)[1] robot bartender project [35]. We survey recent work in the interactive systems community in the form of toolkits used for constructing interactive dialogue systems. We then describe how we use knowledge engineering techniques to perform similar tasks with an epistemic automated planning system. In the specific context of the JAMES robot system, we show how social states are inferred from low-level sensors, using vision and speech as input modalities; how planning domains and problems are modelled for the bartending scenario; and how an epistemic planner is used to construct plans with task, dialogue, and social actions, as an alternative to other methods of interaction management.

[1]http://james-project.eu/.

2 Interaction Management

Since both interaction management and automated planning deal with goal-directed action selection, in principle interaction management presents an opportunity for showcasing planning tools and demonstrating how different approaches can be applied, benchmarked, and compared. Although early applications of planning in this area can be traced back to the 1980s [2, 7, 15, 32, 39], the planning approach has for the most part been largely overlooked more recently. Instead, interactive systems researchers tend to use purpose-built toolkits for constructing end-to-end dialogue systems. Foster and Petrick [9] present a survey of such toolkits; we summarise the main features of some of these toolkits below.

An interaction management toolkit generally incorporates three main features. First, it provides a representational formalism for specifying states and actions. Second, the state/action representation is usually tightly linked to a reasoning strategy that is used to carry out action selection. Finally, most toolkits also include a set of infrastructure building tools designed to support modular system development. While these three features can clearly simplify the task of implementing an individual end-to-end system, the fact that the features are so tightly connected does complicate the task of comparing representational formalisms or reasoning strategies: in general, to carry out such a comparison, there is no alternative but to re-implement the entire system in multiple frameworks [28, 31].

Historically, one of the most widely used approaches to dialogue management was the information state update (ISU) approach, which is exemplified by the TrindiKit toolkit [22]. The core of this approach is the use of an information state which represents the state of the dialogue and which is updated by applying update rules following a given update strategy. A similar ISU approach has also been taken in more recent dialogue systems, but using other infrastructures [18, 19]. A more recent approach is exemplified by OpenDial [23], an open-source toolkit designed to support robust dialogue management, using a hybrid framework that combines logical and statistical approaches through probabilistic rules to represent the internal models of the framework. OpenDial also includes a Java-based blackboard architecture where all modules are connected to a central information hub which represents the dialogue state, along with a plugin framework allowing new modules to be integrated.

Many modern interactive systems are built with online toolkits such as the Amazon Alexa Skills Kit [1] or Dialogflow [14]—these toolkits generally use machine learning to learn the correct responses to user actions given sample inputs. One current interactive system which does incorporate aspects of automated planning is the MuMMER social robot [29], which combines a planner used for action selection with a more traditional dialogue manager. Other approaches [4, 5, 21] have also explored the use of planning for dialogue and interaction, while recent work on explainable planning [12] has also highlighted the links between planning and user interaction.

3 Task-Based Social Interaction: A Robot Bartender Scenario

The goal of this work is to use domain-independent planning as the high-level decision-making mechanism for action selection in an interactive robot system. In particular, the target domain for this work is a task-based human–robot interaction scenario involving a bartending robot, as shown in Fig. 14.1 (left). In this setting, the robot acts as a bartender that serves customers that approach the bar area seeking attention. The robot hardware itself consists of two 6 degrees-of-freedom industrial manipulator arms with grippers, mounted to resemble human arms. Sitting on the main robot torso is an animatronic talking head capable of producing facial expressions, rigid head motion, and lip-synchronised synthesised speech.

A sample bartender interaction is shown in Fig. 14.2. In this example, two people enter the bar area and attempt to order drinks from the robot. During the interaction, a third person approaches the bar and attempts to attract the attention of the bartender. When the third customer appears while the bartender is engaged with the first two customers, the bartender reacts appropriately by telling the third customer to wait, finishing the transaction with the first two people, and then serving the third customer.

Even this simple interaction presents challenges to the robot system tasked with the role of the bartender (see Fig. 14.1, right): the **visual processor** system must track the locations and body postures of the agents; the **speech recogniser** must detect and deal with speech in an open setting and, using the **parser**, transform the input into a logical form representing the speech; the **state manager** must make sense of the social scene using the processed input modalities; the **planner** and **execution monitor** must determine who requires attention and should ensure that they are served correctly, while appropriately responding to unexpected outcomes as they arise; while the **output planner** must select and execute concrete behaviours for each output channel to correctly realise high-level actions, communicating with the **talking-head controller** and **robot motion planner**.

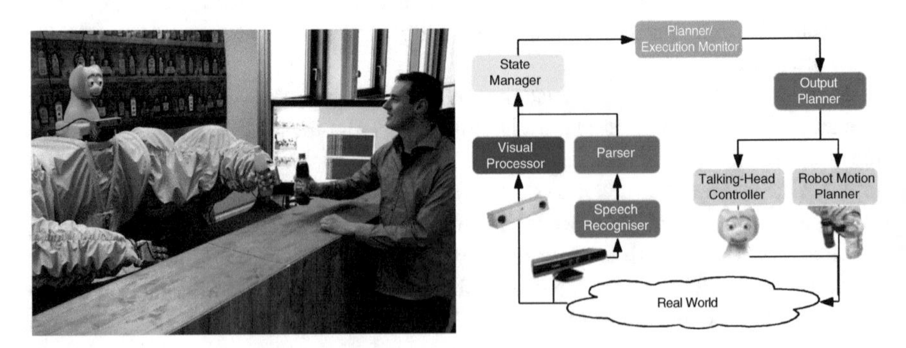

Fig. 14.1 The robot bartender and bar setting (left) and the software architecture for the robot (right)

Interaction		Action type
Two people, A and B, each individually approach and look at the robot bartender		
ROBOT:	[Looks at Person A] How can I help you?	Sensing action
PERSON A:	A pint of cider, please.	
Person C approaches the bar and tries to attract the attention of the bartender		
ROBOT:	[Looks at Person C] One moment, please.	Social action
ROBOT:	[Serves Person A]	Physical action
ROBOT:	[Looks at Person B] What will you have?	Sensing action
PERSON B:	A glass of red wine.	
ROBOT:	[Serves Person B]	Physical action
ROBOT:	[Looks at Person C] Thanks for waiting.	Social action
	How can I help you?	Sensing action
PERSON C:	I'd like a pint of beer.	
ROBOT:	[Serves Person C]	Physical action

Fig. 14.2 An example interaction in the robot bartending scenario

From a high-level planning perspective, the task of applying planning in this scenario centres around the knowledge engineering task of accurately modelling the states, actions, and goals that reflect the types of activities the robot is expected to perform. Considering the sample interaction, this includes a mixture of physical actions in the underlying task domain (e.g., serving the actual drinks), sensing actions that acquire new information (e.g., asking a customer for a drink order), and social actions that help facilitate the interactive context (e.g., thanking a customer). As a result, we also require a suitably expressive representation that enables such actions to co-exist within a planning domain. This task is further complicated by the fact that the planner is a single component situated in a much larger architecture, with the representation of states, actions, and goals having connections to the input and output modalities processed by other system components.

4 Modelling Social Human–Robot Interaction for Planning

In this section, we describe how planning techniques are applied to the problem of social human–robot interaction in the robot bartender scenario by considering how states, actions, and goals are modelled. We begin by presenting an overview of the particular planner used in this work, the epistemic PKS planner; we then describe how states are inferred from the low-level sensor data and how those states are translated into the representations used by PKS in the context of user interaction.

4.1 Planning with Knowledge and Sensing

The high-level planner is responsible for selecting robot actions to respond appropriately in the current scenario state. Since the activities of the robot include a mix of physical, dialogue, and social behaviours, the representation language of the planner must be able to support such action models. In this work, we use PKS (Planning with Knowledge and Sensing) [33, 34], a contingent planner that works with incomplete information and sensing actions. PKS is an **epistemic planner** that operates at the **knowledge level** and reasons about how its knowledge state, rather than the world state, changes due to action. To do this, PKS works with a restricted first-order language with limited inference. While features such as functions and run-time variables are supported, these restrictions mean that some types of knowledge (e.g., general disjunctive information) cannot be modelled.

PKS is based on a generalisation of STRIPS [8]. In STRIPS, the state of the world is modelled by a single database. Actions update this database and, by doing so, update the planner's world model. In PKS, the planner's knowledge state is represented by a set of five databases, each of which models a particular type of knowledge, and can be understood in terms of a modal logic of knowledge. Actions can modify any of the databases, which update the planner's knowledge state. To ensure efficient inference, PKS restricts the type of knowledge it can represent:

`Kf`: This database is like a STRIPS database except that both positive and negative facts are permitted and the closed world assumption is not applied. `Kf` is used to model action effects that change the world and can include any ground literal or function (in)equality mapping ℓ, where $\ell \in$ `Kf` means "the planner knows ℓ".

`Kw`: This database models the plan-time effects of sensing actions that have one of two possible outcomes. $\phi \in$ `Kw` means that at plan time the planner either "knows ϕ or knows $\neg\phi$", and that at run time this disjunction will be resolved. PKS uses such information to build contingent branches in a plan, where each branch assumes one of the possible outcomes is true.

`Kv`: This database stores information about function values that will become known at execution time. `Kv` can model the plan-time effects of sensing actions that return a range of possible constants, where any unnested function term $f \in$ `Kv` means that at plan time the planner "knows the value of f". At execution time, the planner will have definite information about f's value. As a result, PKS can use `Kv` terms as run-time variables in its plans, and can build conditional plan branches when the set of possible mappings for a function is restricted.

`Kx`: This database models the planner's "exclusive or" knowledge. Entries in `Kx` have the form $(\ell_1 \mid \ell_2 \mid \ldots \mid \ell_n)$, where each ℓ_i is a ground literal. Such formulae represent a type of disjunctive knowledge common in planning domains, namely that "exactly one of the ℓ_i is true".

(A fifth database modelling local closed world (LCW) information is not used.) Questions about the knowledge state are answered using a set of **primitive queries**:

`K(`$\boldsymbol{\phi}$**`)`**:	is ϕ known to be true?
`Kw(`$\boldsymbol{\phi}$**`)`**:	does the planner know whether ϕ is true or not?
`Kv(t)`:	does the planner know the value of t?

The negation of the queries is also permitted. An inference procedure evaluates the queries by checking the database contents and applying a set of reasoning rules.

Actions in PKS are modelled by a set of **preconditions** that query the knowledge state and a set of **effects** that update the knowledge state. Preconditions are simply a list of primitive queries. Effects are described by a collection of STRIPS-style "add" and "delete" operations that modify the contents of individual databases. For example, `add(Kf, ϕ)` adds ϕ to the `Kf` database, while `del(Kw, ϕ)` removes ϕ from `Kw`.

PKS builds plans by reasoning about actions in a forward-chaining manner: if the preconditions of a chosen action are satisfied by the knowledge state, then the action's effects are applied to produce a new knowledge state. Planning then continues from the resulting state. PKS can also build plans with branches, by considering the possible outcomes of its `Kw` and `Kv` knowledge. Planning continues along each branch until it satisfies the **goal** conditions, also specified as a list of primitive queries.

4.2 State Management

For PKS to operate successfully in the context of the larger robot system, it requires a discrete representation of the world, the robot, and all entities in the scene, integrating social, interaction-based, and task-based properties. Converting the continuous, low-level sensor information into the discrete states is the job of the state manager.

The social state is represented as a list of properties and their values, where every relation in the state has an associated **confidence** value, represented as a number between 0 and 1. In addition, every relation in the state can potentially have **multiple values**, with each possible value having its own confidence. Table 14.1 shows a sample social state using this representation, including multiple possible values for the `drinkOrder(A1)` relation. Social state properties fall into two main categories: properties that are directly transferred from the input sensors such as `headPos` (which tracks the 3D position of each customer's head), as well as derived properties such as `lastSpeaker` and `seeksAttention` which are computed by the state manager based on the input data.

Table 14.1 Excerpt of a social state identified by the state manager

Property	Value	Confidence
`seeksAttention(A1)`	`true`	0.75
`seeksAttention(A2)`	`false`	0.45
`lastSpeaker()`	`A1`	1.00
`lastEvent()`	`userSpeech(A1)`	1.00
`drinkOrder(A1)`	`green lemonade`	0.68
	`blue lemonade`	0.32
`lastAct(A1)`	`greet`	0.25

For speech, the speech recogniser produces an n-best list of recognition hypotheses, each with an estimated confidence score, along with an estimate of the sound source angle and the angle confidence. The recognised hypotheses are parsed to extract the syntactic and semantic information using a grammar implemented in OpenCCG [38], while the source angle is used together with the location information from vision to estimate which of the customers in the scene is most likely to have been speaking (`lastSpeaker`). If a possible speaker is found, the semantic information from speech is used to update `lastAct`. In the case that the customer says something regarding their drink order, we also update the value of `drinkOrder`, using a generic belief tracking procedure proposed by [37], which maintains beliefs over user goals based on a small number of domain-independent rules using basic probability operations. This enables us to maintain a dynamically updated list of the possible drink orders made by each customer, with an associated confidence value for each.

Information from the robot bartender's vision system [30] provides a continuous estimate of the location, gaze behaviour, and body language of all people in the scene in real time. Every feature reported by the vision system includes an estimated confidence value, which is incorporated into the state and also used for further processing. The information from the vision system contributes to the processing of speech as outlined above; it is also used to estimate which customer(s) are currently seeking attention (`seeksAttention`). `seeksAttention` is one of the most important properties required for the bartender scenario, and we have experimented with several methods of estimating it, including a rule based on the observation of customers in a real bar [24] and a set of classifiers trained on annotated robot bartender interactions [11].

4.3 Representing Properties, Actions, Objects, and Goals

The properties, actions, and goals that make up the planning domain definition are built on the state properties defined by the state manager but exist at a higher level of representation local to the planning system. All of the robot's high-level actions in the bartending scenario (physical, dialogue, and social) are modelled as part of the same planning domain, rather than using specialised tools for certain aspects of the problem (e.g., separating task and dialogue) as is common practice in many modern interactive systems. As a result, the planning domain representation must capture the dynamics of the task, the world, the agents, and the available objects.

Planning domain properties in the bartender scenario are shown in Table 14.2 (left). These properties are defined at a high level of abstraction and in many cases are based on the properties defined by the state manager. For instance, the planning property `seeksAttn(?a)` corresponds to the state manager property `seeksAttention`, while a property like `badASR(?a)` is extracted from the confidence values of other properties maintained by the state manager. Other planning properties like `greeted(?a)` or `served(?a)` do not have a direct analogue in the state manager but are instead derived from a set of properties being tracked at that level.

Table 14.2 Example of properties and action in the robot bartender domain

Properties		Actions	
`seeksAttn(?a)`	?a is seeking attention	`greet(?a)`	Greet ?a
`greeted(?a)`	?a has been greeted	`ask-drink(?a)`	Ask ?a for a drink order
`ordered(?a)`	?a has ordered	`ack-order(?a)`	Acknowledge ?a's order
`ackOrder(?a)`	?a's order has been acknowledged	`serve(?a,?d)`	Serve drink ?d to ?a
`served(?a)`	?a has been served	`wait(?a)`	Tell ?a to wait
`otherAttnReq`	Other agents are seeking attention	`ack-wait(?a)`	Thank ?a for waiting
`badASR(?a)`	?a was not understood	`inform(?a,?d)`	Tell ?a about drink ?d
`transEnd(?a)`	The transaction with ?a has ended	`bye(?a)`	End interaction with ?a
`inTrans=?a`	The robot is interacting with ?a	`not-understood(?a)`	alert that ?a was not understood
`request(?a)=?d`	?a has requested drink ?d		

```
action greet(?a : agent)
    preconds: K(inTrans = nil)
              !K(greeted(?a))
              K(seekAttn(?a))
              !K(ordered(?a))
              !K(otherAttnReq)
              !K(badASR(?a))
    effects:  add(Kf,greeted(?a))
              add(Kf,inTrans = ?a)

action bye(?a : agent)
    preconds: K(inTrans = ?a)
              K(served(?a))
              !K(otherAttnReq)
              !K(badASR(?a))
    effects:  add(Kf,inTrans = nil)

action ask-drink(?a : agent)
    preconds: K(inTrans = ?a)
              !K(ordered(?a))
              !K(otherAttnReq)
              !K(badASR(?a))
    effects:  add(Kf,ordered(?a))
              add(Kv,request(?a))

action serve(?a : agent, ?d : drink)
    preconds: K(inTrans = ?a)
              K(ordered(?a))
              Kv(request(?a))
              K(request(?a) = ?d)
              !K(otherAttnReq)
              !K(badASR(?a))
    effects:  add(Kf,served(?a))
```

Fig. 14.3 Example encoding of PKS actions in the robot bartender domain

Actions in the bartending domain are also described at a high level of abstraction, and are inspired by studies of human customers ordering drinks from real bartenders in real bars [24]. A list of the available actions is given in Table 14.2 (right) with the PKS encoding for a selection of actions shown in Fig. 14.3. The available list includes a mix of physical, dialogue, and social actions to reflect some of the behaviours that arise in typical interactions (e.g., as in Fig. 14.2). For instance, `serve` is a standard planning action with a deterministic effect (i.e., it adds definite knowledge to PKS's `Kf` database so the planner comes to know particular facts like the customer has been served); however, when executed it causes the robot to hand over a drink to an agent and confirm the drink order through speech. Actions like `greet`, `ack-order`, and `bye` are modelled in a similar way, but only map to speech output at run time (e.g., "hello", "okay", and "good-bye"). The `inform` action is used to supply information about specific drinks in response to a customer query. The most interesting action is `ask-drink` which is modelled as a sensing or knowledge-producing action: the function term `request` is added to the planner's `Kv` database as an effect, indicating that this piece of information will become known at execution time. In other words, the planner will come to know the value of the drink the customer requested. The `not-understand` action is used as a directive to the speech output system to produce an utterance that (hopefully) causes the agent to repeat its last response. The `wait` and `ack-wait` actions control interactions when multiple agents are seeking the attention of the bartender.

The planning domain model also includes a list of the objects (drinks) and agents (customers) in the bar. This information is not hard-coded but is instead provided to the planner dynamically by the state manager, based on real-time observations provided by the input sensors, and defined as part of the planning problem's initial state (denoted in PKS syntax using two defined types, `drink` and `agent`). Changes in the object or agent list, when identified by the state manager, are also sent to the planner, causing it to update its domain model. Initially, the `inTrans` function is initially set to `nil` to indicate that the robot is not interacting with any customers. The planner's goal is simply to serve each agent seeking attention, i.e.,

```
forallK(?a : agent) K(seeksAttn(?a)) ⇒ K(transEnd(?a)).
```

This goal is viewed as a rolling target which is reassessed each time a state update is received from the state manager.

5 Planning for Social Human–Robot Interaction

Using the planning model defined above, plans can now be generated to respond to many common interactive situations that arise in the bartender domain. This process is triggered by the appearance of agents (customers) in the scene which are reported to be seeking attention by the state manager. The planner responds by attempting to generate a plan to achieve the goal of serving all agents in the bar. Here we consider the generated behaviour in a number of common scenarios.

5.1 Ordering a Drink

The simplest interactive situation in the bartender domain is the case where a single agent `A1` is seeking attention in the bar, represented by the state manager adding a new fact `seeksAttn(A1)` to the initial `Kf` database. Initially, the robot is not interacting with any agent (inTrans = nil ∈ Kf). In response, the planner can build the following plan to achieve the goal:

greet(A1)	Greet agent A1
ask-drink(A1)	Ask A1 for drink order
ack-order(A1)	Acknowledge A1's drink order
serve(A1,request(A1))	Serve A1 the drink they requested
bye(A1)	End the transaction

Initially, the planner can choose greet(A1) since no transaction is taking place (inTrans = nil ∈ Kf) and A1 is seeking attention (seeksAttn(A1) ∈ Kf). The other preconditions of greet(A1) are trivially satisfied (i.e., none of greeted(A1), ordered(A1), otherAttnReq, or badASR(A1) are in Kf). After greeting A1, the ask-drink(A1) action is then chosen, updating the planner's knowledge state so that ordered(A1) ∈ Kf and request(A1) ∈ Kv, i.e., the planner knows that A1 has ordered and knows the value of the drink that was requested. The ack-order(A1) is then selected to acknowledge the drink order to the customer. The most interesting action in the plan is serve(A1,request(A1)) which, intuitively, has the effect of "serving A1 the drink that A1 requested". This follows as a consequence of the planner knowing the value of request(A1), which is recorded in the planner's Kv database. Thus, request(A1) acts as a run-time variable whose definite value (A1's actual drink order) will become known at run time. Finally, after serving the drink the bye(A1) action can be selected, resulting in inTrans = nil ∈ Kf and transEnd(A1) ∈ Kf, thereby ending the transaction and satisfying the goal.

5.2 Ordering Drinks with Multiple Agents

The planning domain model also enables more than one agent to be served if multiple customers are reported as seeking attention. For instance, in the case of two agents, A1 and A2, the following plan might be built:

wait(A2)	Tell agent A2 to wait
greet(A1)	Greet agent A1
ask-drink(A1)	Ask A1 for drink order
ack-order(A1)	Acknowledge A1's drink order
serve(A1,request(A1))	Give the drink to A1
bye(A1)	End A1's transaction
ack-wait(A2)	Thank A2 for waiting
ask-drink(A2)	Ask A2 for drink order
ack-order(A2)	Acknowledge A2's drink order
serve(A2,request(A2))	Give the drink to A2
bye(A2)	End A2's transaction

Thus, `A1`'s drink order is taken and processed, followed by `A2`'s order. The `wait` and `ack-wait` actions (which are not needed in the single-agent plan) act as social actions that are used to defer a transaction with `A2` until `A1`'s transaction has finished. (The `otherAttnReq` property, whose value depends on `seeksAttn`, ensures that other agents seeking attention are told to wait before an agent is served.)

Larger number of customers result in plans with the same general structure. For example, a plan for three agents, `A1`, `A2`, and `A3`, would look like the following:

`wait(A2)`	Tell agent `A2` to wait
`wait(A3)`	Tell agent `A3` to wait
`greet(A1)`	Greet agent `A1`
...	Transact with `A1`
`bye(A1)`	End `A1`'s transaction
`ack-wait(A2)`	Thank `A2` for waiting
...	Transact with `A2`
`bye(A2)`	End `A2`'s transaction
`ack-wait(A3)`	Thank `A3` for waiting
...	Transact with `A3`
`bye(A3)`	End `A3`'s transaction

Similarly, if a new customer appears, it is dynamically reported to the planner, possibly triggering a replanning operation: the newly built plan might result in the extension of an existing plan (which might reflect a transaction currently in progress) to include actions for interacting with the new agent if they are seeking attention. However, it is important to note that we are not just stitching together single-agent plans to account for the number of agents in the scenario. Instead, the planner generates a plan appropriate to the social context in response to the state information reported to it by the state manager.

5.3 *Ordering a Drink with Restricted Drink Choices*

From a planning point of view, the above plans rely on the planner's ability to reason about particular types of knowledge (e.g., functions like `request(A1)`) which act as variables in parameterised plans. However, an alternative type of plan can also be built in the case that the possible set of drinks is explicitly restricted. For instance, consider a single-agent `A1` seeking attention, where the planner also told there are three possible drinks that can be ordered: juice, water, and beer. This information is represented in the planner as a type of "exclusive or" knowledge in the `Kx` database:

$$\texttt{request(A1)} = \texttt{juice} \mid \texttt{request(A1)} = \texttt{water} \mid \texttt{request(A1)} = \texttt{beer}$$

The planner can now build a plan of the following form to serve the customer:

`greet(A1)`	Greet agent `A1`
`ask-drink(A1)`	Ask `A1` for drink order
`ack-order(A1)`	Acknowledge `A1`'s order
`branch(request(A1))`	*Form conditional plan*
`  K(request(A1) = juice):`	*If* `juice` *was requested*
`    ...`	
`    serve(A1,juice)`	Serve juice to `A1`
`  K(request(A1) = water):`	*If* `water` *was requested*
`    ...`	
`    serve(A1,water)`	Serve water to `A1`
`  K(request(A1) = beer):`	*If* `beer` *was requested*
`    ...`	
`    serve(A1,beer)`	Serve beer to `A1`
`bye(A1)`	End the transaction

In this case, a contingent plan is built with branches for each possible mapping of `request(A1)`. For example, in the first branch `request(A1) = juice` is assumed to be true; in the second branch `request(A1) = water` is true; and so on. Planning continues along each branch under the given assumption. (We note that this type of branching is only possible here because the planner had initial `Kx` knowledge that restricted `request(A1)`, combined with `Kv` knowledge provided by the `ask-drink` action.) Along each branch, an appropriate `serve` action is added to deliver the appropriate drink. The places in the plan indicated by "`...`" indicate places where drink-specific interactions (subdialogues) could be inserted. For instance, each branch may require different actions to serve a drink, such as putting the drink in a special glass, or requesting additional information from the customer (i.e., "would you like ice in your water?").

6 Plan Execution, Monitoring, and Recovery

Once a plan is built, it is executed by the robot one action at a time. A plan execution monitor tracks the plan, comparing the expected plan states against sensed states provided by the state manager, to determine whether a plan should continue to be executed. To do this, it tries to ensure that a state still permits the next action (or set of actions) in the plan to be executed and that effects needed by actions or goals later in the plan have been achieved as expected. In the case of a mismatch, the planner is directed to build a new plan, using the sensed state as a new initial state.

The execution of individual actions is handled by dividing each high-level planned action into specific output modalities—speech, head motions, and arm manipulation behaviour—that can be executed by the robot. This mapping is specified by a simple rule-based structure containing specifications of each output [17]. The resulting structure is then passed to the multimodal output planner, which mediates execution to each output channel. Language output is specified in terms of communicative acts based on rhetorical structure theory (RST) [25], using a generation module that translates RST into speech by the robot's animatronic head. The robot also expresses itself through facial expressions, gaze, and arm motions. The animatronic head can express a number of predefined expressions, while the robot arm can perform tasks like grasping objects (e.g., to hand over a drink to a customer). Multimodal behaviour is coordinated across the various output channels to ensure they are synchronised temporally and spatially. For instance, an action `serve(?a,?d)` to serve an agent a drink might be transformed into multimodal outputs that result in the robot smiling at `?a` (an animatronic head facial expression) while physically handing over drink `?d` (a robot arm manipulation action) and saying to the customer "here is your drink" (speech output).

If the plan execution monitor detects a situation where a plan has failed, for instance, due to unexpected outcomes like action failure, the planner is invoked to construct a new plan. This method is particularly useful for responding to unexpected responses by agents interacting with the robot. For example, if the planner receives a report that an agent `A1`'s response to `ask-drink(A1)` was not understood due to low-confidence automatic speech recognition, the state report sent to the planner will have no value for `request(A1)`, and `badASR(A1)` will be set to true. This situation will be detected by the plan execution monitor and the planner will be directed to build a new plan. One possible result is a modified version of the original plan that first informs `A1` they were not understood before repeating the `ask-drink` action and continuing the plan:

`ask-drink(A1)`	Ask `A1` for drink order
`???`	`A1` was not understood
`[Replan]`	Replan
`not-understood(A1)`	Alert `A1` it was not understood
`ask-drink(A1)`	Ask `A1` again for drink order
`...`	

Another consequence of this approach is that certain types of overanswering can be handled through plan execution monitoring and replanning. For instance, a `greet(A1)` action by the robot might cause the customer to respond with an utterance that includes a drink order:

`greet(A1)`	Greet `A1`
`???`	`A1` says "I'd like a beer"
`[Replan]`	Replan
`ack-order(A1)`	Acknowledge `A1`'s drink order
`serve(A1,request(A1))`	Serve `A1` their drink
`...`	

In this case, the state manager would include `request(A1)=beer` in its state report, along with `ordered(A1)`. The execution monitor would detect that the preconditions of `ask-drink(A1)` are not met and direct the planner to replan. A new plan could then omit `ask-drink` and proceed to acknowledge and serve the requested drink.

7 Discussion and Conclusions

This chapter has described how automated planning can be applied to the problem of social human–robot interaction in the JAMES robot bartending domain, as an alternative to mainstream approaches to interaction management. In particular, we have shown how the planning representation is engineered from social states induced from different input modalities, and how plans are built incorporating a mix of task, dialogue, and social actions, with execution involving various output modalities on the robot. The use of the epistemic PKS planner has also provided certain benefits during the work, such as enabling sensing actions to be used to model certain types of dialogue actions, generating parameterised high-level plans, and considering subdialogues with contingent branches.

The planning approach has also presented certain technical advantages. For instance, the JAMES robot system has been evaluated through a series of user studies aimed at exploring socially appropriate interaction in the bartender scenario, where participants successfully ordered and received drinks from the bartender [10]. An interesting variant of the study compared the full planning domain described above with a domain that dealt with task-based actions only. From a representation point of view, this was done by simply removing the social actions from the domain model. Results showed that the social version led to more efficient dialogues [13]. Another variant of multiple customer drink ordering also considered different ordering strategies when agents arrive in groups (e.g., interacting with a group representative versus transacting with all agents in a single group before moving to another group). Again, the changes required to support planning in this new setting resulted from modifications to the domain model: in this case adding a new property to track agents in groups, and introducing another type of drink ordering action to accommodate multiple agents ordering drinks in a group.

More generally, interactive systems also offer several opportunities for the automated planning community to showcase their tools and techniques. For instance, interaction problems could form the basis for new challenge domains in planning, and the standard planning representation languages offer an approach to modelling problems that break the tight link between representation and reasoning that is often found in interaction toolkits. There are lessons that the planning community can also learn from the interactive systems community. For example, the issue of user evaluation is at the heart of interactive systems research, with a focus on (non-expert) users interacting with the developed tools. The fact that interactive systems are also inherently application driven means that planning must be situated in the context of larger, more complex systems, requiring a degree of maturity and robustness in development that often goes beyond lab settings, but which could facilitate the wider adoption of planning approaches in such settings. Our ongoing research aims to address some of these issues by adapting our planning techniques to other types of service robots and scenarios that involve interacting with humans in public spaces.

Acknowledgements The authors thank their JAMES colleagues who helped implement the bartender system: Andre Gaschler, Manuel Giuliani, Amy Isard, Maria Pateraki, and Richard Tobin. This research has received funding from the European Union's 7th Framework Programme under grant No. 270435 (JAMES, http://james-project.eu/).

References

1. Amazon (2020) Alexa Skills Kit Official Site. https://developer.amazon.com/en-GB/alexa/alexa-skills-kit, accessed: 2020-02-09
2. Appelt D (1985) Planning English Sentences. Cambridge University Press
3. Asher N, Lascarides A (2003) Logics of Conversation. Cambridge University Press
4. Benotti L (2008) Accommodation through tacit sensing. In: Proceedings of LONDIAL 2008, London, United Kingdom, pp 75–82
5. Brenner M, Kruijff-Korbayová I (2008) A continual multiagent planning approach to situated dialogue. In: Proceedings of LONDIAL 2008, pp 67–74
6. Bui TH (2006) Multimodal dialogue management - state of the art. Tech. Rep. 06–01, University of Twente (UT), Enschede, The Netherlands
7. Cohen P, Levesque H (1990) Rational interaction as the basis for communication. In: Intentions in Communication, MIT Press, Cambridge, MA, pp 221–255
8. Fikes RE, Nilsson NJ (1971) STRIPS: A new approach to the application of theorem proving to problem solving. Artificial Intelligence 2:189–208
9. Foster ME, Petrick RPA (2017) Separating representation, reasoning, and implementation for interaction management: Lessons from automated planning. In: Dialogues with Social Robots: Enablements, Analyses, and Evaluation, Springer Singapore, Singapore, pp 93–107, https://doi.org/10.1007/978-981-10-2585-3_7
10. Foster ME, Gaschler A, Giuliani M, Isard A, Pateraki M, Petrick RPA (2012) Two people walk into a bar: Dynamic multi-party social interaction with a robot agent. In: Proceedings of ICMI 2012, pp 3–10, https://doi.org/10.1145/2388676.2388680
11. Foster ME, Gaschler A, Giuliani M (2017) Automatically classifying user engagement for dynamic multi-party human–robot interaction. International Journal of Social Robotics 9(5):659–674, https://doi.org/10.1007/s12369-017-0414-y
12. Fox M, Long D, Magazzeni D (2017) Explainable planning. In: Proceedings of the IJCAI Workshop on Explainable AI
13. Giuliani M, Petrick RPA, Foster ME, Gaschler A, Isard A, Pateraki M, Sigalas M (2013) Comparing task-based and socially intelligent behaviour in a robot bartender. In: Proceedings of ICMI 2013, https://doi.org/10.1145/2522848.2522869
14. Google (2020) Dialogflow. https://dialogflow.com/, accessed: 2020-02-09
15. Hovy E (1988) Generating natural language under pragmatic constraints. Lawrence Erlbaum Associates, Hillsdale, NJ, USA
16. ICAPS (2019) ICAPS Competitions. http://www.icaps-conference.org/index.php/Main/Competitions, accessed: 2019-08-01

17. Isard A, Matheson C (2012) Rhetorical structure for natural language generation in dialogue. In: Proceedings of SemDial-2012 (SeineDial), pp 161–162
18. Janarthanam S, Hastie H, Deshmukh A, Aylett R, Foster ME (2015) A reusable interaction management module: Use case for empathic robotic tutoring. In: Proceedings of goDIAL 2015, Gothenburg, Sweden
19. Johnston M, Bangalore S, Vasireddy G, Stent A, Ehlen P, Walker M, Whittaker S, Maloor P (2002) MATCH: An architecture for multimodal dialogue systems. In: Proceedings of ACL 2002, Philadelphia, Pennsylvania, USA, pp 376–383
20. Jokinen K, McTear M (2009) Spoken dialogue systems. Synthesis Lectures on Human Language Technologies 2(1):1–151
21. Koller A, Stone M (2007) Sentence generation as planning. In: Proceedings of ACL 2007, Prague, Czech Republic, pp 336–343
22. Larsson S, Traum DR (2000) Information state and dialogue management in the TRINDI dialogue move engine toolkit. Natural Language Engineering 6(3&4):323–340, https://doi.org/10.1017/S1351324900002539
23. Lison P (2015) A hybrid approach to dialogue management based on probabilistic rules. Computer Speech & Language https://doi.org/10.1016/j.csl.2015.01.001
24. Loth S, Huth K, De Ruiter JP (2013) Automatic detection of service initiation signals used in bars. Frontiers in Psychology 4(557), https://doi.org/10.3389/fpsyg.2013.00557
25. Mann WC, Thompson SA (1988) Rhetorical structure theory: Toward a functional theory of text organization. Text 8(3):243–281
26. McDermott D, Ghallab M, Howe A, Knoblock C, Ram A, Veloso M, Weld D, Wilkins D (1998) PDDL – The Planning Domain Definition Language (Version 1.2). Technical Report CVC TR-98-003/DCS TR-1165, Yale Center for Computational Vision and Control
27. McTear M, Callejas Z, Griol D (2016) The Conversational Interface. Springer International Publishing, https://doi.org/10.1007/978-3-319-32967-3
28. Olaso JM, Milhorat P, Himmelsbach J, Boudy J, Chollet G, Schlögl S, Torres MIT (2016) A multi-lingual evaluation of the vAssist spoken dialog system: Comparing Disco and RavenClaw. In: Proceedings of IWSDS 2016, Saariselkä, Finland
29. Papaioannou I, Dondrup C, Lemon O (2018) Human-robot interaction requires more than slot filling - multi-threaded dialogue for collaborative tasks and social conversation. In: Proceedings of the FAIM/ISCA Workshop on Artificial Intelligence for Multimodal Human Robot Interaction, pp 61–64
30. Pateraki M, Sigalas M, Chliveros G, Trahanias P (2013) Visual human-robot communication in social settings. In: Proceedings of ICRA Workshop on Semantics, Identification and Control of Robot-Human-Environment Interaction
31. Peltason J, Wrede B (2011) The curious robot as a case-study for comparing dialog systems. AI Magazine 32(4):85–99, https://doi.org/10.1609/aimag.v32i4.2382
32. Perrault CR, Allen JF (1980) A plan-based analysis of indirect speech acts. American Journal of Computational Linguistics 6(3–4):167–182
33. Petrick RPA, Bacchus F (2002) A knowledge-based approach to planning with incomplete information and sensing. In: Proceedings of AIPS 2002, pp 212–221
34. Petrick RPA, Bacchus F (2004) Extending the knowledge-based approach to planning with incomplete information and sensing. In: Proceedings of ICAPS 2004, pp 2–11
35. Petrick RPA, Foster ME (2013) Planning for social interaction in a robot bartender domain. In: Proceedings of ICAPS 2013, Rome, Italy
36. Rintanen J (2004) Complexity of planning with partial observability. In: Proceedings of ICAPS 2004, pp 345–354
37. Wang Z, Lemon O (2013) A simple and generic belief tracking mechanism for the dialog state tracking challenge: On the believability of observed information. In: Proceedings of SIGDIAL 2013
38. White M (2006) Efficient realization of coordinate structures in Combinatory Categorial Grammar. Research on Language and Computation 4(1):39–75
39. Young RM, Moore JD (1994) DPOCL: a principled approach to discourse planning. In: Proceedings of INLG 2004, Kennebunkport, Maine, USA, pp 13–20

Printed in the United States
by Baker & Taylor Publisher Services